IEE TELECOMMUNICATIONS SERIES 9

Series Editors: Professor J. E. Flood
                Professor C. J. Hughes
                Professor J. D. Parsons

PHASE NOISE IN SIGNAL SOURCES

(Theory and Applications)

Other volumes in this series:

Volume 1 **Telecommunications networks** J. E. Flood (Editor)
Volume 2 **Principles of telecommunication-traffic engineering** D. Bear
Volume 3 **Programming electronic switching systems** M. T. Hills and
          S. Kano
Volume 4 **Digital transmission systems** P. Bylanski and D. G. W. Ingram
Volume 5 **Angle modulation: the theory of system assessment**
          J. H. Roberts
Volume 6 **Signalling in telecommunications networks** S. Welch
Volume 7 **Elements of telecommunications economics** S. C. Littlechild
Volume 8 **Software design for electronic switching systems** S. Takamura,
          H. Kawashima, N. Nakajima
Volume 9 **Phase noise in signal sources** W. P. Robins
Volume 10 **Local telecommunications** J. M. Griffiths (Editor)
Volume 11 **Principles and practices of multi-frequency telegraphy**
          J. D. Ralphs
Volume 12 **Spread spectrum in communications** R. Skaug and
          J. F. Hjelmstad
Volume 13 **Advanced signal processing** D. J. Creasey (Editor)
Volume 14 **Land mobile radio systems** R. J. Holbeche (Editor)
Volume 15 **Radio receivers** W. Gosling (Editor)
Volume 16 **Data communications and networks** R. L. Brewster (Editor)
Volume 17 **Local telecommunications 2** J. M. Griffiths (Editor)
Volume 18 **Satellite communication systems** B. G. Evans (Editor)
Volume 19 **Telecommunications traffic, tariffs and costs** R. E. Farr
Volume 20 **An introduction to satellite communications** D. I. Dalgleish
Volume 21 **SPC digital telephone exchanges** F. J. Redmill and
          A. R. Valdar
Volume 22 **Data communications and networks II** R. L. Brewster (Editor)

PHASE NOISE IN SIGNAL SOURCES

(Theory and Applications)

W. P. Robins

Peter Peregrinus Ltd. on behalf of the Institution of Electrical Engineers

Published by: Peter Peregrinus Ltd. on behalf of the Institution of
Electrical Engineers, London, United Kingdom

First published 1982
© 1982: Peter Peregrinus Ltd.
Paperback edition (with minor corrections) 1984
Reprinted 1998

Peter Peregrinus Ltd.
The Institution of Electrical Engineers,
Michael Faraday House,
Six Hills Way, Stevenage,
Herts. SG1 2AY, United Kingdom

British Library Cataloguing in Publication Data

A CIP catalogue record for this book
is available from the British Library

ISBN 0 86341 026 X

ACKNOWLEDGEMENTS

The writing of this book would not have been possible without the forbearance and understanding of my wife Doreen over an extended period.

My gratitude is also due to Mr. S. H. Moss, Chief Mathematician Marconi Space and Defence Systems Ltd., for numerous lunch-time discussions and for illuminating comments on parts of my draft manuscript. His comments have enabled me to avoid a number of errors which I would otherwise have made.

The analysis carried out in Section 9.4.5 was originally formulated in a Technical Note by Dr. H. Smith, formerly one of my collegues. His permission to publish this work is hereby acknowledged with thanks.

Microwave Associates Inc., of Burlington Mass. USA and Microwave Associates Ltd., of Dunstable have given me permission to describe two of their synthesisers and to quote measured results in Section 8.3 of this book. I am pleased to acknowledge this.

CONTENTS

                                                                              page
ACKNOWLEDGEMENTS   i

LIST OF MAJOR SYMBOLS USED   viii

1 INTRODUCTION   1
   1.1   The Importance of Spectral Purity   1
   1.2   Towards a Physical Understanding   3
   1.3   The Concept of Noise Density   4
   1.4   Sinusoidal Representation of Narrow Band Noise   6

2 REVIEW OF MODULATION THEORY   9
   2.1   Amplitude Modulation   9
   2.2   Phase Modulation   9
   2.3   Frequency Modulation   11
   2.4   The Addition of an Arbitrary Phase Angle   12
   2.5   Linear Approximation   13
   2.6   Phasor Representation   13
   2.7   The Concept of Conformability   14
   2.8   Coherent Demodulation of Conformable Signals   16

3 THE RELATIONSHIP BETWEEN PHASE JITTER
AND NOISE DENSITY   18
   3.1   Sinusoidal Phase Jitter   18
   3.2   The Representation of Narrow Band Noise   19
   3.3   Phase Jitter due to Superposed SSB Noise   20
   3.4   Simplified Treatment of Phase Jitter due to Superposed
       DSB White Noise   24
   3.5   Fuller Treatment of DSB Superposed White Noise   26
       3.5.1   Phase Modulation Index   26
       3.5.2   Amplitude Modulation Index   29
       3.5.3   Sideband Power Relationships 29
       3.5.4   Phase Jitter   30
       3.5.5   Phasor Representation   31
       3.5.6   Real and Potential Sidebands 33
       3.5.7   Philosophic Difficulties   36

3.6   A Note on Notation   37
   3.6.1   Parameters to be Distinguished   37
   3.6.2   The Concept of Phase Noise Density   37
   3.6.3   The Notation Used   37
3.7   Integration Over a Frequency Band   37
   3.7.1   Pure Noise   37
   3.7.2   Phase Noise   38
3.8   Summary of the Relationships between Phase Jitter and Noise   38
   3.8.1   Pure Phase Noise   38
      3.8.1.1   In Terms of Noise Density   39
      3.8.1.2   Integrated Over a Baseband Bandwidth b 39
   3.8.2   Superposed Thermal Noise   39
      3.8.2.1   SSB Noise   39
      3.8.2.2   DSB Noise   39

4   NOISE INDUCED FREQUENCY MODULATION   **41**
4.1   Basic Relationships   41
4.2   Restricted Value of the Concept of Frequency Deviation   44
4.3   Comparison of Phase Noise and Frequency Deviation   45

5   NOISE IN OSCILLATORS   **47**
5.1   The Effects of Non-Linearity   47
5.2   Analysis   48
5.3   Phase Noise Density as a Function of Offset Frequency   54
5.4   The Feedback Q 55
5.5   The Loaded Q 56
5.6   The Choice of Parameters   60
5.7   Examples   61
5.8   Integration of Phase Noise of an Oscillator   65
5.9   Other Factors which Affect Oscillator Phase Noise   65
   5.9.1   Disturbances on the Tuning Voltage of a VCO   66
   5.9.2   Disturbances on Other Power Supply Rails   68
      5.9.2.1   Due to Non-Linearities   68
      5.9.2.2   Due to Transistor Capacitance Changes   69
   5.9.3   Disturbances at the Output   69
      5.9.3.1   Added Noise   69
      5.9.3.2   Mismatch Effects   69
   5.9.4   Vibration   70
   5.9.5   Manual Tuning Mechanisms   72
   5.9.6   Dual Resonators   73
   5.9.7   Subsequent Filtering   73
5.10 A Simple Misunderstanding   74

6   FREQUENCY MULTIPLIER CHAINS   **75**
6.1   The Need for Frequency Multiplication   75
6.2   The Effect of Frequency Multiplication   75
   6.2.1   An Amplitude Modulated Wave   75
   6.2.2   A Phase Modulated Wave   76
   6.2.3   Effect of Frequency Multiplication on Phase Noise   77
   6.2.4   Effect of Frequency Multiplication on Thermal Noise   78
      6.2.4.1   SSB Thermal Noise   78
      6.2.4.2   DSB Thermal Noise   80

6.3		Added Amplifier Noise	81
	6.3.1	A Fundamental Frequency Source	81
	6.3.2	A Source with Amplification Prior to Frequency Multiplication	83
	6.3.3	A Source with Amplification After Frequency Multiplication	84
6.4		Example—An X-Band Xtal Oscillator/Multiplier Source	85
6.5		Limitations of Simple Oscillator and Oscillator/Multiplier Sources	87
6.6		The Performance of some X-Band Sources	90
6.7		Spurious Outputs	92

7 THE USE OF PHASE LOCK LOOPS   **93**

7.1		Introduction	93
7.2		Phase Lock Loop Configuration	94
	7.2.1	VCO	96
	7.2.2	Phase Detector	97
	7.2.3	Loop Filter	100
	7.2.4	The Open Loop Gain	104
	7.2.5	The Closed Loop Gain	105
	7.2.6	Loop Natural Frequency and Damping Factor	105
7.3		Phase Lock Loop Characteristics	109
	7.3.1	Frequency Transmission Characteristics	109
	7.3.2	Residual Phase Error	109
	7.3.3	Mathematical Convergence of $1/f^3$ Noise Operated on by a PLL	114
7.4		The Need for Frequency Division	115
7.5		Programmable Dividers	116
7.6		The Effects of Frequency Division on Correction Ratio and PLL Parameters	119
7.7		Phase Detector Noise Floors	122
	7.7.1	Analogue	122
	7.7.2	Digital using 'Exclusive Or' Gate PD	127
	7.7.3	Digital using Edge Triggered PD	128
	7.7.4	The Use of a Saturating Amplifier	128
	7.7.5	The Effect of Frequency Division	130
7.8		Frequency Conversion within a PLL	131

8 FREQUENCY SYNTHESISERS   **134**

8.1		Types of Synthesiser	134
	8.1.1	Direct Synthesis	134
	8.1.2	Synthesis using PLLs	135
8.2		Factors Affecting Choice of Configuration	135
8.3		Analysis of some Examples	140
	8.3.1	A UHF Synthesiser	140
	8.3.2	An X-Band Synthesiser	147
8.4		A TI 58/59 Program to Integrate Phase Jitter over Specific Frequency Bands	152
	8.4.1	General	152
	8.4.2	Theory	153
	8.4.3	The Program	153

8.5   Discrete Spurious Outputs   154
       8.5.1   Production of Discrete Spurious Outputs   154
       8.5.2   Phase Jitter due to Discrete Spurious Outputs   154
               8.5.2.1   Contributions due to a Single Sinusoid   154
               8.5.2.2   Integration of Contributions due to a Number
                         of Spurious Signals   155
       8.5.3   Enhancement of MO Spurious Outputs   156
       8.5.4   Breakthrough of PD Signals   157
       8.5.5   Unwanted Harmonics of Frequency Multipliers   161
       8.5.6   Harmonics of the Output Frequency   161
       8.5.7   Discrete Signals due to Power Supply Ripple or
               Vibration   161
8.6   Engineering Design Requirements   162
       8.6.1   Power Supplies   162
       8.6.2   Earthing   163
       8.6.3   Screening and Filtering   163
       8.6.4   Vibration   164
8.7   Some Important Aspects of Phase Noise Measurement   164
       8.7.1   A Simple Method of Measuring N_{op}/C or Integrated
               Phase Jitter   164
       8.7.2   Methods of Calibration   165
       8.7.3   Derivation of Calibration Constants   166
       8.7.4   Statistical Aspects   168
8.8   A Generalisation of the Concept of Phase Noise Density   171

9   THE RECIPROCAL RELATIONSHIPS BETWEEN PHASE
    NOISE AND FREQUENCY STABILITY (FREQUENCY
    DOMAIN TO TIME DOMAIN TRANSFORMATIONS
    AND THEIR INVERSES)   172
9.1   Introduction   172
9.2   Reciprocal Density Relationships   173
9.3   The Calculation of Integrated Frequency Jitter from
       Phase Noise Density Characteristics   173
       9.3.1   The Theoretical Relationship   173
       9.3.2   A TI 58/59 Program   174
       9.3.3   The Relevant offset Frequency Range   177
       9.3.4   Statistical Aspects   180
9.4   The Measurement of Integrated Frequency Jitter   180
       9.4.1   Initial Review of Concepts   180
       9.4.2   The Allan Variance   184
       9.4.3   An Important Relationship   185
       9.4.4   'Transfer Function' to a Frequency Jitter Input   186
       9.4.5   Transfer Function to a Phase Noise Input   189
       9.4.6   Statistical Aspects   193
9.5   Measurement of Density Using Time Domain Methods   195
       9.5.1   Measurement of Frequency Jitter Density   195
       9.5.2   Measurement of Phase Noise Density   199
       9.5.3   Statistical Aspects   199
       9.5.4   The Hadamard Variance   200

9.6   Calculation of Phase Noise Density Knowing Integrated
      Frequency Jitter   200
      9.6.1   When $(N_{op}/C)_f$ follows a $1/f^3$ law   200
      9.6.2   For Other Known Laws   202

10 SYSTEM PHASE NOISE REQUIREMENTS   **203**
      10.1   A Single Dish CW Radar System   204
             10.1.1   General Description   204
             10.1.2   Delayed Reflections of a Signal with Spurious AM
                      and PM   207
             10.1.3   Permissible AM   211
             10.1.4   Permissible PM   217
             10.1.5   Single Sideband Signals and Double Sideband
                      Interference   222
      10.2   A Radar using High Deviation FM   224
             10.2.1   Types of FM Radar   224
             10.2.2   Required Frequency Deviation   226
             10.2.3   Spurious AM and PM   229
      10.3   Communication Systems using Single Channel per Carrier FM 232
             10.3.1   Introduction   232
             10.3.2   Phase Noise Requirements for a Simple FM System   233
             10.3.3   The Addition of Pre-emphasis/De-emphasis   239
      10.4   FDM/FM Communication Systems   241
             10.4.1   Introduction   241
             10.4.2   The Required Value of C/N_0 242
             10.4.3   Phase Noise Requirements   244
      10.5   Communication Systems Using PSK   246
             10.5.1   Introduction   246
             10.5.2   Carrier Recovery   249
             10.5.3   Signal Error Rate   254
             10.5.4   Clock Recovery   260
             10.5.5   Differential PSK   262
             10.5.6   The Effects of Phase Transients   268
      10.6   Antenna Tracking   272

REFERENCES   **277**

APPENDIX I   SUMMARY OF IMPORTANT FORMULAE   **279**

**APPENDIX II   NOISE TEMPERATURE AND NOISE
                FIGURE REVIEW**   **285**
      AII.1   Background   285
      AII.2   Noise Temperature and Noise Figure   286
      AII.3   Waveguide Losses at Different Temperatures   288
      AII.4   The Reasons for a Mismatch at the Input to a Low
              Noise Amplifier   290
      AII.5   The Noise Figure of an FET Amplifier   292
      AII.6   The Noise Bandwidth of a Single Tuned Circuit   293
      AII.7   Mixer Noise Figure, Noise Temperature and NTR   294
      AII.8   The Suppression of Local Oscillator AM Noise by a
              Balanced Mixer   296

APPENDIX III   **THE QUADRATURE REPRESENTATION OF**
NARROW BAND NOISE   **300**

APPENDIX IV   **THE Q OF VARACTOR TUNED**
OSCILLATORS   **304**

APPENDIX V   **THE PHASE NOISE PERFORMANCE**
OF GUNN OSCILLATORS   **309**

INDEX   **311**

LIST OF MAJOR SYMBOLS USED

f_0	the carrier frequency.
ω (or w)	the angular frequency of the carrier $= 2\pi f_0$.
f (or f_m)	the frequency offset from the carrier.
p	the angular offset frequency $= 2\pi f_m$.
$N_0(f)$	the *single sideband* noise density at an offset frequency f. This represents a long term statistical average. $N_0 = kT$ where k is Boltzmann's Constant and T is the effective noise temperature in degrees Kelvin.
$n_0(f)$	the single sideband noise power density at an offset frequency f, during a specific second of time.
N	the total noise power integrated over some RF frequency band of interest. Whether it corresponds to SSB or DSB noise is made clear in the context in which it is used. It represents a long term statistical value (LTSV).
C	the carrier power in watts.
θ	the peak phase jitter of a carrier in radians (LTSV).
$\theta_0(f)$	the *peak* phase jitter in radians per $\sqrt{\text{Hz}}$ of baseband bandwidth (LTSV).
ϕ	the *rms* phase jitter of the carrier in radians integrated over some baseband frequency range (LTSV).
$\phi_0(f)$	the *rms* phase jitter of the carrier in radians $\sqrt{\text{Hz}}$ of baseband bandwidth (LTSV).
$\overline{\phi^2}$	the phase jitter variance in (radians)^2 integrated over some baseband frequency range (LTSV).
$\overline{\phi_0^2}(f)$	the phase jitter variance in (radians)^2 per Hz of baseband bandwidth (LTSV).

N_p	the *single sideband* phase noise power integrated over some offset frequency band of interest (LTSV).
$N_{op}(f)$	the *single sideband* phase noise density per Hz of RF bandwidth corresponding to the offset frequency f (LTSV).
N_a	the *single sideband* amplitude noise power integrated over some offset frequency band of interest (LTSV).
$N_{oa}(f)$	the single sideband amplitude noise power density per Hz of RF bandwidth corresponding to the offset frequency f (LTSV).
$\dfrac{N_p}{C}$	the *single sideband* phase noise to carrier ratio (LTSV).
$\left(\dfrac{N_{op}}{C}\right)_f$	the *single sideband* phase noise density to carrier ratio at an offset frequency f (LTSV).
$\dfrac{N_a}{C}$	the *single sideband* amplitude noise to carrier ratio (LTSV).
$\left(\dfrac{N_{oa}}{C}\right)_f$	the *single sideband* amplitude noise density to carrier ratio at an offset frequence f (LTSV).
$(n_{op})(f)$	the single sideband phase noise density at the offset frequency f during the specific second under consideration.
$\left(\dfrac{n_{op}}{C}\right)_f$	the single sideband phase noise density to carrier ratio at an offset frequency f during the specific second under consideration.
b	the baseband bandwidth (in a digital system this is often made equal to the bit rate).
$B = 2b(= \pm b)$	the RF bandwidth, i.e. between $\pm b$.
δf	the *rms* frequency deviation of a carrier in Hz. If the frequency deviation is due to noise it is also a long term statistical value. In FDM/FM systems it also represents the rms deviation of the Test Tone per channel.
Δf	the *peak* frequency deviation of a carrier in Hz. For both deterministic and noise modulation $\Delta f = \sqrt{2}\,\delta f$.
δf_0	the *rms* frequency deviation per $\sqrt{\text{Hz}}$. It is a long term statistical value. It may also be defined as the standard deviation of the result of a series of measurements of frequency difference from the average frequency when referred to a 1 Hz bandwidth.
Δf_0	the *peak* frequency deviation density per $\sqrt{\text{Hz}}$. It is a long term statistical value. Apart from typographical convenience it could be written $(\langle \Delta f_0^2 \rangle)^{1/2}$. In the latter form it also represents the standard deviation of the difference between two measurements of average frequency after normalisation to a 1 Hz bandwidth.

Q_u	the unloaded quality factor of a resonant circuit.
Q_L	the quality factor of a resonant circuit with external loading.
Q'	the (very high value) effective Q of an oscillator tuned circuit due to the effect of positive feedback.
f_c	the highest offset frequency below which the phase noise density of an oscillator follows a $1/f^3$ law.
K_d	a phase detector gain constant in volts per radian.
K_0	the gain constant of a VCO in radians per second per volt.
f_n	the natural frequency of a phase lock loop in Hz.
ζ	the damping factor of a phase lock loop.
n	usually a frequency multiplication or division ratio.
y_k	one sample of the normalized frequency i.e. f_k/f_0.
ΔF	peak or peak to peak frequency deviation of a linear FM radar.
F_m	modulation rate for a linear FM radar.
S/N	baseband signal to noise ratio.
x	a numerical ratio by which the phase noise component of thermal noise is required to exceed local oscillator phase noise.
E	the received energy per bit in a digital communication system.
wrt	with respect to
SMO	Station Master Oscillator
MO	Master Oscillator

CHAPTER 1

INTRODUCTION

1.1 The Importance of Spectral Purity

In communications, radar and similar systems, signal sources are required as carriers for increasingly sophisticated baseband information. They are also required as local oscillators or pump sources for achieving frequency conversions. A simple source might consist of a crystal oscillator alone or perhaps a manually tuned oscillator. With the increased usage of the frequency spectrum the permissible tolerance on the nominal frequency of signal sources is continually reduced. This has necessitated the development of more complex signal sources. For single frequency operation these usually consist of crystal oscillators followed by amplifiers and frequency multiplying stages. Where manual or remote accurate selection of frequencies is required, frequency synthesisers are normally specified.

A naive view of a signal source might be to regard it as having an output at a single frequency. This is to say that its spectrum is an infinitely narrow line (delta function) in the frequency domain. In practice, if only because the oscillator is not switched on for an infinite time, the resultant spectral line will have a finite width. Real signal sources, even those of the highest quality, have other imperfections also which result in unwanted amplitude and phase modulation of the carrier.

If a carrier with unwanted amplitude or phase modulation is modulated, either in amplitude or phase, by an information carrying baseband signal, then the signal quality or error rate of the communications link will be worsened as a result of the unwanted carrier modulation. In practice spurious phase modulation is more important technically and economically than spurious amplitude modulation. This is partly because the majority of high capacity, and hence expensive, communication systems use angle modulation, that is frequency modulation or phase modulation, but also because spurious amplitude modulation in complex signal sources is usually of a much lower level than spurious phase modulation. Frequency multiplication which is necessary in the majority of stable high frequency signal sources, results in a degradation of the unwanted angle modulation but not inherently in the unwanted AM, as will become apparent in the course of the analysis, (Section 6.2). A further reason for the greater importance of spurious angle modulation is that frequency or phase modulation of a local oscillator in a transmitter or receiver will be directly transferred to the signal which is being subjected to frequency conversion. The effects of local oscillator AM may be greatly mitigated by the use of balanced mixers.

The economic importance of a small degradation in link performance is well illustrated, admittedly in a very crude fashion, by considering an international satellite communication system. The total cost of such a system over a decade, including the cost of the space segment and of all ground stations together with staffing costs might well exceed £300 million. A 3 dB degradation in performance would halve the traffic capacity. One way to restore the originally required traffic capacity, although admittedly not the most economic, would be to duplicate the complete system at a cost of a further £300 million. Thus, in a very crude sense indeed, we might say that in such a system 3 dB is equivalent to £300 million and hence 0·1 dB is very roughly equivalent to £10 million. This rather absurd example is given merely to emphasise that in such a system a small signal degradation is expensive and that therefore relatively complex steps to reduce link degradation are economically justified.

As satellite communication systems develop, and the radio frequency power and antenna gain available from satellites increase, there is a growing tendency to devise systems which use small cheap ground stations, probably in large numbers, thus economising in land line connections between the ultimate user and the earth terminal (References 1 and 2). The use of small ground stations with low figures of merit implies that the traffic capacity of each station is limited. To avoid the necessity of complex signal processing at the earth terminals, the signal or signals directed to a single earth terminal, which may be a telephony channel or one or more data channels, are usually modulated on to a carrier specifically directed to that earth terminal. Thus, there is a tendency for such a system to adopt single channel per carrier (SCPC) operation. The carrier directed to a very small earth terminal may in fact carry only a single low data rate channel. If this is angle modulated on to the carrier concerned, using either frequency shift keying or phase shift keying, then any spurious phase modulation of the carrier at a rate comparable to that of the data stream will produce signal degradation. With the use of low bit rates it is becoming more and more important to achieve a low level of spurious phase modulation at low modulation rates. Spurious phase modulation at relatively low rates in the time domain, is equivalent in the frequency domain to spurious phase sidebands very close to the carrier.

For reasons which will become apparent it is much more difficult to achieve spectral purity close to the carrier, that is at small offset frequencies, than it is for large offset frequencies. In the time domain this implies that spurious modulation at high rates is more readily controlled than such modulation at low rates. For this reason it becomes increasingly important to understand the mechanism of phase noise in signal sources as a first step towards its control and reduction.

1.2 Towards a Physical Understanding

The primary orientation of this treatment will be aimed towards achieving a physical understanding, using only sufficient mathematical analysis to enable the quantitative contributions of each factor to be assessed. The symbols used in the inevitable mathematical treatment will be chosen for their physical meaning and their relationship to measurable parameters. Conventional symbols will be used as far as possible but, where the use of such symbols conflicts with the directness of physical understanding, novel symbols will be used and their relationship to the conventional symbols stated.

With the exceptions mentioned in a subsequent paragraph, every formula used will be derived from first principles. In addition, the analysis will be carried out by a process of successive refinement: the simpler, and usually major, effects being treated first.

It is the author's opinion that the majority of textbooks and technical papers cannot be effectively understood if the reader is sitting in an armchair without pencil and paper available to carry out the intermediate calculations which are normally left to the reader. This approach is advantageous in so far as it certainly results in brevity, but it is a major disadvantage for any novice to the field except those with great facility in mathematical manipulation. Thus the present intention is to show every line of any mathematical argument and to use the lowest level of mathematical sophistication that is compatible with a full analysis.

The majority of communication and radar engineers are familiar with the noise performance of amplifiers and frequency converters, but usually less familiar with the problems relating to phase noise in signal sources and the phase noise requirements which must be met in any specific system. For this reason the calculation of system noise temperatures, amplifier noise and converter noise figures will be assumed to be understood by the reader, although a brief review will be given in Appendix II. By assuming such an understanding it is hoped to achieve brevity and a detailed step by step treatment of phase noise, simultaneously.

Thus, the main subjects treated in this book will be as follows:

(*a*) A review of modulation theory, which is necessary for a full understanding of phase noise.
(*b*) A consideration of the relationship between phase noise, phase jitter and thermal noise.

(*c*) A discussion of noise in oscillators.

(*d*) A consideration of frequency multiplier chains and their effects on source noise.

(*e*) The use of phase lock loops to control source noise over specific frequency ranges.

(*f*) The 'noise floors' of phase sensitive detectors and the modification to the effective noise floor resulting from frequency division in a phase lock loop.

(*g*) The calculation of the noise performance of some typical frequency synthesisers. This will include discussion of the factors which influence the choice of synthesiser block diagram given, output frequency ranges, the size of frequency step required and the phase noise requirements over various offset frequency bands.

(*h*) The relationship between frequency stability and phase noise.

(*i*) The calculation of phase noise requirements for a number of radar and communication systems.

Measurement of the spectral purity of high quality signal sources is a large subject; any complete treatment would probably double the size of this book. Hence it has only been treated to the extent necessary for a full understanding of the concepts and parameters associated with phase noise and frequency stability. The results of measurements are given in a number of instances in an attempt to correlate theory and practice.

1.3 The Concept of Noise Density

The mean available noise power per Hz of bandwidth from a resistor at a temperature of $T°$K is:

$$N = kT \text{ watts per Hz}$$

where k is Boltzmann's Constant $= 1.38 \times 10^{-23}$ joules per $°$K.

The mean noise per Hz of bandwidth will hereafter always be designated N_0. Thus N_0 is equal to the noise power spectral density in watts per Hz of bandwidth or alternatively it may be expressed in dB watts per Hz of bandwidth.

In the case of a receiving system (such as a satellite ground station) with an overall effective noise temperature T_s, referred to the low noise amplifier (LNA) input:

$$N_0 = kT_s \text{ watts per Hz.}$$

Note that for

$$T_s = 1°\text{K}, \qquad N_0 = k = -228.6 \text{ dBW/Hz};$$

and for

$$T_s = 290°\text{K}, \qquad N_0 = -204 \text{ dBW/Hz} = -174 \text{ dBm/Hz} \qquad (1.1)$$

Consider a broad band receiving system such as a satellite ground station, with an overall noise temperature (T_s) referred to the LNA input. In the absence of a signal the mean noise power density will be independent of the specific frequency within the frequency coverage of the station (assuming the station noise temperature to be constant over the useable bandwidth). Consider an idealised case where the station receives a pure CW signal from the satellite (that is a single frequency signal uncontaminated by any noise components of its own), at a frequency in the centre of the station pass band. The total input to the LNA will consist of this carrier with white noise of spectral density N_0 (due to T_s) on either side of this carrier. Let the carrier frequency be f_0 and consider a noise component in a 1 Hz bandwidth at ($f_0 + \delta f$) and also another 1 Hz wide noise component at ($f_0 - \delta f$). The total noise power in the two sidebands together equals $2N_0$. This will be referred to as double sideband (DSB) noise power.

Thus:

$$N_0 = \text{power density (SSB)} = kT_s.$$

$$2N_0 = \text{power density (DSB)} = 2kT_s.$$

$N_0 G$ is the mean power which would be measured at the output of a 1 Hz wide RF or IF filter placed in a receiver chain with a total power gain G between the LNA input and the filter output. 1 Figure 1 illustrates these ideas.

It should be noted that in some mathematical treatments a different definition of noise power density is adopted. In carrying out Fourier transformations it is necessary to integrate over both positive and negative frequencies. To allow for the contribution from negative frequencies the 'two sided' power density is defined as $N_0/2$. N_0 is then described as the 'single sided' noise density; that is the noise density resulting from the combined contributions from positive and negative frequencies. We shall not use the concept of 'two sided' noise density at all and it is only mentioned here in case the reader should have encountered the concept from other treatments.

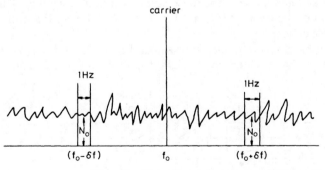

Illustrating Noise Power Density   **1 Figure 1**

To avoid confusion it is important to remember that N_0 as used in this book is equivalent to the *'single sided'* noise density used in such mathematical treatments. Our double sideband noise power equals $2N_0$ and should not be confused with the 'two sided' noise density (equals $N_0/2$) of the formal mathematical treatment. In this book we are nowhere concerned with negative frequencies; all the frequencies in 1 Figure 1 are positive, regardless of the fact that some are above the carrier frequency and others below.

1.4 Sinusoidal Representation of Narrow Band Noise

The characteristics of narrow band Gaussian noise may readily be observed on an oscilloscope by carrying out the following experiment. Take a high gain bandpass amplifier with a centre frequency within the frequency range of the oscilloscope to be used. This amplifier should have a bandwidth which is small compared with its centre frequency. Connect this amplifier to a suitable oscilloscope and adjust the oscilloscope gain until a noise-like display is visible. Set the oscilloscope timebase to give a total trace time of approximately $1/B$ seconds where B is the bandwidth of the bandpass amplifier in Hz. Switch the timebase to the single shot triggered mode of operation and arrange that it should be triggered by a peak of the incoming noise waveform. The waveform displayed will appear very similar to a perfect sinusoid at the centre frequency of the bandpass amplifier. If the duration of the timebase is increased a few times and the experiment repeated it will be noted that the early part of the trace (immediately following the trigger) is sinusoidal and the latter part of the trace loses its sinusoidal appearance. A block diagram for such an experiment is given in 1 Figure 2. In this figure the bandwidth of the 1 MHz amplifier chosen is 10 kHz. The reason why the output noise appears to be sinusoidal for approximately 0·1 ms after the timebase triggers, is that neither the amplitude nor the 'phase' of a signal passing a bandpass filter can change appreciably in a time which is short compared with the reciprocal of the bandwidth. Thus, in the example given, the 'phase' and amplitude of the noise at the instant of triggering is maintained for a period of the order of 0·1 ms. For times much longer than this there will be negligible coherence.

If (in the bandpass amplifier shown in 1 Figure 2) we insert an additional very narrow band filter of bandwidth 1 Hz at a frequency of (say)

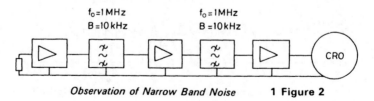

Observation of Narrow Band Noise   **1 Figure 2**

1 MHz + 200 Hz then the output will remain coherent for the order of 1 second. Thus, for periods of time up to 1 second, noise in a 1 Hz bandwidth may be represented by a sinusoid if the amplitude and phase are suitably chosen. As the signal is actually Gaussian noise, both the amplitude and the phase will vary from one second to another and can only be specified statistically. If the noise figure (F) and the gain of the amplifier up to the output of the 1 Hz filter are known then it is relatively simple to calculate the rms value of the noise voltage at the filter output. Alternatively the available noise power output may be calculated. This is given by:

$$N_0 = GFkT \quad \text{(where } G \text{ is the available power gain)}$$

Knowing the output impedance, the rms noise voltage (in a 1 Hz bandwidth) may readily be calculated. This is a statistical value and is the square root of the long time mean square of the voltage. If for a period of less than approximately 1 second we represent the noise voltage by a sinusoid this too may be allotted an rms value (and even a 'peak' value $= \sqrt{2}\,V_{rms}$) but it is important in many calculations that the rms value specified for the equivalent sinusoid should be the long term standard deviation or rms value. Thus, if a narrow band noise component is represented in sinusoidal form it must be remembered that the rms value attributed to this sinusoid is not necessarily that which would be observed in any 1 second period. It is purely a statistical value. As long as these limitations are borne in mind it is valid and extremely useful to represent narrow band noise components as sinusoids of peak amplitude $= \sqrt{2}$ times the rms value.

This approach is particularly illuminating in the case where narrow band noise is superposed upon a carrier passing through an amplifier. Consider a carrier at a frequency f_0 and a noise component in a 1 Hz bandwidth at frequency $(f_0 + \delta f)$. For periods short compared with 1 second the noise voltage may be considered to be a sinusoid at a frequency $(f_0 + \delta f)$. We then use well known phasor methods to calculate the effect of combining these two different frequency sinusoids. This may be done by considering the carrier to be stationary and the noise component to be rotating at an angular velocity $\delta p = 2\pi\delta f$. This is shown in 1 Figure 3. It will be noted that as the noise vector

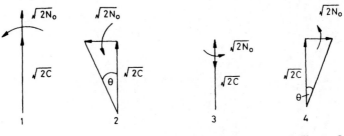

Carrier Plus Narrow Band Noise   **1 Figure 3**

rotates, it produces amplitude modulation of the carrier when in positions 1 and 3, and phase modulation of the carrier when in positions 2 and 4. The carrier amplitude is shown as $\sqrt{2C}$ and the noise sinusoid is shown with an amplitude $\sqrt{2N_0}$. In the usual way these values are the peak values of the sinusoids. It would seem from this diagram that the peak value of the amplitude modulation is given by:

$$M' = \sqrt{\frac{N_0}{C}}$$

Similarly the peak phase modulation (phase jitter) is given by:

$$\theta = \tan^{-1}\sqrt{\frac{N_0}{C}}$$

Given the peak values as calculated above, the rms values of the amplitude and phase modulation indices may then readily be calculated if required. More detailed analysis supplements this simple picture but also shows it to be correct as far as it goes.

CHAPTER 2

REVIEW OF MODULATION THEORY

2.1 Amplitude Modulation (AM)

An amplitude modulated signal is given by

$$V(t) = \sqrt{2C}\,[1 + M'(t)]\,\sin\,wt$$

where C is the carrier power and (without loss of generality) the impedance level is 1 ohm. We shall deal with the simplest case and write

$$M'(t) = M'\,\sin\,pt\ (M'\ \text{normally}\ < 1)$$

$$\therefore\quad V(t) = \sqrt{2C}\,[1 + M'\,\sin\,pt]\,\sin\,wt$$

(2.1)

Now $\sin A \sin B = \frac{1}{2}\cos(A - B) - \frac{1}{2}\cos(A + B)$

$$\therefore\quad V(t) = \sqrt{2C}\left[\underset{\underset{\text{Carrier}}{\uparrow}}{\sin\,wt} + \underset{\underset{\text{Lower Sideband}}{\uparrow}}{\frac{M'}{2}\cos(w - p)t} \quad - \underset{\underset{\substack{\text{Upper}\\\text{Sideband}}}{\uparrow}}{\frac{M'}{2}\cos(w + p)t}\right]$$

(2.2)

This shows that modulation with a single tone $p/2\pi$ produces just a single sideband on each side of the carrier (together with the carrier). The ratio of the amplitude of each sideband to that of the carrier (normalised sideband amplitude) is $M'/2$. M' is the modulation depth and $100M'$ is the percentage modulation.

2.2 Phase Modulation (PM)

A phase modulated wave may be written

$$V(t) = \sqrt{2C}\,\sin\,[wt + \theta(t)]$$

Again taking the simplest case we write

$$\theta(t) = \theta \sin pt \tag{2.3}$$

where θ is the peak angular deviation in radians.

$$\therefore \quad V(t) = \sqrt{2C} \sin [wt + \theta \sin pt] \tag{2.4}$$

$$\therefore \quad V(t) = \sqrt{2C} [\sin wt \cos (\theta \sin pt) + \cos wt \sin (\theta \sin pt)]$$

Using the identities

$$\cos (x \sin \phi) \equiv J_0(x) + 2[J_2(x) \cos 2\phi + J_4(x) \cos 4\phi \cdots] \tag{2.5}$$

and

$$\sin (x \sin \phi) \equiv 2[J_1(x) \sin \phi + J_3(x) \sin 3\phi + \cdots] \tag{2.6}$$

where $J_0(x)$, $J_1(x)$, ..., are Bessel Functions of the first kind of argument (x) and order 0, 1, ..., respectively.

$$V(t) = \sqrt{2C} \sin wt[J_0(\theta) + 2J_2(\theta) \cos 2pt + 2J_4(\theta) \cos 4pt + \cdots]$$
$$+ \sqrt{2C} \cos wt[2J_1(\theta) \sin pt + 2J_3(\theta) \sin 3pt + \cdots] \tag{2.7}$$

$$\therefore \quad V(t) = \sqrt{2C} [J_0(\theta) \sin wt + 2J_2(\theta) \sin wt \cos 2pt$$
$$+ 2J_4(\theta) \sin wt \cos 4pt \cdots]$$
$$+ \sqrt{2C} [2J_1(\theta) \cos wt \sin pt + 2J_3(\theta) \cos wt \sin 3pt + \cdots]$$

$$\therefore \quad V(t) = \sqrt{2C} [J_0(\theta) \sin wt + J_2(\theta) \sin (w + 2p)t$$
$$+ J_2(\theta) \sin (w - 2p)t$$
$$+ J_4(\theta) \sin (w + 4p)t + J_4(\theta) \sin (w - 4p)t + \cdots]$$
$$+ \sqrt{2C} [J_1(\theta) \sin (w + p)t - J_1(\theta) \sin (w - p)t$$
$$+ J_3(\theta) \sin (w + 3p)t - J_3(\theta) \sin (w - 3p)t + \cdots]$$

$$\therefore \quad V(t) = \sqrt{2C} [J_0(\theta) \sin wt + J_1(\theta) \sin (w + p)t - J_1(\theta) \sin (w - p)t$$
$$+ J_2(\theta) \sin (w + 2p)t + J_2(\theta) \sin (w - 2p)t + \cdots]$$

$$\tag{2.8}$$

The carrier amplitude is $\sqrt{2C} J_0(\theta)$

The first sideband amplitude is $\quad + \sqrt{2C} J_1(\theta)$ (Upper)

The first sideband amplitude is $\quad - \sqrt{2C} J_1(\theta)$ (Lower)

The second sideband amplitude is $+ \sqrt{2C} J_2(\theta)$ (Upper)

The second sideband amplitude is $+ \sqrt{2C} J_2(\theta)$ (Lower)

etc. ...

The Bessel Functions may be evaluated from tables but for small values of θ such as occur in random noise modulation

$$J_0(\theta) \simeq 1 \tag{2.9}$$

$$J_1(\theta) \simeq \frac{\theta}{2} \tag{2.9}$$

$$J_2(\theta) \simeq J_3(\theta) \simeq J_4(\theta) \cdots \simeq 0 \tag{2.9}$$

This follows from the identity:

$$J_n(x) \equiv \left(\frac{x}{2}\right)^n\left[\frac{1}{n!} - \left(\frac{x}{2}\right)^2 \frac{1}{1!\,(n+1)!} + \left(\frac{x}{2}\right)^4 \frac{1}{2!\,(n+2)!} - \cdots\right] \tag{2.10}$$

$\therefore$ for small values of θ (small sideband to carrier ratios) we have:

$$V(t) = \sqrt{2C}\left[\sin wt + \frac{\theta}{2}\sin (w+p)t - \frac{\theta}{2}\sin (w-p)t\right] \tag{2.11}$$

Within the approximations of (2.9) above, this is identical to equation (2.2) above for the AM case if θ is replaced by M', sin by cos and the $\pm$ signs in front of the sidebands are reversed.

2.3 Frequency Modulation (FM)

The equation for a frequency modulated wave is:

$$V(t) = \sqrt{2C}\,\sin\left[wt + \frac{\Delta f}{f_m}\sin 2\pi f_m t\right] \tag{2.12}$$

where Δf is the peak frequency deviation and f_m is the modulating frequency. For convenience $2\pi f_m$ will be written as p.

$\Delta f/f_m$ is usually written M and is known as the Modulation Index.

$$M = \frac{\Delta f}{f_m} \tag{2.13}$$

$\therefore$ Equation (2.12) may be written

$$V(t) = \sqrt{2C}\,\sin\,[wt + M\sin pt] \tag{2.14}$$

(2.14) is identical with (2.4) above if M, the Modulation Index, is replaced by θ, the peak phase deviation in radians.

$\therefore$ the whole analysis of Section 2.2 above may be applied to (2.14) to give

$$\begin{aligned}
V(t) = \sqrt{2C}\,[&J_0(M)\sin wt + J_1(M)\sin (w+p)t - J_1(M)\sin (w-p)t \\
&+ J_2(M)\sin (w+2p)t + J_2(M)\sin (w-2p)t \\
&+ J_3(M)\sin (w+3p)t - J_3(M)\sin (w-3p)t \\
&+ \cdots]
\end{aligned}$$

For the case of small values of M (see (2.9)) we have:

$$V(t) \simeq \sqrt{2C} \left[\sin wt + \frac{M}{2} \sin (w + p)t - \frac{M}{2} \sin (w - p)t \right] \tag{2.15}$$

and

$$\Delta f = M f_m \tag{2.16}$$

where Δf is the peak frequency deviation at a rate f_m, producing a sideband distant f_m from the carrier.

2.4 The Addition of an Arbitrary Phase Angle

For the two basic cases of amplitude modulation and angle modulation the modulating waveform is given by equations (2.1) and (2.3) repeated below:

$$M'(t) = M' \sin pt \tag{2.1}$$

$$\theta(t) = \theta \sin pt \tag{2.3}$$

If we add an arbitrary phase angle ψ to the modulation waveform, in order to achieve greater generality, (2.1) and (2.3) become:

$$M'(t) = M' \sin (pt + \psi) \tag{2.1.1}$$

$$\theta(t) = \theta \sin (pt + \psi) \tag{2.3.1}$$

For the AM case, substituting $(pt + \psi)$ for (pt) results in equation (2.2) being modified as follows:

$$V(t) = \sqrt{2C} \left\{ \sin wt + \frac{M'}{2} \cos [(w - p)t - \psi] \right.$$

$$\left. - \frac{M'}{2} \cos [(w + p)t + \psi] \right\} \tag{2.2.1}$$

This constitutes a more general expression for the carrier and sidebands of an amplitude modulated signal.

For the angle modulated case, substituting $(pt + \psi)$ for (pt) results in equations (2.11) and (2.15) becoming, for the PM case:

$$V(t) = \sqrt{2C} \left\{ \sin wt + \frac{\theta}{2} \sin [(w + p)t + \psi] \right.$$

$$\left. - \frac{\theta}{2} \sin [(w - p)t - \psi] \right\} \tag{2.11.1}$$

and for the FM case:

$$V(t) = \sqrt{2C} \left\{ \sin wt + \frac{M}{2} \sin [(w + p)t + \psi] \right.$$

$$\left. - \frac{M}{2} \sin [(w - p)t - \psi] \right\} \qquad (2.15.1)$$

2.5 Linear Approximation

In general angle modulation (PM or FM) is non-linear: if we double the modulation index when modulating by a single sinusoid, we do not double the sideband amplitude. Hence, if we modulate simultaneously with two baseband sinusoids of different frequencies, using the same modulation index for each, the amplitudes of each of the sidebands are different from the amplitudes observed when we modulate with either separately.

However, for very low modulation indices, such as usually apply in the case of spurious noise modulation of a carrier, equation (2.9) may be satisfied to an adequate approximation. In this case $J_1(\theta) \simeq \theta/2$ and $J_1(2\theta) \simeq 2\theta/2$. The angle modulation may then be considered to be a linear process to the degree of approximation involved.

For example: if

$$\theta = 0.05 \text{ radians;} \qquad J_1(\theta) = 0.024992$$

$$2\theta = 0.1 \text{ radians;} \qquad J_1(2\theta) = 0.04994$$

This is accurately linear to 0.12%.

Thus, for most purposes, if the total modulation index due to all noise components does not exceed 0.1 radians, we may consider that we are dealing with a linear modulation process and that we may use the Superposition Theorem when convenient. In doing this we must of course not forget that independent noise powers add and that independent noise voltages are combined by taking the square root of the sum of the squares.

2.6 Phasor Representation

Whilst the phasor representation of equations (2.2) and (2.11) is almost self evident there is a point of interpretation which is important if confusion is to be avoided.

The phasor diagrams are shown in 2 Figure 1 with exaggerated sideband amplitudes.

We will take the PM case for purposes of illustration, although a similar point arises in the AM case. Each of the sidebands has a peak amplitude $\sqrt{2C}\,\theta/2$. When the two PM 'sidebands' come into 'phase' as shown in 2

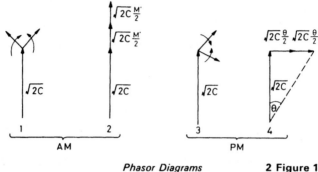

Phasor Diagrams   **2 Figure 1**

Figure 1-4 the combined amplitude is doubled and becomes $\sqrt{2C}\,\theta$. A careless interpretation of the phasor diagram might then lead to the supposition that the combined power of the two sidebands is four times that of one of the sidebands. On the other hand, the upper and lower sidebands are at different frequencies and it is well known that the combined power of two sinusoids at different frequencies is equal to the sum of the individual powers.

The puzzle may be resolved by remembering that in a phasor diagram the carrier has been made stationary, in one sense it has been translated to DC, and the sideband phases are thus at relative angular frequencies of $\pm p$. Thus the phasor diagram is, from one point of view, a baseband diagram rather than a bandpass representation. Hence, although in the bandpass case the combined power of the two sidebands is the sum of the individual powers; in the lowpass case, for example after coherent demodulation, the combined power from the two contributions is equal to four times that of a single contribution.

A detailed analysis of this effect is given in Section 2.8.

2.7 The Concept of Conformability

The two phase modulation sidebands given by equation (2.11) are, in a strict mathematical sense, uncorrelated; as may be seen by taking the square of the sum, to arrive at the power and finding that the average value of the cross product term is zero. However, they have equal amplitudes, a definite 'phase' relationship and each has a constant frequency. Addition of a sufficiently large purely sinusoidal carrier at the mean frequency (that is half way between the two frequencies) will produce, pure AM, pure PM or a mixture of AM and PM, depending on the phase of the added carrier. This is shown in 2 Figure 2.

Consider now a more complicated case where a carrier is phase modulated by two baseband sinusoids at different angular frequencies (p_1 and p_2) simultaneously. As long as the overall modulation index is small we may invoke the Superposition Theorem and consider that two upper sidebands are

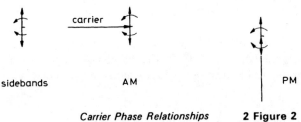

sidebands   AM   PM

Carrier Phase Relationships   **2 Figure 2**

produced at angular frequencies $(w + p_1)$ and $(w + p_2)$. Similarly there will be two lower sidebands at angular frequencies $(w - p_1)$ and $(w - p_2)$. For this to be true it is not necessary that the amplitudes of the two baseband sinusoids should be equal and hence the amplitudes of the two upper sidebands may not be equal. It remains true, however, that the amplitudes of corresponding upper and lower sidebands will be equal: that is the amplitude of sidebands at $(w \pm p_2)$ will be equal. In fact it may readily be seen that the upper and lower sidebands will be symmetrical but inverted with respect to frequency. In addition, reference to (2.11) or to (2.2) shows that the mean phase differs from wt by a constant at all times; i.e. their phases are equal and opposite *wrt* $(wt + \text{constant})$. The two sidebands are mirror images *wrt* the carrier.

Such sidebands, considered on their own without a carrier, do not constitute a phase or amplitude modulated signal. They might be regarded as constituting suppressed carrier PM or AM signals. Reintroduction of a carrier of sufficient amplitude and the correct frequency would produce a PM signal, an AM signal or a mixture of the two depending on the phase of the added carrier.

Signals possessing all these properties are not mathematically correlated, being at different frequencies, but they do have a very special relationship. It is useful to have a single word to refer to two signals which are related in this way. A search of the literature does not reveal any useage of a single word for this concept.

Two signals which satisfy all these relationships will be said to be CONFORMABLE. Two signals which do not satisfy all these relationships will be said to be NON-CONFORMABLE.

Definition

Two signals are said to be conformable when they satisfy all the following conditions:

(a) Their mean frequency $(w/2\pi)$ is large compared with their maximum frequency difference.

(b) Their amplitudes are equal at the same offset frequency with respect to their mean frequency.

(c) Their phases are equal and opposite *wrt* $(wt + \text{constant})$, and their amplitudes are equal at all times, (hence Conformable).

It will be shown in the next section that, if two signals are conformable, coherent addition of a carrier of a suitable amplitude, frequency and phase followed by coherent demodulation will produce complete correlation between the two sideband contributions to the baseband signal. In a sense this is equivalent to the well known 'Low Pass Analogue' concept.

2.8 Coherent Demodulation of Conformable Signals

Take a simple case: consider the sum of two conformable signals given by:

$$e' = \sqrt{2C} \left[\frac{\theta}{2} \sin (w + p)t - \frac{\theta}{2} \sin (w - p)t \right] \tag{2.17}$$

Add a suitably phased carrier:

$$e'_c = \sqrt{2C} \sin wt \tag{2.18}$$

The resultant signal is given by:

$$e = \sqrt{2C} \left[\sin wt + \frac{\theta}{2} \sin (w + p)t - \frac{\theta}{2} \sin (w - p)t \right] \tag{2.19}$$

(This is merely a phase modulated signal see (2.11))

Coherently demodulate the signal by beating it in a phase detector with a 'clean' correctly phased carrier $\sqrt{H} \cos wt$. Pass the output signal from the phase detector through a low pass or baseband filter which will reject all terms involving the carrier frequency $(w/2\pi)$ and its harmonics. The rejected carrier frequency terms will be dropped from any expressions derived during the analysis at any convenient stage of the argument. The phase detector may be considered to be a perfect multiplier. Writing e_b for the baseband output voltage and multiplying (2.19) by $\sqrt{H} \cos wt$ we get:

$$e_b = \sqrt{2CH} \left[\sin wt \cos wt + \frac{\theta}{2} \sin (w + p)t \cos wt \right.$$

$$\left. - \frac{\theta}{2} \sin (w - p)t \cos wt \right]$$

Now $\sin A \cos A = \frac{1}{2}(\sin 2A + \sin 0) = \frac{1}{2} \sin 2A$

$$\therefore \quad e_b = \sqrt{2CH} \left\{ \frac{\sin 2wt}{2} + \frac{\theta}{2} [\sin wt \cos pt \cos wt + \cos^2 wt \sin pt] \right.$$

$$\left. - \frac{\theta}{2} [\sin wt \cos wt \cos pt - \cos^2 wt \sin pt] \right\} \tag{2.20}$$

$$\therefore \quad e_b = \sqrt{2CH} \left[\frac{\theta}{2} \cos^2 wt \sin pt + \frac{\theta}{2} \cos^2 wt \sin pt \right] \tag{2.21}$$

It will be noted that, of the two terms inside the bracket, the first is the contribution of the upper sideband and the second that of the lower sideband. These two contributions are in phase and therefore add on a *voltage* basis. Thus the baseband power due to two conformable sidebands will be 4 times that due to a single sideband.

Now

$$\cos^2 A = \cos A \cos A = \tfrac{1}{2}(\cos 2A + \cos 0) \tag{2.22}$$

$$\therefore \quad e_b = \sqrt{2CH} \, \frac{\theta}{2} \frac{2}{2} (\cos 2wt + 1) \sin pt \tag{2.23}$$

$$\therefore \quad e_b = \sqrt{2CH} \, \frac{\theta}{2} \frac{2}{2} \sin pt \tag{2.24}$$

This is a baseband sinusoid with contributions:

$$e_{bssB} = \sqrt{2CH} \, \frac{\theta}{2} \frac{1}{2} \sin pt \tag{2.25}$$

from each of the two sidebands which are in phase (fully correlated) and add directly to give (2.24). From (2.24)

the peak amplitude of the DSB produced baseband signal is $\sqrt{2CH} \, \dfrac{\theta}{2}$

and its rms amplitude is $\sqrt{CH} \, \dfrac{\theta}{2}$ $\qquad\qquad\qquad\qquad\qquad\qquad$ (2.26)

Without loss of generality we may work in voltage squared or normalised power. This merely implies an impedance level of 1 ohm. Unless otherwise stated, when the word 'power' is used in this book, it will refer to normalised power.

From (2.26) the baseband power due to conformable DSB signals (P_b DSB) is given by:

$$P_b \text{ DSB} = CH \, \frac{\theta^2}{4} \tag{2.27}$$

An analysis of the coherent demodulation of an AM signal, such as that given by (2.2), by beating it with a clean carrier ($\sqrt{H} \sin wt$) may be carried out in a similar way. The result shows that the two AM sidebands again add on an amplitude basis (rather than on a power basis) at baseband. However the carrier power is effectively modified by the fact that it is converted to DC, and the DC power is the square of the *peak* carrier amplitude $\sqrt{2C}$ rather than $1/\sqrt{2}$ of the peak value.

Thus the ratio of 'carrier' power to modulation power at baseband is the same as the ratio of the carrier to the sum of the two sideband powers prior to demodulation.

CHAPTER 3

THE RELATIONSHIP BETWEEN PHASE JITTER AND NOISE DENSITY

3.1 Sinusoidal Phase Jitter

Before considering the various relationships between noise power and phase jitter it is useful to review the corresponding relationships in the case of modulation by a simple sinusoidal (and hence deterministic) waveform.

Using equation (2.11) we may deduce the sideband characteristics from the modulating waveform: alternatively we may know the characteristics of the carrier and superposed sidebands and wish to calculate the characteristics of the equivalent modulating waveform. In so far as we may represent superposed noise by simple sinusoids, the argument may then be extended to determine relationships between phase jitter and noise density.

Assume that an otherwise clean carrier has a sinusoidal phase jitter with a peak deviation of θ radians at a rate f_m. That is θ is the peak modulation index.

Assume that $\theta < 0.1$ radians.

From equation (2.11) the peak amplitude of each resultant sideband at angular frequencies $(w + p)$ and $(w - p)$, where $p = 2\pi f_m$ is given by:

$$V = \sqrt{2C}\left(\frac{\theta}{2}\right)$$

$$\therefore \quad V_{rms} = \sqrt{C}\frac{\theta}{2}$$

$\therefore$ power in one sideband is given by:

$$P_{SSB} = \frac{C\theta^2}{4} \tag{3.1}$$

The sidebands are at different frequencies and thus orthogonal, hence total power in both sidebands is given by:

$$P_{DSB} = \frac{C\theta^2}{2} \tag{3.2}$$

Due to the small value of the modulation index, and the resultant linearity of the modulation process, no amplitude modulation (AM) will be produced. Thus, the sideband power given by equations (3.1) and (3.2) is purely phase modulation power. It should be noted also that the two sidebands are of equal amplitude and are conformable.

3.2 The Representation of Narrow Band Noise

Remembering the experiment described in Section 1.4 it is apparent that we may represent narrow band noise by a sinusoid as long as we observe essential time and bandwidth relationships. The noise waveform will only remain approximately sinusoidal for a period $(1/B)$ seconds where B is the RF bandwidth under consideration.

Consider a single noise sideband in a 1 Hz bandwidth at an angular frequency $(w + p)$. This may be adequately represented by a sinusoid for periods up to approximately 1 second. The amplitude of this sinusoid will vary from one second to another (as will its reference phase). Consider the 1 Hz bandwidth noise during a specific second which is selected at random.

The noise power during this second may be very different from the long term mean value which will be the noise density N_0 (see 1 Figure 1). Nevertheless (assuming an impedance level of 1 ohm, which does not affect the generality of the argument) we may take the square root of the actual power during this specific second and regard the result as the rms value of the voltage of the sinusoid during that second. The equivalent 'peak' voltage will be obtained by multiplying by $\sqrt{2}$.

It is important for many, but not all, purposes to bear clearly in mind the distinction between the rms value of the waveform for any specific second as just described, and the long term mean rms value which is $\sqrt{N_0}$.

Thus an equation for a noise signal in 1 Hz bandwidth at an angular frequency $(w + p)$ may be written:

$$v_n(t) = \sqrt{2n_0} \sin\left[(w + p)t + \psi_0\right] \tag{3.3}$$

where $v_n(t)$ is the noise voltage, n_0 is the noise power during the specific second chosen and ψ_0 is the relative 'phase' angle during this specific second.

When our interest is in the long term mean power contribution, which is usually the case, then we may write:

$$V_n(t) = \sqrt{2N_0} \sin\left[(w + p)t + \psi\right] \tag{3.4}$$

where $V_n(t)$ is the instantaneous value of the sinusoid corresponding to the well defined long term rms value $\sqrt{N_0}$ and ψ is an indeterminate, and ultimately irrelevant reference phase.

This is the basic mathematical representation of a noise waveform and its use will form our main concern. N_0 is the statistical mean value of the power density and does not represent the value which might be measured during any specific second.

In the case of equation (3.3) n_0 and ψ_0 are statistical variables which vary from second to second. n_0 has a Rayleigh distribution and ψ_0 a uniform distribution over $\pm\pi$.

For the sake of completeness it is worth mentioning that alternative expressions for the noise voltage may be written in terms of two suitably chosen noise components in quadrature. This is treated from a rather more rigorous point of view in Appendix III.

For most purposes equation (3.4) is much more useful than equation (3.3), which contains an unpredictable term 'n_0', which varies from second to second. There are some cases, however, such as the superposition of DSB noise on a carrier, where the concept 'n_0' is of assistance in the initial stages of analysis. The fact that, during any specific second, although the values are unknown, the value for the upper sideband will almost always be different from that for the lower sideband, is very important.

3.3 Phase Jitter due to Superposed Single Sideband Noise

Consider a perfectly 'clean' carrier, of power level C and angular frequency w, with a superposed single noise sideband in a 1 Hz bandwidth at an angular frequency $(w + p)$. See 3 Figure 1.

We might approach the analysis of this situation in either of two ways. Firstly we might take the long term mean value of the sideband power (i.e. N_0), find the 'peak' value $\sqrt{2N_0}$ corresponding to the long term rms value and arrive at equation (3.4) to represent the noise sideband. The addition of a carrier term $(\sqrt{2C} \sin wt)$, would result in a mathematical expression for the carrier and noise sideband as follows:

$$V(t) = \sqrt{2C} \sin wt + \sqrt{2N_0} \sin [(w + p)t + \psi] \tag{3.5}$$

Superposed Single Sideband Noise   **3 Figure 1**

The generality would not be increased by the addition of an arbitrary phase angle to the carrier. It must be remembered that ψ is considered to be a constant during any one second, but that its value is not constant from second to second.

Secondly we might consider the performance during a single arbitrarily chosen second of time. As the noise is statistical in nature, the noise power in the sideband during a specific second, cannot be predicted; although it must have a value (which we have called n_0). Adding equation (3.3), for the noise sideband, to the carrier $\sqrt{2C} \sin wt$, gives an equation for the noise voltage during this particular second:

$$v(t) = \sqrt{2C} \sin wt + \sqrt{2n_0} \sin [(w + p)t + \psi_0] \tag{3.6}$$

In many respects (3.6) is an inconvenient equation to use as it contains a term 'n_0' which cannot be predicted. For this reason the long term noise power density will be used in the analysis. It is however important to note that n_0 may be substituted for N_0 in every step in the equations which follow, including the final results. This shows that, in this case (that of superposed *single* sideband noise), the results are true not only for the long term mean power situation, but also during each and every second.

Re-writing equation (3.5)

$$V(t) = \sqrt{2C} \sin wt + \sqrt{2N_0} \sin [(w + p)t + \psi] \tag{3.5}$$

$$V(t) = \sqrt{2C} \sin wt + \sqrt{2N_0}[\sin wt \cos (pt + \psi) + \cos wt \sin (pt + \psi)]$$

$$V(t) = \sin wt[\sqrt{2C} + \sqrt{2N_0} \cos (pt + \psi)]$$

$$+ \cos wt \,[\sqrt{2N_0} \sin (pt + \psi)] \tag{3.7}$$

Equation (3.7) is of the form:

$$A \sin wt + B \cos wt$$

The resultant amplitude is given by:

$$R^2 = A^2 + B^2$$

The phase angle with respect to $\sin wt$ is given by

$$\tan \theta = \frac{B}{A}$$

$$\therefore \quad \tan \theta(t) = \frac{\sqrt{2N_0} \sin (pt + \psi)}{\sqrt{2C} + \sqrt{2N_0} \cos (pt + \psi)} \tag{3.8}$$

Now $\cos (pt + \psi)$ cannot exceed 1 and $\sqrt{2C}$ is very much greater than $\sqrt{2N_0}$.

$$\therefore \quad \tan \theta(t) = \sqrt{\frac{N_0}{C}} \sin (pt + \psi) \tag{3.9}$$

$$\text{For} \quad \theta \ll 1, \quad \tan \theta \simeq \theta \tag{3.10}$$

$$\therefore \quad \theta(t) = \sqrt{\frac{N_0}{C}} \sin (pt + \psi) \tag{3.11}$$

This represents a sinusoidal variation in phase where $\sqrt{N_0/C}$ is equal to the peak value (θ) of the sinusoid.

Thus the modulation index θ is given by:

$$\theta \simeq \sqrt{\frac{N_0}{C}} \tag{3.12}$$

Let the rms modulation index be denoted by ϕ

$$\therefore \quad \phi = \frac{\theta}{\sqrt{2}} \simeq \sqrt{\frac{N_0}{2C}} \tag{3.13}$$

For simplicity the approximation sign in equation (3.13) will be replaced by a sign of equality. Thus the equations which follow are only approximately true as long as the conditions of equations (2.9) and (3.10) are satisfied.

Equation (3.5) is illustrated in 3 Figure 2 (with exaggerated noise amplitudes).

The resultant AM modulation index may be calculated from the phasor diagram. It is:

$$M' = \sqrt{\frac{N_0}{C}} \tag{3.14}$$

From the phasor diagram it is also apparent that the AM modulating waveform leads the PM modulating waveform by $\pi/2$ radians. Thus

$$\theta(t) = \sqrt{\frac{N_0}{C}} \sin (pt + \psi) \tag{3.15.1}$$

and

$$M'(t) = \sqrt{\frac{N_0}{C}} \sin \left(pt + \frac{\pi}{2} + \psi \right) \tag{3.15.2}$$

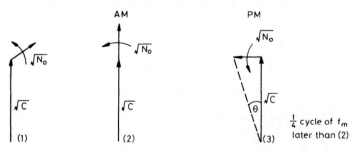

Phase Jitter due to SSB Noise   **3 Figure 2**

Substituting (3.15.1) and (3.15.2) for (2.3.1) and (2.1.1) and remembering that $\cos(A \pm \pi/2) = \mp \sin A$, we find from (2.11.1) and (2.2.1) that the various sidebands are as follows:

$$\text{Upper PM sideband} = + \sqrt{2C}\, \frac{1}{2} \sqrt{\frac{N_0}{C}} \sin\left[(w + p)t + \psi\right]$$

$$= + \sqrt{\frac{N_0}{2}} \sin\left[(w + p)t + \psi\right] \tag{3.16.1}$$

$$\text{Lower PM sideband} = - \sqrt{\frac{N_0}{2}} \sin\left[(w - p)t - \psi\right] \tag{3.16.2}$$

$$\text{Upper AM sideband} = - \sqrt{\frac{N_0}{2}} \cos\left(wt + pt + \frac{\pi}{2} + \psi\right)$$

$$= + \sqrt{\frac{N_0}{2}} \sin\left[(w + p)t + \psi\right] \tag{3.16.3}$$

$$\text{Lower AM sideband} = + \sqrt{\frac{N_0}{C}} \cos\left(wt - pt - \frac{\pi}{2} - \psi\right)$$

$$= + \sqrt{\frac{N_0}{2}} \sin\left[(w - p)t - \psi\right] \tag{3.16.4}$$

It is illuminating initially to pursue the following argument. The peak absolute magnitude of each of the four sidebands is $\sqrt{N_0/2}$.

Thus the rms amplitude of each sideband $= \frac{1}{2}\sqrt{N_0}$.

$\therefore$ the power in each sideband $= \frac{1}{4}N_0$.

The original power of the single noise sideband was N_0, and this has been equally shared between the four sidebands. The two PM sidebands together take half the total power. Thus half the total power is converted to AM and half to PM. The phase modulation index $\sqrt{N_0/C}$ is in fact the peak phase jitter and this will be unchanged by the simultaneous presence of AM.

It should be noted that these statements are not only statistically true, but are also true for every second if n_0 is substituted for N_0.

Closer examination of equations (3.16) shows that the lower sidebands cancel (3.16.2) and (3.16.4) as is to be expected, as we started only with an upper sideband. Thus the simultaneous presence of AM with the PM in which we are primarily interested, cancels the lower PM sideband without affecting the phase jitter. Thus if the AM were removed, say by a perfect limiter, which absorbs the AM power, pure PM would be left with a total double sideband power of $N_0/2$, consisting of an upper and a lower sideband each of power $N_0/4$.

It remains to consider the power balance if a limiter is not provided and the two lower sidebands are suppressed by their mutual cancellation. The two upper sidebands (3.16.1) and (3.16.3) are at the same angular frequency and are in phase. The resultant is obtained by adding (3.16.1) and (3.16.3) which gives:

$$\text{Combined upper sideband} = +2 \sqrt{\frac{N_0}{2}} \sin \left[(w + p)t + \psi\right] \qquad (3.17.1)$$

$$\text{Peak amplitude of upper sideband} = \sqrt{2N_0} \qquad (3.17.2)$$

$$\text{rms amplitude of upper sideband} = \sqrt{N_0}$$

$$\therefore \quad \text{power in upper sideband} = N_0$$

This is identical to the SSB power with which we started.

Equation (3.17.1) is in fact completely identical to the superposed upper sideband shown in (3.5) from which we started.

Thus, except for the 'Logical Positivist', there is a significant sense in which the lower sideband, although suppressed, is potentially present. Initially people holding such views are likely to dismiss the concept of single sideband phase noise as meaningless and to state that only phase jitter can be considered. We shall use the concept of single sideband phase noise power, and single sideband phase noise power density, as it is of great theoretical, and even great practical value. The fact that it is more easily measured than phase jitter (using a limiter and a spectrum analyser) might, if emphasised, result in a belated recognition of the concept by the 'Logical Positivist' or 'Verificationalist'. To distinguish between noise power density and 'phase noise' power density we shall use the symbol N_{op} to refer to 'phase noise' power density. As only half the noise power is converted, or potentially converted, into DSB PM:

$$N_{op} = \frac{N_0}{4} \qquad (3.18.1)$$

Similarly

$$N_{oa} = \frac{N_0}{4} \qquad (3.18.2)$$

where N_{oa} is the AM noise power density.

3.4 Simplified Treatment of Phase Jitter due to Superposed Double Sideband White Noise

Consider a carrier of power C watts and angular frequency w, together with two 1 Hz noise sidebands at angular frequencies $(w + p)$ and $(w - p)$. As we are considering pure additive noise the two noise sidebands will be incoherent. The total noise power will be $2N_0$. The 'peak' noise voltage will be $\sqrt{2}\sqrt{2N_0}$. Due to the linearity of the modulation process for low modulation indices

($\theta < 0.1$ rads) the phase jitter due to the two sidebands is equivalent in amplitude to that produced by a single sinusoidal modulating signal of amplitude $2\sqrt{N_0}$.

However it differs from the case of a carrier and an additive single sideband of peak amplitude $2\sqrt{N_0}$ in that real, rather than potential, AM and PM sidebands will be present on both sides of the carrier.

Because the originally superposed upper and lower sidebands are due to (white) Gaussian noise, they will be incoherent. That is, although their long term rms values will be equal, their values during any particular one second period are most unlikely to be equal. Similarly their 'phases' will be uniformly distributed and their instantaneous phase difference may have any value from 0 to 2π radians. The analysis is, of course, carried out using the sinusoidal representation of narrow band noise where the 'peak' values of the sinusoids are $\sqrt{2}$ times the long term rms values and the phase constants in the two sideband equations are different.

One further very important point which will be mentioned here, but fully explained later (Section 3.5), is that the 'phase noise' sidebands will be conformable although they are due to the superposition of two uncorrelated noise sidebands and a pure carrier. Similarly the pure 'AM noise' sidebands will be conformable with one another. Taking the case of pure 'phase noise', "conformable" means that the phase noise sidebands at $\pm f_m$ will instantaneously have equal amplitudes and satisfy the other requirements defined in Section 2.7. Similarly, and simultaneously the 'AM noise' sidebands will satisfy equation (2.2). It is the interaction of the AM and 'phase noise' sidebands which produces resultant overall sidebands which are non-conformable.

That the 'phase noise' sidebands must be conformable can be most simply understood by remembering that the superposed noise at frequencies $\pm f_m$ produces phase jitter of the carrier at a rate f_m. The analyses of Sections 2.2 and 2.4 are then directly applicable.

We will now carry out an initial simple calculation of the phase jitter resulting from superposition of DSB noise from two sidebands 1 Hz wide at $f_0 \pm f_m$ (where f_0 is the carrier frequency $= w/2\pi$ and $f_m = p/2\pi$).

Substituting $2N_0$ for N_0 in (3.12) we have:

$$\theta = \sqrt{\frac{2N_0}{C}} \tag{3.19}$$

$$\therefore \quad \phi = \sqrt{\frac{N_0}{C}} \text{ rads rms per } \sqrt{\text{Hz}} \tag{3.20}$$

Now $N_0 = 2N_{op}$

$$\therefore \quad \phi = \sqrt{\frac{2N_{op}}{C}} \text{ rads rms per } \sqrt{\text{Hz}} \tag{3.21}$$

As ϕ is the rms phase deviation per $\sqrt{Hz}$ (of DSB noise) it is convenient hereafter to write it ϕ_0 by analogy with N_0 which denotes a density.

$$\phi_0 = \sqrt{\frac{N_0}{C}} \text{ rads rms per } \sqrt{Hz} \tag{3.22.1}$$

and

$$\phi_0 = \sqrt{\frac{2N_{op}}{C}} \text{ rads rms per } \sqrt{Hz} \tag{3.22.2}$$

$$\text{and } \overline{\phi_0^2} = \frac{2N_{op}}{C} \text{ rads}^2 \text{ per Hz} \tag{3.22.3}$$

where $\overline{\phi_0^2}$ is the phase jitter variance.

Equations (3.22) express very important relationships. Equation (3.22.3) states that the mean square phase jitter density is equal to the DSB phase noise density to carrier ratio.

For completeness it is interesting to correlate (3.22.3) and (3.1). From (3.1):

$$\frac{P_{SSB}}{C} = \frac{\theta^2}{4}$$

Now

$$\theta = \sqrt{2}\,\phi$$

$$\therefore \quad \frac{P_{SSB}}{C} = \frac{2\phi^2}{4} = \frac{\phi^2}{2}$$

The two sidebands are at different frequencies.

$$\therefore \quad P_{DSB} = 2P_{SSB}$$

$$\therefore \quad \frac{P_{DSB}}{C} = \phi^2$$

This is analogous in form to (3.22.3) apart from the fact that it refers to modulation by a sinusoid rather than by narrow band noise.

3.5 Fuller Treatment of DSB Superposed White Noise

3.5.1 Phase Modulation Index
Consider a 'clean' carrier of angular frequency w with superposed white noise, and in particular, two 1 Hz white noise sidebands equally spaced on either side of the carrier at angular frequencies $(w \pm p)$ or frequencies $(f_0 \pm f_m)$.

Although, during any one second period, the two sidebands will have different amplitudes their long term rms values will be equal as the noise is white. Confusion is most easily avoided if we use the voltage values for each of the sidebands which are the current values for one specific, arbitrarily chosen, second of time. For this second, the values remain constant.

Let n_{o1} be the power in the upper sideband for this second.

Let n_{o2} be the power in the lower sideband for the same second.

The mathematical expressions for the sinusoidal representation of each of the two noise sidebands must contain different phase constants ψ_1 and ψ_2 for the upper and lower sidebands respectively.

We may now write an equation for the carrier with superposed noise sidebands as follows:

$$v(t) = \sqrt{2C}\,\sin wt + \sqrt{2n_{o1}}\,\sin\left[(w+p)t + \psi_1\right]$$
$$+ \sqrt{2n_{o2}}\,\sin\left[(w-p)t + \psi_2\right] \tag{3.23}$$

$$\therefore \quad v(t) = \sqrt{2C}\,\sin wt + \sqrt{2n_{o1}}\,\sin wt \cos(pt + \psi_1)$$
$$+ \sqrt{2n_{o1}}\,\cos wt \sin(pt + \psi_1)$$
$$+ \sqrt{2n_{o2}}\,\sin wt \cos(pt - \psi_2)$$
$$- \sqrt{2n_{o2}}\,\cos wt \sin(pt - \psi_2)$$

$$\therefore \quad v(t) = \sin wt\left[\sqrt{2C} + \sqrt{2n_{o1}}\,\cos(pt + \psi_1) + \sqrt{2n_{o2}}\,\cos(pt - \psi_2)\right]$$
$$+ \cos wt\left[\sqrt{2n_{o1}}\,\sin(pt + \psi_1) - \sqrt{2n_{o2}}\,\sin(pt - \psi_2)\right]$$

This is of the form

$$A \sin wt + B \cos wt$$

(see 3 Figure 3; where for clarity, B has been greatly exaggerated)

$$A = \sqrt{2C} + \sqrt{2n_{o1}}\,\cos(pt + \psi_1) + \sqrt{2n_{o2}}\,\cos(pt - \psi_2)$$

and

$$B = \sqrt{2n_{o1}}\,\sin(pt + \psi_1) - \sqrt{2n_{o2}}\,\sin(pt - \psi_2)$$

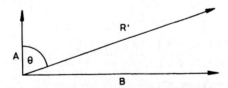

3 Figure 3

Phase Relationships for Carrier with
Superposed DSB Noise

Tan $\theta = B/A$ where θ is the resultant phase with respect to the carrier.

$$\therefore \quad \tan\theta = \frac{\sqrt{2n_{o1}}\,\sin(pt + \psi_1) - \sqrt{2n_{o2}}\,\sin(pt - \psi_2)}{\sqrt{2C} + \sqrt{2n_{o1}}\,\cos(pt + \psi_1) + \sqrt{2n_{o2}}\,\cos(pt - \psi_2)}$$

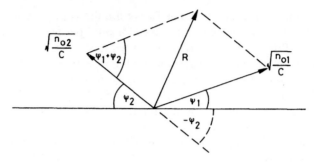

Baseband Phase Relationships   **3 Figure 4**

Now $\cos (pt \mp \psi)$ cannot exceed 1 and $\sqrt{2C} \gg \sqrt{2n_{o1}}$ or $\sqrt{2n_{o2}}$

$$\therefore \quad \tan \theta \simeq \frac{\sqrt{2n_{o1}} \sin (pt + \psi_1) - \sqrt{2n_{o2}} \sin (pt - \psi_2)}{\sqrt{2C}}$$

For small values of $\tan \theta$, $\tan \theta \simeq \theta$

$$\therefore \quad \theta(t) = \sqrt{\frac{n_{o1}}{C}} \sin (pt + \psi_1) + \sqrt{\frac{n_{o2}}{C}} \sin (pt - \psi_2 + \pi) \qquad (3.24)$$

$\theta(t)$ is the resultant phase modulation index as it represents the instantaneous phase of the combined signal due to carrier and sidebands.

The phase relationships between the baseband modulation terms due to the two sidebands are illustrated in 3 Figure 4.

Put

$$\theta(t) = R \sin (pt + \psi)$$

Remember the cosine rule

$$a^2 = b^2 + c^2 - 2bc \cos A$$

$\therefore$ from 3 Figure 4

$$R^2 = \frac{n_{o1}}{C} + \frac{n_{o2}}{C} - 2 \sqrt{\frac{n_{o1}}{C}} \sqrt{\frac{n_{o2}}{C}} \cos (\psi_1 + \psi_2) \qquad (3.25)$$

Now ψ_1 and ψ_2 are random phases with a uniform distribution over the range $0-2\pi$. $\cos (\psi_1 + \psi_2)$ is therefore uniformly distributed over the range -1 to $+1$. Its long term mean value will be zero.

The long term mean value of R^2 (i.e. $\overline{R^2}$) is determined by the sum of the long term mean values of:

$$\frac{n_{o1}}{C}, \quad \frac{n_{o2}}{C} \quad \text{and} \quad -2 \sqrt{\frac{n_{o1}}{C}} \sqrt{\frac{n_{o2}}{C}} \cos (\psi_1 + \psi_2).$$

The variance of the third term is zero and the variances of the first two terms are both equal to N_0/C.

$$\therefore \quad \overline{R^2} = \frac{2N_0}{C} \tag{3.26}$$

$$\therefore \quad \theta(t) = \sqrt{\frac{2N_0}{C}} \sin (pt + \psi) \tag{3.27}$$

This checks with equation (3.19).

3.5.2 Amplitude Modulation Index

At times t' when $\sin (pt' + \psi) = 0$ (see 3.27) $\theta(t) = 0$. At such times the modulation waveform is in phase with the carrier, as may be realised by remembering that θ is defined with respect to the carrier (see 3 Figure 3).

Thus amplitude modulation is also produced with a peak amplitude $\sqrt{2N_0/C}$. This is the long term mean value of the peak amplitude. Thus the amplitude modulation index is given by:

$$M' = \sqrt{\frac{2N_0}{C}} \tag{3.28}$$

which is identical to the phase modulation index.

3.5.3 Sideband Power Relationships

From equations (2.11) and (2.2) the peak amplitude of the PM sidebands (for small modulation index) is equal to $\sqrt{2C}\,\theta/2$ and the peak amplitude of the AM sideband is $\sqrt{2C}\,M'/2$. Thus the amplitude of each of the four sidebands (upper and lower PM and upper and lower AM) is

$$\tfrac{1}{2}\sqrt{2C}\,\sqrt{\frac{2N_0}{C}} = \sqrt{N_0}\,.$$

Thus the rms value of each sideband $= \sqrt{N_0/2}$ and the power in each sideband $= N_0/2$.

Of the total superposed DSB noise power $2N_0$, half has been converted into AM and half into PM: a power $N_0/2$ appearing in each of the four sidebands.

Thus the phase noise power density (N_{op}) and the amplitude noise power density (N_{oa}) due to superposed DSB white noise of power density N_0 are given by:

$$N_{op} = \frac{N_0}{2} \tag{3.29.1}$$

$$N_{oa} = \frac{N_0}{2} \tag{3.29.2}$$

Compare with equations (3.18) which were derived for the case of superposed SSB noise.

It must be remembered that these relationships are only true on a long term statistical basis. Referring to (3.24) and (3.25), during any particular second, n_{o1} is unlikely to be equal to n_{o2} and $\cos(\psi_1 + \psi_2)$ is unlikely to be simultaneously zero. There is a small probability that during a specific second *all* the superposed noise would be converted to PM or all to AM. In this respect the characteristics for superposed DSB noise are different from those due to SSB superposed noise. In that case the AM and PM sideband power were equal during every second and not merely on a statistical basis. In that case the AM and PM sideband powers, although always equal to one another, varied from second to second.

3.5.4 Phase Jitter
From equation (3.27)

$$\theta = \sqrt{\frac{2N_0}{C}}$$

θ is the peak phase jitter. The rms phase jitter is given by:

$$\phi = \sqrt{\frac{N_0}{C}}$$

To show that this is the phase jitter due to DSB superposed white noise in 1 Hz wide bands equispaced above and below the carrier ϕ will be written ϕ_0.

$$\therefore \quad \phi_0 = \sqrt{\frac{N_0}{C}} \tag{3.30.1}$$

But (3.29.1) $N_0 = 2N_{op}$

$$\therefore \quad \phi_0 = \sqrt{\frac{2N_{op}}{C}} \tag{3.30.2}$$

$$\text{and} \quad \overline{\phi_0^2} = \frac{2N_{op}}{C} \tag{3.30.3}$$

Equations (3.30) are of fundamental importance. They confirm equations (3.22) derived using a simpler treatment in Section 3.4.

It should be noted that (3.30.3) states that the phase jitter variance in a 1 Hz bandwidth is equal to the double sideband phase noise density to carrier ratio. The fact that this is only true on a statistical basis is emphasised by the bar over ϕ_0^2.

3.5.5 *Phasor Representation*

The phasor representation of a carrier plus DSB noise is helpful in visualising the relationships. This situation using the relevant values for a specific second is given by (3.23), which is repeated below for convenience.

$$v(t) = \sqrt{2C} \sin wt + \sqrt{2n_{o1}} \sin [(w + p)t + \psi_1]$$
$$+ \sqrt{2n_{o2}} \sin [(w - p)t + \psi_2]$$

If we wish to represent the long term average situation on a phasor diagram we may first substitute the long term mean value N_0 for n_{o1} and n_{o2}. It remains to consider how we should represent the phase constants ψ_1 and ψ_2. The long term mean value of the last term in equation (3.25)

$$2 \sqrt{\frac{n_{o1}}{C}} \sqrt{\frac{n_{o2}}{C}} \cos (\psi_1 + \psi_2)$$

is zero because the two noise sidebands are incoherent with one another. As the mean value of $\cos (\psi_1 + \psi_2) = 0$ the actual values of ψ_1 and ψ_2 are ultimately irrelevant when our final interest is in the mean power. If we put

$$\psi_2 = -\psi_1 \pm \frac{\pi}{2}$$

we satisfy the requirement that

$$\overline{\cos (\psi_1 + \psi_2)} = 0$$

Thus in a useful sense we may write

$$V(t) = \sqrt{2C} \sin wt + \sqrt{2N_0} \sin [(w + p)t + \psi_1]$$
$$+ \sqrt{2N_0} \sin \left[(w - p)t - \psi_1 \pm \frac{\pi}{2} \right] \qquad (3.31)$$

A series of phasor diagrams for this case is given in 3 Figure 5. Due to our assumptions about the phase constants these diagrams are somewhat arbitrary but they do help in visualising a complicated situation whilst implying the right power relationships. The phasor diagrams for a particular second might of course look very different; assuming the amplitudes and phase constants for that particular second to be known.

At a time half way between diagrams 1 and 2 the two noise phasors will come into phase with one another giving a resultant of $2\sqrt{2N_0}$. Projecting this on to the horizontal axis gives a component in quadrature with the carrier of magnitude

$$\frac{1}{\sqrt{2}} 2\sqrt{2N_0} = 2\sqrt{N_0}.$$

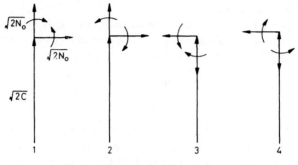

Phasor Diagrams for Carrier with   **3 Figure 5**
Superposed DSB Noise

The large magnitude may be puzzling if the warning given in Section 2.6 above is not remembered. Because the carrier on a phasor diagram is represented as stationary, the square of the sum of the two sideband amplitudes represents the peak baseband power due to both sidebands.

This is twice as great as the sum of the two sideband peak powers.

We may verify that the diagram is correct by evaluating the phase modulation index.

From 3 Figure 6:

$$\tan \theta \simeq \theta = \sqrt{\frac{2N_0}{C}}$$

This value checks with that given by equation (3.27).

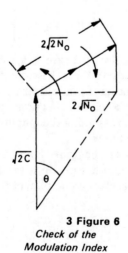

3 Figure 6
Check of the
Modulation Index

3.5.6 Real and Potential Sidebands

When DSB noise (of 1 Hz bandwidth) is superposed upon a carrier the phase modulation index is given by equation (3.24) i.e.

$$\theta(t) = \sqrt{\frac{n_{o1}}{C}} \sin(pt + \psi_1) - \sqrt{\frac{n_{o2}}{C}} \sin(pt - \psi_2)$$

$\theta(t)$ is the algebraic sum of two terms, one due to each of the superposed noise sidebands. For reasonable carrier to noise ratios, which includes most cases of practical interest, we may assume linearity and invoke the Superposition Theorem considering the effects of upper and lower superposed noise sidebands separately when convenient, summing the two effects at the end of the discussion.

Due to Upper Sideband Superposed Noise
In this case the modulation indices are:

$$\text{PM:} \quad \theta_U(t) = \sqrt{\frac{n_{o1}}{C}} \sin(pt + \psi_1) \tag{3.32.1}$$

$$\text{AM:} \quad M'_U(t) = \sqrt{\frac{n_{o1}}{C}} \sin\left(pt + \psi_1 + \frac{\pi}{2}\right) \tag{3.32.2}$$

The relative phases of these two modulation indices are illustrated in 3 Figure 2.

Due to Lower Sideband Superposed Noise
In this case the phase modulation index is:

$$\text{PM:} \quad \theta_L(t) = -\sqrt{\frac{n_{o2}}{C}} \sin(pt - \psi_2) \tag{3.32.3}$$

The amplitude modulation index will lag the PM index by $\pi/2$ radians (see 3 Figure 7)

$$\therefore \quad M'_L(t) = -\sqrt{\frac{n_{o2}}{C}} \sin\left(pt - \psi_2 - \frac{\pi}{2}\right) \tag{3.32.4}$$

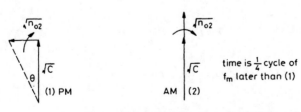

time is $\frac{1}{4}$ cycle of f_m later than (1)

Phasor Diagram for Superposed   3 Figure 7
Lower Sideband

The phase and amplitude modulation indices are given by equation (3.32). If these values are compared with equations (2.3.1) and (2.1.1), the resultant PM and AM sidebands may be written down by inspection using equations (2.11.1) and (2.2.1).

PM and AM Sidebands due to Upper Superposed Sideband
The resultant sidebands will be distinguished by using UPM for the upper PM sideband and LPM for the lower sideband. Similarly the AM sidebands will be distinguished by using UAM and LAM.

$$\text{UPM} = \tfrac{1}{2}\sqrt{2C}\,\sqrt{\frac{n_{o1}}{C}}\,\sin\left[(w+p)t+\psi_1\right]$$

$$\therefore\quad \text{UPM} = \sqrt{\frac{n_{o1}}{2}}\,\sin\left[(w+p)t+\psi_1\right] \tag{3.33.1}$$

$$\text{LPM} = -\sqrt{\frac{n_{o1}}{2}}\,\sin\left[(w-p)t-\psi_1\right] \tag{3.33.2}$$

$$\text{UAM} = -\sqrt{\frac{n_{o1}}{2}}\,\cos\left[(w+p)t+\psi_1+\frac{\pi}{2}\right]$$

$$\therefore\quad \text{UAM} = +\sqrt{\frac{n_{o1}}{2}}\,\sin\left[(w+p)t+\psi_1\right] \tag{3.33.3}$$

$$\text{LAM} = \sqrt{\frac{n_{o1}}{2}}\,\cos\left[(w-p)t-\psi_1-\frac{\pi}{2}\right]$$

$$\therefore\quad \text{LAM} = \sqrt{\frac{n_{o1}}{2}}\,\sin\left[(w-p)t-\psi_1\right] \tag{3.33.4}$$

PM and AM Sidebands due to Lower Superposed Sideband

$$\text{UPM} = -\sqrt{\frac{n_{o2}}{2}}\,\sin\left[(w+p)t-\psi_2\right] \tag{3.33.5}$$

$$\text{LPM} = \sqrt{\frac{n_{o2}}{2}}\,\sin\left[(w-p)t+\psi_2\right] \tag{3.33.6}$$

$$\text{UAM} = \sqrt{\frac{n_{o2}}{2}}\,\cos\left[(w+p)t-\psi_2-\frac{\pi}{2}\right]$$

$$\therefore\quad \text{UAM} = \sqrt{\frac{n_{o2}}{2}}\,\sin\left[(w+p)t-\psi_2\right] \tag{3.33.7}$$

$$\mathrm{LAM} = -\sqrt{\frac{n_{o2}}{2}}\, \cos\left[(w - p)t + \psi_2 + \frac{\pi}{2}\right]$$

$$\therefore \quad \mathrm{LAM} = \sqrt{\frac{n_{o2}}{2}}\, \sin\left[(w - p)t + \psi_2\right] \tag{3.33.8}$$

The eight real or potential sidebands are listed in 3 Table 1.
It will be noted that items (2) and (4) cancel, whilst items (1) and (3) add, to give:

$$2\sqrt{\frac{n_{o1}}{2}}\, \sin\left[(w + p)t + \psi_1\right]$$

which represents the upper superposed noise sideband (3.23) from which we started.

Items (5) and (7) cancel, whilst items (6) and (8) add to give:

$$+2\sqrt{\frac{n_{o2}}{2}}\, \sin\left[(w - p)t + \psi_2\right]$$

This is identical to the lower superposed noise sideband (3.23) from which we started.

If the AM is suppressed by a limiter, or becomes negligible due to high order frequency multiplication, then only items 1, 2, 5 and 6 of 3 Table 1 remain.
The combined UPM sideband consists of items 1 + 5 giving:

$$v(t) = +\sqrt{\frac{n_{o1}}{2}}\, \sin\left[(w + p)t + \psi_1\right] - \sqrt{\frac{n_{o2}}{2}}\, \sin\left[(w + p)t - \psi_2\right]$$

3 Table 1   *Noise Components*

	Superposed Noise at $(w + p)$	Superposed Noise at $(w - p)$
Upper PM	(1) $+\sqrt{\dfrac{n_{o1}}{2}}\, \sin\left[(w + p)t + \psi_1\right]$	(5) $-\sqrt{\dfrac{n_{o2}}{2}}\, \sin\left[(w + p)t - \psi_2\right]$
Lower PM	(2) $-\sqrt{\dfrac{n_{o1}}{2}}\, \sin\left[(w - p)t - \psi_1\right]$	(6) $+\sqrt{\dfrac{n_{o2}}{2}}\, \sin\left[(w - p)t + \psi_2\right]$
Upper AM	(3) $+\sqrt{\dfrac{n_{o1}}{2}}\, \sin\left[(w + p)t + \psi_1\right]$	(7) $+\sqrt{\dfrac{n_{o2}}{2}}\, \sin\left[(w + p)t - \psi_2\right]$
Lower AM	(4) $+\sqrt{\dfrac{n_{o1}}{2}}\, \sin\left[(w - p)t - \psi_1\right]$	(8) $+\sqrt{\dfrac{n_{o2}}{2}}\, \sin\left[(w - p)t + \psi_2\right]$

From second to second peak $v(t)$ will vary over the range

$$\left| \sqrt{\frac{n_{o1}}{2}} - \sqrt{\frac{n_{o2}}{2}} \right| \quad \text{to} \quad \left| \sqrt{\frac{n_{o1}}{2}} + \sqrt{\frac{n_{o2}}{2}} \right|$$

n_{o1} and n_{o2} themselves varying.

Similarly the combined LPM sideband is given by:

$$v(t) = - \sqrt{\frac{n_{o1}}{2}} \sin \left[(w - p)t - \psi_1 \right] + \sqrt{\frac{n_{o2}}{2}} \sin \left[(w - p)t + \psi_2 \right]$$

3.5.7 Philosophic 'Difficulties'

In the foregoing fairly detailed analysis we have introduced two concepts which are likely to cause disagreement in some circles. These are the concepts of 'phase noise power' and 'potential sidebands'. Firstly the disagreement might be expressed by saying that phase modulation is essentially something that happens to a carrier and the sidebands cannot be considered apart from the carrier: phase jitter of a carrier is a meaningful concept, phase sidebands are not. Secondly it will be said that 'potential sidebands' have no real existence and are therefore 'meaningless'.

Two replies can be made to these criticisms. Extreme empiricism of this type which regards theoretical concepts as 'meaningless' unless they can be directly 'verified' experimentally would, if accepted, destroy the whole of theoretical physics. In philosophic circles this view is known as 'Logical Positivism' or 'Verificationalism'. This philosophy was conclusively refuted by K. R. Popper (Reference 3) many years ago. Even its leading British exponent, A. J. Ayer, has had to admit that (in 40 years) he has failed to produce a satisfactory formulation of the 'Verification Principle' (Reference 4).

The second reply is that any part of either of the phase noise sidebands of a carrier can be separately observed using a limiter and a spectrum analyser or separated from the carrier by a narrow band filter. In fact a 'potential' lower PM sideband due to upper sideband superposed noise, becomes 'real' after the combined signal is passed through a limiter.

It provides some additional understanding to consider the case of a 'phase noise' upper sideband, of power P which has been separated from the carrier by a narrow band filter. If this separated sideband is superposed upon another 'clean' carrier, of the same frequency as the original carrier, then both amplitude and phase modulation (AM and PM) of the inserted carrier will result. Actually each of the four 'potential' or real AM or PM sidebands will have equal powers ($P/4$) but the real upper AM and PM sidebands will be in phase with one another to give an effective upper sideband power P, and the lower 'potential' AM and PM sidebands will be in antiphase with one another and will cancel.

3.6 A Note on Notation

3.6.1 Parameters to be Distinguished
In considering the magnitudes of phase noise or phase jitter there are four separate possibilities of making 3 dB errors in the calculations. These are:

(a) Confusing peak and rms values.
(b) Confusing phase noise with pure noise.
(c) Confusing baseband and RF bandwidths.
(d) Confusing SSB and DSB superposed gaussian noise.

The notation adopted in this book has been chosen to minimise the likelihood of such errors.

3.6.2 The Concept of Phase Noise Power Density
As 'phase noise' has different characteristics from pure noise it is considered essential to use a different, if analogous symbol to distinguish the two, if confusion is to be avoided. Thus N_{op} is used for SSB 'phase noise' power density by analogy with N_0 which is used for the corresponding power density of pure noise. Similarly N_p is used to distinguish 'phase noise' from pure noise N. In the literature 'script $\mathscr{L}$' is sometimes used to denote (N_{op}/C). Phase noise differs from pure noise in the following respects:

(a) The upper and lower sidebands are conformable.
(b) A carrier may have 'phase noise' sidebands and relatively negligible AM sidebands either due to limiting or to frequency multiplication (as will be seen later). In fact frequency multiplication has an entirely different effect on 'PM noise' from its (negligible) effect on 'AM noise'.
(c) The PM noise component due to superposed pure noise is given by:

$$N_{op} = \frac{N_0}{2}$$

(d) Some methods of measurement and some systems are primarily responsive to PM noise.

3.6.3 The Notation Used (see Reference 5)
The symbols used are listed at the front of the book, immediately preceding Chapter 1.

3.7 Integration over a Frequency Band

3.7.1 Pure Noise
For pure noise of noise power density $N_0(f)$ the total noise power in an RF bandwidth $\pm b$, centred on f_0, that is in an RF bandwidth $2b$ is:

$$N_{(DSB)} = \int_{-b}^{+b} [N_0(f)] \, df \tag{3.34}$$

If the noise is white, with noise density N_0, over the offset frequency band $\pm b$ and zero or irrelevant outside this band then:

$$N_{\text{DSB}} = 2N_0 b \qquad (3.35)$$

The integrated phase jitter variance over this band is given by (3.13):

$$\overline{\phi^2} = \int_{-b}^{+b} \left(\frac{N_0}{2C}\right)_f \, df \text{ rads}^2 \qquad (3.36)$$

If the noise is white over this offset frequency band then:

$$\overline{\phi^2} = \frac{N_0 b}{C} \text{ rads}^2 \qquad (3.37)$$

and

$$\phi = \sqrt{\frac{N_0 b}{C}} \text{ radians} \qquad (3.38)$$

3.7.2 Phase Noise

If AM noise is negligible, or alternatively if we know the phase noise power in a certain bandwidth, and we wish to determine the phase jitter over an offset frequency range $\pm b$, that is the phase jitter up to a maximum rate b, we get:

$$N_p = \int_{-b}^{+b} [N_{op}(f)] \, df \qquad \cdot \qquad (3.39)$$

from (3.30.3)

$$\overline{\phi^2} = \int_0^b \left(\frac{2N_{op}}{C}\right)_f \, df \text{ rads}^2 \qquad (3.40)$$

If the phase noise is white over the offset frequency range $\pm b$ then:

$$\overline{\phi^2} = \frac{2N_{op}}{C} b \text{ rads}^2 \qquad (3.41)$$

3.8 Summary of the Relationships Between Phase Jitter and Noise

In order to avoid confusion it is necessary to distinguish clearly between the following cases.

3.8.1 Pure Phase Noise

Only 'phase noise' is present: 'AM noise' being negligible. As will be seen later (Chapter 6) this might well apply in practice at the output of a source consisting of an oscillator followed by a frequency multiplier chain. Alternatively only the 'phase noise' power has been measured or is of interest.

3.8.1.1 In Terms of Noise Density

$$\overline{\phi_0^2}(f) = \left(\frac{2N_{op}}{C}\right)_f \text{ rads}^2 \text{ per Hz} \tag{3.42}$$

Thus $\overline{\phi_0^2}(f)$ is the DSB 'phase noise' power density to carrier ratio at an offset frequency f. Conversely:

$$\left(\frac{N_{op}}{C}\right)_f = \left[\frac{\overline{\phi_0^2}(f)}{2}\right] \equiv 10 \log_{10}\left[\frac{\overline{\phi_0^2}(f)}{2}\right] \text{ dB/Hz} \tag{3.43}$$

3.8.1.2 Integrated over a Baseband Bandwidth b

$$\overline{\phi^2} = \int_{-b}^{+b} \left(\frac{N_{op}}{C}\right)_f df = \int_0^b \left(\frac{2N_{op}}{C}\right)_f df \text{ rads}^2 \tag{3.44}$$

This equality is valid because the two sidebands are conformable and hence symmetrical.

For white 'phase noise' over an offset frequency range $\pm b$.

$$\overline{\phi^2} = \frac{2N_{op}}{C} b \text{ rads}^2 \tag{3.45}$$

$\overline{\phi^2}$ is the DSB 'phase noise' power to carrier ratio.

3.8.2 Superposed Thermal Noise

3.8.2.1 Single Sideband Noise

$$\overline{\phi_0^2}(f) = \left(\frac{N_0}{2C}\right)_f \text{ rads}^2 \text{ per Hz} \tag{3.46}$$

$$\overline{\phi^2} = \int_0^b \left(\frac{N_0}{2C}\right)_f df \text{ rads}^2 \tag{3.47}$$

3.8.2.2 Double Sideband Noise
For white noise

$$\overline{\phi_0^2}(f) = \left(\frac{N_0}{C}\right)_f \text{ rads}^2 \text{ per Hz} \tag{3.48}$$

For the general case

$$\overline{\phi^2} = \int_{-b}^{+b} \left(\frac{N_0}{2C}\right)_f df \text{ rads}^2 \tag{3.49}$$

For white noise only $N_0(-b) = N_0(+b)$ and

$$\overline{\phi^2} = \int_0^b \left(\frac{N_0}{C}\right)_f df \text{ rads}^2 \tag{3.50}$$

and also for white noise only

$$\overline{\phi^2} = \frac{N_0}{C} b \tag{3.51}$$

All the above formulae are only valid if ϕ is sufficiently small that the conditions given by equation (2.9) and by $\tan \theta \simeq \theta$ are simultaneously satisfied.

CHAPTER 4

NOISE INDUCED FREQUENCY DEVIATION

4.1 Basic Relationships

As is apparent from the analysis in Chapter 2, there is a close relationship between frequency modulation and phase modulation. For the case of sinusoidal FM, equation 2.16 gives:

$$\Delta f = M f_m \tag{4.1}$$

Δf is the peak frequency deviation and M is the peak phase modulation in radians. f_m is the baseband modulating frequency or the rate at which the carrier is modulated. The frequency offset of the first order sidebands from the carrier will also be equal to f_m.

The phase modulation index (M) may also be written as θ, where θ is the peak modulation index in radians.

$$\therefore \quad \Delta f = \theta f_m \tag{4.2}$$

Writing δf for the rms frequency deviation and ϕ for the rms phase modulation index:

$$\delta f = \phi f_m \tag{4.3}$$

This equivalence may be extended to the case where we adopt the sinusoidal representation of narrow band noise. For white noise, the phase jitter density ϕ_0 is independent of the offset frequency f_m. We write $\delta f_0(f_m)$ for the rms frequency deviation measured at the output of a discriminator with a baseband filter 1 Hz wide centred on a frequency f_m. Thus $\delta f_0(f_m)$ is the density of the rms frequency deviation at an offset frequency f_m and it is essentially a DSB concept as it is measured at baseband. Thus for white phase noise:

$$\delta f_0(f_m) = \phi_0 f_m \tag{4.4}$$

4 Table 1 δf Hz (rms)/$\sqrt{\text{Hz}}$ Versus N_{op}/C(dB/Hz) and f_m

$\dfrac{N_{op}}{C}$ (dB/Hz) \ f_m	10 Hz	100 Hz	1 kHz	10 kHz	100 kHz	1 MHz	10 MHz
−20	1·41	14·1	141	1414	14140	$1\cdot4 \times 10^5$	$1\cdot4 \times 10^6$
−25	0·795	7·95	79·5	795	7950	$7\cdot95 \times 10^4$	$7\cdot95 \times 10^5$
−30	0·447	4·47	44·7	447	4470	$4\cdot47 \times 10^4$	$4\cdot47 \times 10^5$
−35	0·251	2·51	25·1	251	2510	$2\cdot51 \times 10^4$	$2\cdot51 \times 10^5$
−40	0·141	1·41	14·1	141	1410	$1\cdot41 \times 10^4$	$1\cdot41 \times 10^5$
−45	0·0795	0·795	7·95	79·5	795	7950	79500
−50	0·0447	0·447	4·47	44·7	447	4470	44700
−55	0·0251	0·251	2·51	25·1	251	2510	25100
−60	0·0141	0·141	1·41	14·1	141	1410	14100
−65	0·00795	0·0795	0·795	7·95	79·5	795	7950
−70	0·00447	0·0447	0·447	4·47	44·7	447	4470
−75	0·00251	0·0251	0·251	2·51	25·1	251	2510

	1410	141	14·1	1·41	0·141	0·0141	0·00141
−80	1410	141	14·1	1·41	0·141	0·0141	0·00141
−85	795	79·5	7·95	0·795	0·0795	0·00795	0·000795
−90	447	44·7	4·47	0·447	0·0447	0·00447	0·000447
−95	251	25·1	2·51	0·251	0·0251	0·00251	0·000251
−100	141	14·1	1·41	0·141	0·0141	0·00141	0·000141
−105	79·5	7·95	0·795	0·0795	$7·95 \times 10^{-3}$	$7·95 \times 10^{-4}$	$7·95 \times 10^{-5}$
−110	44·7	4·47	0·447	0·0447	$4·47 \times 10^{-3}$	$4·47 \times 10^{-4}$	$4·47 \times 10^{-5}$
−115	25·1	2·51	0·251	0·0251	$2·51 \times 10^{-3}$	$2·51 \times 10^{-4}$	$2·51 \times 10^{-5}$
−120	14·1	1·41	0·141	0·0141	$1·41 \times 10^{-3}$	$1·41 \times 10^{-4}$	$1·41 \times 10^{-5}$
−125	7·95	0·795	0·0795	$7·95 \times 10^{-3}$	$7·95 \times 10^{-4}$	$7·95 \times 10^{-5}$	$7·95 \times 10^{-6}$
−130	4·47	0·447	0·0447	$4·47 \times 10^{-3}$	$4·47 \times 10^{-4}$	$4·47 \times 10^{-5}$	$4·47 \times 10^{-6}$
−135	2·51	0·251	0·0251	$2·51 \times 10^{-3}$	$2·51 \times 10^{-4}$	$2·51 \times 10^{-5}$	$2·51 \times 10^{-6}$
−140	1·41	0·141	0·0141	$1·41 \times 10^{-3}$	$1·41 \times 10^{-4}$	$1·41 \times 10^{-5}$	$1·41 \times 10^{-6}$
−145	0·795	0·0795	$7·95 \times 10^{-3}$	$7·95 \times 10^{-4}$	$7·95 \times 10^{-5}$	$7·95 \times 10^{-6}$	$7·95 \times 10^{-7}$
−150	0·447	0·0447	$4·47 \times 10^{-3}$	$4·47 \times 10^{-4}$	$4·47 \times 10^{-5}$	$4·47 \times 10^{-6}$	$4·47 \times 10^{-7}$

In the general case where ϕ_0 is also a function of f_m:

$$\delta f_0(f_m) = \phi_0(f_m) f_m \qquad (4.5)$$

Now (3.42)

$$\phi_0(f_m) = \left(\frac{2N_{op}}{C}\right)^{1/2}_{f_m}$$

where $(N_{op}/C)_{f_m}$ is the value of the phase noise density to carrier ratio at an offset frequency f_m.

$$\therefore \quad \delta f_0(f_m) = \left(\frac{2N_{op}}{C}\right)^{1/2}_{f_m} f_m \qquad (4.6)$$

and

$$(\delta f_0)^2(f_m) = \left(\frac{2N_{op}}{C}\right)_{f_m} f_m^2 \qquad (4.7)$$

Thus if δf_0 is measured at a baseband frequency f_m, the corresponding DSB or SSB phase noise density to carrier ratios $(2N_{op}/C)_{f_m}$ and $(N_{op}/C)_{f_m}$ at an offset frequency f_m may readily be calculated using (4.7). Values of δf (Hz rms)$/\sqrt{\text{Hz}}$ have been calculated over a range of values of N_{op}/C(dB/Hz) and offset frequency f_m. The results are given in 4 Table 1.

4.2 Restricted Value of the Concept of Frequency Deviation

In a radar or communication system using frequency modulation it is natural to express any spurious angle modulation in terms of unwanted frequency deviation, as it is this which directly causes performance degradation.

Apart from cases of this type and the assessment of short term frequency stability (see Chapter 9), it is usually greatly preferable to work in terms of noise density, phase noise density or phase jitter. For example, the relationship between the AM noise and PM noise of a signal source is unnecessarily obscured by plotting spurious angle modulation in terms of frequency deviation, whilst plotting AM noise sidebands in terms of noise density to carrier ratio, as is often done (c.f. References 19 and 20).

The use of the expression 'FM noise' is deprecated. A carrier with noise-like FM sidebands has sideband powers proportional to the phase modulation index and not to the frequency deviation. The operation of an FM demodulator is to produce an output power proportional to the square of the frequency deviation. It is only this operation which converts the random 'FM sidebands' into baseband noise power.

In fact, even for signal sources for use in FM systems, it is usually preferable to work in terms of phase noise and only convert to frequency deviation as a last step using equation (4.7) or 4 Table 1.

In the case where it is necessary to calculate the effective frequency deviation over some finite baseband bandwidth say $(f_{m1} - f_{m2})$ then:

$$(\delta f)^2 = \int_{f_{m1}}^{f_{m2}} \left(\frac{2N_{op}}{C}\right)_{f_m} f_m^2 \, df_m \tag{4.8}$$

Except for very small percentage bandwidths when $f_{m2}/f_{m1} \simeq 1$, this integral must be rigorously evaluated. Even for white noise, for which $(N_{op}/C)_{f_m}$ is constant over the band

$$\delta f^2 = \left(\frac{2N_{op}}{C}\right) \int_{f_{m1}}^{f_{m2}} f_m^2 \, df_m \tag{4.9}$$

$$\therefore \quad \delta f = \left[\left(\frac{2N_{op}}{C}\right)\left(\frac{f_{m2}^3 - f_{m1}^3}{3}\right)\right]^{1/2} \tag{4.10}$$

This should be compared with the very much simpler expression for phase jitter in this case (3.45):

$$\overline{\phi^2} = \left(\frac{2N_{op}}{C}\right) b$$

where $b = f_{m2} - f_{m1}$.

In FDM/FM communication systems the spurious frequency deviation in a single telephone channel of 3·1 kHz bandwidth is of importance. For this case the baseband frequency of the channel covers the range $(f_{m2} - f_{m1}) = 3\cdot1$ kHz. The centre frequency of the channel (say f_m) is normally large compared with the bandwidth. For this case the following approximation is normally used:

$$\delta f^2 = \phi_0^2 f_m^2 b \tag{4.11}$$

where b is the baseband bandwidth of 3·1 kHz. This assumes that the phase jitter density (and hence the phase noise density) may be considered to be constant over the bandwidth and that simultaneously f_m is constant over the bandwidth.

4.3 Comparison of Phase Noise and Frequency Deviation

It is interesting to compare graphical plots of phase noise density and frequency deviation density of the same signal source in order to assist in visualising the differences previously discussed mathematically. Such plots for a simple idealised case are shown in 4 Figure 1. Curves (a) and (b) are plots of phase noise density/carrier ratio and frequency deviation density respectively, both expressed in dB, at different offset frequencies (f_m). Curve (a) represents a constant value of N_{op}/C (-60 dB/Hz) over the offset frequency band. It will be

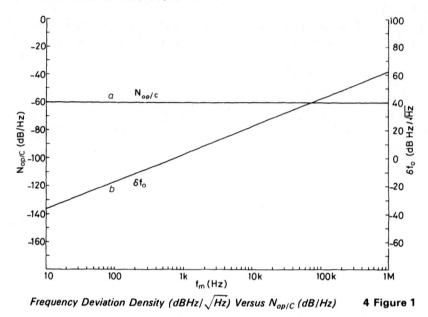

Frequency Deviation Density (dBHz/$\sqrt{Hz}$) Versus $N_{op/C}$ (dB/Hz)   **4 Figure 1**

noted that curve (*b*) for the consequent frequency deviation has a rising slope of 6 d/octave or 20 dB per decade with increasing frequency.

For a much more detailed discussion of the relationship between frequency deviation and phase noise see Chapter 9.

CHAPTER 5

NOISE IN OSCILLATORS

5.1 The Effects of Non-Linearity

An oscillator is an inherently non-linear device. If it had an effective loop gain of exactly unity for all signal levels, an initial low level oscillation would never build up in amplitude. Some non-linearity must be present to reduce the effective loop gain to unity at the full output level. During each RF cycle the instantaneous voltage will sweep over the full range and thus will normally experience this non-linearity. The effects of such non-linearity will include the generation of harmonics and intermodulation between any noise components which may be present and also between such noise components and the carrier.

If we consider the oscillator to have been switched on for some time so that the output has reached its normal level we might, as a first approximation, consider that within the oscillator loop we had a large, relatively clean carrier accompanied by various forms of noise. The noise may initially be divided into two classes, noise at frequencies of the same order as the carrier and noise at low frequencies relatively close to DC. If the oscillator is of high quality and its non-linearity is relatively small, cross products of small noise components will become negligibly small. However, cross product terms involving the large carrier may not be negligible. A carrier at a frequency f_0 and noise components of the same order of frequency (say $f_0 + \delta f$) will only produce additional unwanted frequencies within the pass band of the oscillator as a result of third order, or higher, intermodulation products. In a relatively linear oscillator these will be very small.

Due to Flicker effects in the active device used, noise close to DC will tend to follow a $1/f$ power law with frequency. For example a noise component in a 1 Hz bandwidth at a frequency of 5 Hz will be relatively large and due to the non-linearity will beat with the carrier to produce noise components at $(f_0 \pm 5)$ Hz which will be of significant amplitude.

The only other significant contribution to the noise output of a good oscillator would be expected to be the noise components existing initially at frequencies adjacent to f_0. These would be identical in an ideal linear oscillator to those in a real somewhat non-linear oscillator.

Thus, it is to be expected that, if a linear model were used to predict the carrier to noise ratio of an oscillator, the main error would be due to the neglect of the effects of $1/f$ noise transposed to small offset frequencies. Such a model should, therefore, be adequate to determine the performance of high quality oscillators except for very small offset frequencies.

5.2 Analysis

In setting up a model for analysis, four assumptions will be made. These are:

(*a*) That the oscillator is linear. This is reasonably valid for a high quality oscillator except in predicting carrier/noise ratios for small offset frequencies. For low level high quality crystal oscillators at carrier frequencies of a few MHz, using modern transistors or FETs, it gives good results for offset frequencies as low as 100 Hz and in some cases considerably lower.

(*b*) That, apart from the noise figure of the equivalent amplifier, and the tuned circuit damping, the exact circuit configuration is unimportant as far as the derivation of the carrier/noise performance equations is concerned. The validity of this assumption will be more readily granted when the analysis is complete. Basically it is dependent upon the fact that the loop gain of a linear oscillator is very close to unity.

(*c*) That the output power of an oscillator consists only of noise amplified by the positive feedback and filtered by the effective Q' achieved by positive feedback.

(*d*) For the initial analysis it is assumed that the feedback is as tight as possible: that is that the total output admittance, including the effects of the final load, suffers an impedance transformation so that at the input to the active device there is an admittance match at resonance.

In the absence of a detailed non-linear analysis (see Reference 6), which is beyond the scope of this treatment, the reader who doubts the validity of assumption (*a*), even for a high quality oscillator, may be somewhat reassured by the following physical reasoning. Consider a crystal oscillator at a frequency of 5 MHz. Assume it is to be provided with an automatic gain control loop with a time constant of the order of 1 second. Assume also that this loop controls the level so that the peak to peak swing over an RF cycle is well within the linear working range of the active device. Thus, over many RF cycles the performance will be linear and in particular no single RF cycle will be distorted, as changes in loop gain take place very slowly. It might be expected that other gain control mechanisms could be provided which would

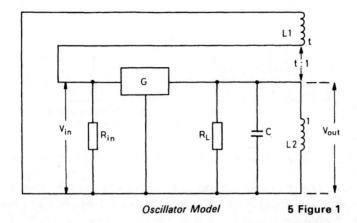

Oscillator Model   **5 Figure 1**

approximate this performance in the sense that the non-linearity during an RF cycle could be made small.

The reader who doubts the validity of assumption (*c*) is reminded that the noise due to the noise figure F is present for the whole time that the oscillator is switched on. Its total energy is therefore large compared with any 'switch on' transient.

As with any hypothesis, the validity of these assumptions rests ultimately on agreement between measured results and results calculated using a theory based on these assumptions. (See Section 5.7.1.)

For simplicity the circuit chosen for analysis is as given in 5 Figure 1.

The boxed item is an 'ideal amplifier' with a transconductance G. R_{in} includes all input circuit losses including those of the real transconductance amplifier. Reactive components at the input are assumed to have been tuned out by a tuned circuit with a Q which is low compared with that of the output tuned circuit. L_1 and L_2 form an ideal transformer with a voltage transformation ratio *t*. The output conductance of the amplifier, the tuned circuit losses and the coupling to the final output are all included in R_L.

Under the operating impedance, gain and current conditions, but with the loop broken at L_1, the noise figure of the equivalent amplifier is *F*. (To achieve the unchanged impedance conditions experimentally, after disconnecting L_1, it would be necessary to simulate the loading at the input normally transferred via L_1). See the last paragraph of Section 5.9.3.1 for further discussion of the value of *F*. In the analysis all noise sources are referred to the input as an available input noise power $FkTB$.

Assume that *t* has been chosen to give impedance matching, i.e.,

$$t^2 = \frac{R_{in}}{R_L}$$

Consider the transformer equivalent circuit referred to the output as shown in 5 Figure 2 (b).

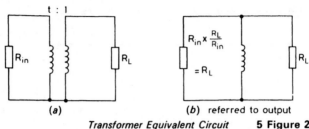

Transformer Equivalent Circuit   **5 Figure 2**

Thus $V_{out} = V_{in}G(R_L/2)$ at resonance. Let the available input noise voltage be $V_{N_{in}}$.

If all noise generators are lumped as a voltage generator in series with R_{in}, then half the open circuit EMF will appear across R_{in} (due to the matching to R_L via the transformer).

$$\therefore \quad V_{N_{in}} = \frac{\sqrt{4FkTR_{in}}}{2} = \sqrt{FkTR_{in}} \text{ per } \sqrt{\text{Hz}}$$

Now

$$V_{in} = tV_{out} + V_{N_{in}}$$

$$\therefore \quad V_{out} = (tV_{out} + V_{N_{in}})G\frac{R_L}{2} \text{ at resonance}$$

$$\therefore \quad V_{out} - G\frac{R_L}{2}tV_{out} = G\frac{R_L}{2}V_{N_{in}}$$

$$\therefore \quad V_{out} = \frac{G(R_L/2)V_{N_{in}}}{[1 - G(R_L/2)t]} \text{ at resonance} \tag{5.1}$$

Removing the restriction to resonance and assuming the output circuit to be the only high Q circuit

$$V_{out} = \frac{G(R_L/2)V_{N_{in}}}{1 - G(R_L/2)t + 2jQ(\delta f/f_0)} \tag{5.2}$$

The response will be 3 dB down when:

$$2Q\frac{\delta f}{f_0} = 1 - G\frac{R_L}{2}t \tag{5.3}$$

$\therefore$ Due to positive feedback

$$\text{effective } Q = Q' = \frac{Q}{1 - G(R_L/2)t} \tag{5.4}$$

$$\therefore \quad \text{effective bandwidth } B' = B\left(1 - G\frac{R_L}{2}t\right) \tag{5.5}$$

From (5.1) power in R_L at resonance is:

$$P = \frac{[G(R_L/2)V_{N_{in}}]^2}{R_L[1 - G(R_L/2)t]^2} \tag{5.6}$$

$$\text{Power in} = \left(\frac{V_{N_{in}}}{R_{in}}\right)^2 \tag{5.7}$$

$$\therefore \quad \text{Power Gain} = \frac{G^2(R_L/2)^2 R_{in}}{R_L[1 - G(R_L/2)t]^2} \tag{5.8}$$

Now the output of an oscillator is just amplified noise in a very narrow band and the input noise power is FkT times the effective noise bandwidth. The noise bandwidth of a single tuned circuit of 3 dB bandwidth B' is $(\pi/2)B'$ (see Appendix II). Thus

$$\text{input power} = FkTB' \frac{\pi}{2} \tag{5.9}$$

Let P be the total power generated including output power and all losses.

$$\therefore \quad \text{Power Gain} = \frac{P}{FkTB'(\pi/2)} \tag{5.10}$$

Equating (5.8) and (5.10)

$$\frac{P}{FkTB'(\pi/2)} = \frac{G^2(R_L/2)^2 R_{in}}{R_L[1 - G(R_L/2)t]^2} \tag{5.11}$$

But

$$\frac{R_{in}}{R_L} = t^2 \text{ (impedance matching)} \tag{5.12}$$

$$\therefore \quad \frac{P}{FkTB'(\pi/2)} = \frac{G^2(R_L/2)^2 t^2}{[1 - G(R_L/2)t]^2} \tag{5.13}$$

Substitute the value of B' from (5.5)

$$\frac{P}{(\pi/2)FkTB[1 - G(R_L/2)t]} = \frac{G^2(R_L/2)^2 t^2}{(1 - G(R_L/2)t)^2}$$

$$\therefore \quad \left(1 - G\frac{R_L}{2}t\right) = \frac{\pi}{2}\frac{FkTB}{P}\left(G\frac{R_L}{2}t\right)^2 \tag{5.14}$$

But $[G(R_L/2)t]$ is the loop gain which is closely equal to unity.

$$\therefore \quad \left(1 - G\frac{R_L}{2}t\right) = \frac{\pi}{2}\frac{FkTB}{P} \tag{5.15}$$

Substitute value of

$$\left(1 - G\frac{R_L}{2}t\right)$$

in (5.4) and (5.5)

$$\therefore \quad Q' = Q\frac{2}{\pi}\frac{P}{FkTB} \tag{5.16}$$

$$\therefore \quad B' = B\frac{\pi}{2}\frac{FkTB}{P} \tag{5.17}$$

It will be noted that the effective Q' under positive feedback conditions is very much larger than the normal Q (see Reference 7 which originally provoked this analysis in 1964).

Remember that the output power, and also the somewhat greater power P dissipated in R_L, is just amplified noise in a very narrow band. Initially assume the effective noise bandwidth $(\pi/2)B'$ to be rectangular.

Then total power generated per Hz of bandwidth at peak of response:

$$N_0 = \frac{P}{(\pi/2)B'} \text{ watts/Hz} \tag{5.18}$$

Removing the restrictions to a rectangular noise bandwidth and the peak of the response:

$$N_0 = \frac{2}{\pi}\frac{P}{B'}\frac{1}{|1 + 2jQ'(\delta f/f_0)|^2} \tag{5.19}$$

$\therefore$ Noise density to carrier ratio is:

$$\frac{N_0}{P} = \frac{2}{\pi}\frac{1}{B'}\frac{1}{|1 + 2jQ'(\delta f/f_0)|^2} \tag{5.20}$$

Now

$$\left|1 + 2jQ'\frac{\delta f}{f_0}\right|^2 = \left(1 + 4Q'^2\frac{\delta f^2}{f_0^2}\right)$$

$$\therefore \quad \frac{N_0}{P} = \frac{2}{\pi}\frac{1}{B'}\frac{1}{1 + 4Q'^2(\delta f/f_0)^2} \tag{5.21}$$

Consider points sufficiently far from resonance that:

$$4Q'^2\left(\frac{\delta f}{f_0}\right)^2 \gg 1 \tag{5.22}$$

$$\frac{N_0}{P} = \frac{2}{\pi}\frac{1}{B'}\frac{1}{4Q'^2(\delta f/f_0)^2} \tag{5.23}$$

Substitute values of Q' and B' from (5.16) and (5.17)

$$\therefore \quad \frac{N_0}{P} = \left(\frac{2}{\pi}\right)^2 \frac{1}{B} \frac{P}{FkTB} \frac{1}{4Q^2} \left(\frac{\pi}{2}\right)^2 \left(\frac{FkTB}{P}\right)^2 \left(\frac{f_0}{\delta f}\right)^2$$

$$\therefore \quad \frac{N_0}{P} = \frac{FkT}{P} \frac{1}{4Q^2} \left(\frac{f_0}{\delta f}\right)^2 \tag{5.24}$$

But δf is the frequency offset from resonance. Let $\delta f = f_m$

$$\therefore \quad \frac{N_0}{P} = \frac{FkT}{P} \frac{1}{4Q^2} \left(\frac{f_0}{f_m}\right)^2 \tag{5.25}$$

The present analysis is based on the assumption of linearity which is approximately valid over most of the offset frequency range. The assumption of an ideal amplifier is considered specifically to exclude variation of amplifier input and output capacitances with drive level and any resultant frequency pulling. Except in the case of SHF bipolar or FET oscillators, suitable circuit design can reduce such effects to the point where they become negligible. Under these circumstances, the output noise power density given by equation (5.25) will have all the characteristics of thermal noise filtered by a high Q resonant circuit. Thus, half the noise power density will be AM and half will be PM (3.29).

Thus, the phase noise density to carrier ratio becomes:

$$\frac{N_{op}}{P} = \frac{1}{2} \frac{FkT}{P} \frac{1}{4Q^2} \left(\frac{f_0}{f_m}\right)^2 \tag{5.26}$$

This may, of course, be written:

$$\left(\frac{N_{op}}{C}\right)_{fm} = \frac{FkT}{C} \frac{1}{8Q^2} \left(\frac{f_0}{f_m}\right)^2 \tag{5.27}$$

as long as it is remembered that F refers to the noise figure at a power level P. This is an important formula. (See also Reference 8). P is the total power generated by the oscillator. This power is shared between:

(a) Actual power output C.
(b) Losses in the output tuned circuit.
(c) The output admittance of the active device.
(d) The transformed input admittance of the active device.

This analysis was originally carried out by the author in 1964 and used as a guide in the design of a series of low phase noise crystal oscillator/frequency multiplier X-band sources which were widely used in USA and UK. However, it was not published for commercial reasons. It is believed that Reference 8 represents the first publication of this formula, although in a different notation.

5.3 Phase Noise Density as a Function of Offset Frequency

It will be noted that equation (5.27) gives an output noise density to carrier ratio of the form:

$$\frac{N_{op}}{C} = \frac{K}{(f_m)^2} \qquad (5.28)$$

Where K is independent of the offset frequency. This inverse square law will hold over an offset frequency range from the incidence of a $1/f^3$ term (due to intermodulation of carrier and Flicker Noise) at some low offset frequency, up to a higher offset frequency where flat phase noise in any subsequent electronic devices, begins to dominate. A typical performance for a 120 MHz crystal oscillator is shown in 5 Figure 3.

In some instances there may be a further slope of 6 dB per octave improvement with offset frequency, due to a second lower Q circuit within the oscillator. Due to its lower Q this circuit will only begin to affect the response at a considerably larger offset frequency than that at which the high Q circuit takes effect. If such a circuit is present, and superposed white noise in equipment connected to the output is of a sufficiently low level, then part of the offset region may show a $1/f^4$ law.

After calculating the overall performance of an oscillator using equation (5.27) one important parameter is missing: this is the value of the offset frequency at which the transition between $1/f^2$ and $1/f^3$ laws occurs. This will be called the Transition Frequency which will be designated f_c (to avoid confusion with transistor terminology).

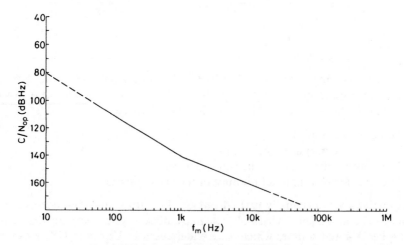

Phase Noise Performance of a 120 MHz Crystal   **5 Figure 3**
Oscillator

In 5 Figure 3, f_c has been (arbitrarily) assumed to occur at 1 kHz. The value of f_c may be determined by measurement on a completed oscillator, or by calculation if the Flicker noise performance of the transistor and the linearity characteristics of the oscillator, are precisely known. This is not usually the case, except when the considerable work involved is justified by the overriding importance of the close to carrier phase noise; as might be true if the oscillator were required as a Station Master Oscillator (SMO). Nevertheless, every effort should be made to choose a transistor and its operating conditions to minimize Flicker Noise.

5.4 The Feedback Q

Consider a high quality 100 MHz crystal oscillator. Low noise crystals are available in this frequency region with a Q of 2×10^4. Assume a power output of 1 mW. If the matching is such that the loaded Q is 5×10^3 and is otherwise optimum then:

$$\frac{P}{C} \simeq 4$$

$$\therefore \quad P = 4\,\text{mW} = 6\,\text{dBm}$$

Assume the use of a rather poorly suited VHF transistor, say type 2N918. For a high degree of linearity the DC working point must be chosen to permit much greater current and voltage swings than those required at a 4 mW RF power level. Assuming a 20 Volt supply and allowing an RF peak voltage of 5 Volts, demands an RF peak current swing of 1·6 mA. For low distortion this might demand a standing current in the region of 7 mA. The DC power consumption will then be 140 mW which is within the transistor rating. At 100 MHz and these operating levels the noise figure will be rather poor even with operation from the optimum source impedence.

Assume $F = 20\,\text{dB}$

·Now $f_0 = 100\,\text{MHz}$

Loaded $Q = 5 \times 10^3$

$$B = \frac{f_0}{Q} = 20\,\text{kHz} = +43\,\text{dB Hz}$$

$kT = -174\,\text{dBm/Hz}$ and $FkTB = -111\,\text{dBm}$

$$\therefore \quad P/FkTB = (+6 + 111)\,\text{dB} = 10^{11\cdot7}$$

From (5.16)

$$Q' = Q \frac{2}{\pi} \frac{P}{FkTB}$$

$$Q' = \frac{2}{\pi} \times 5 \times 10^3 \times 10^{11 \cdot 7}$$

$$Q' = 1 \cdot 6 \times 10^{15}$$

From (5.17)

$$B' = \frac{B\pi}{2} \frac{FkTB}{P}$$

$$B' = 2 \times 10^4 \frac{\pi}{2} \frac{1000}{4} \times FkTB$$

$$B' = 6 \cdot 24 \times 10^{-8} \text{ Hz}$$

If we put $\tau = \dfrac{1}{B'}$, $\tau = 185$ days

At first sight these are surprising figures. They show that the effect of the positive feedback is to produce a fantastically narrow effective bandwidth and a very long effective time constant. It must be remembered that these figures only apply under the steady state conditions when the loop gain has settled at a value very close to unity.

From (5.15) the difference from unity is given by:

$$\left(1 - G\frac{R_L}{2}t\right) = \frac{\pi}{2} \frac{FkTB}{P} = 3 \cdot 12 \times 10^{-12}$$

It is this factor by which the open loop bandwidth B is multiplied to produce the closed loop bandwidth B'. If on first switching on, the loop gain for the small initial signal, were (say) 11 then

$$\left(1 - G\frac{R_L}{2}t\right) = -10$$

The bandwidth would then be $10B = 200$ kHz and τ would become 5 μs.

This shows why the build up in a normal oscillator is in fact rapid, although the steady state time constant is very long.

5.5 The Loaded Q

The loaded Q of the resonant circuit is dependent on a number of factors as follows:

(a) The operating frequency (f_0).

(*b*) The type of resonant circuit selected.

(*c*) The couplings between this resonant circuit and the oscillator including its output load.

For a single frequency oscillator in the frequency range 100 kHz to 150 MHz, a crystal would be the obvious choice to meet any high performance requirement. The choice of crystal type at a given frequency should be determined by aiming for the highest Q, subject to a suitable power rating and freedom from vibration or acoustically induced phase modulation. This implies rugged (ribbon) quartz mounting arrangements (Reference 9). In addition, requirements for absolute frequency accuracy may influence the choice of crystal. If a very narrow tuning range, remotely or electrically controlled, is a requirement; then a Voltage Controlled Crystal Oscillator (VCXO) would probably be specified. Use of quartz crystals in this mode normally involves some sacrifice in Q.

If the required output frequency is higher than about 150 MHz, and frequency multiplication is not acceptable, then a high Q distributed circuit oscillator would be the normal choice. Depending on the space available and the performance and cost limitations, this might be a strip line circuit, a trough line, a coaxial cavity or a waveguide cavity. The values of Q obtainable may readily be estimated. (References 10 and 11).

In many cases it may be essential to provide voltage tuning facilities. This will usually be done with varactor diodes. For large percentage tuning ranges this will normally result in some degradation of resonant circuit Q. With unsophisticated circuit coupling arrangements between the transistor and resonant circuit the provision of a large varactor tuning range may result in a drastic reduction in Q (to a value as low as 10).

We will turn now to consider the Q degradation of the resonant circuit due to the circuit couplings. In doing so we shall only initially observe the restriction imposed by assumption (d) of Section 5.2. This assumption made a trivial difference to the development of the analysis without affecting the final formula derived. It was included so that the physical meaning of the various steps in the argument was more apparent. Matching the admittance of the feedback network to the input admittance of the transistor also happens to be roughly an optimum if a bipolar transistor is used, as will shortly be apparent.

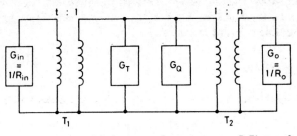

Admittance at Resonance   **5 Figure 4**

The total admittance shunting the resonant circuit is shown in 5 Figure 4, which is derived from 5 Figure 1 with the addition of some detail.

$G_{in} = 1/R_{in}$ is the input admittance of the transistor.

G_T is the output admittance of the transistor plus that due to transformer losses in T_1 and T_2.

G_Q is the shunt admittance of the resonant circuit.

G_0 is the admittance of the final output load.

A simplified equivalent circuit, after performing the impedance transformations, is given in 5 Fig. 5.

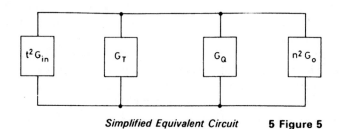

Simplified Equivalent Circuit   5 Figure 5

Let G be the total admittance. Then:

$$G = t^2 G_{in} + G_T + G_Q + n^2 G_0 \qquad (5.29)$$

Assume initially that G_{in} is matched so that:

$$t^2 G_{in} = G_T + G_Q + n^2 G_0 \qquad (5.30)$$

$$\therefore \quad G = 2t^2 G_{in} = 2(G_T + G_Q + n^2 G_0) \qquad (5.31)$$

This implies that half the power generated (P) is dissipated in G_{in}. Let $\frac{1}{4}$ of the power generated ($P/4$) be dissipated in the final output load G_0.

$$n^2 G_0 = \frac{G}{4} \qquad (5.32)$$

$$G = 2\left(G_T + G_Q + \frac{G}{4} \right)$$

$$\frac{G}{2} = 2(G_T + G_Q)$$

$$G = 4(G_T + G_Q) \qquad (5.33)$$

Thus, with half the power dissipated in G_{in} and one quarter in the final output, the resonant circuit Q (Q_{T+Q}) resulting from coupling to the output circuit of the transistor (however this is achieved) will become

$$\frac{Q_{(T+Q)}}{4}.$$

Thus under these conditions the very best that can be achieved is:

$$Q_L \simeq \frac{Q_u}{4}$$

where Q_L is the loaded Q, and Q_u is the unloaded Q. It should be noted also that under these conditions

$$C = \frac{P}{4}. \tag{5.34}$$

Thus the transistor must be capable of generating a total power at least four times the required power output and it must do this with a good noise figure F and, for many applications, have a good flicker noise performance also.

Consider now the assumption of matching to the input admittance of the transistor.

$$t^2 G_{in} = \frac{G}{2} \quad \text{or} \quad t^2 = \frac{R_{in}}{R_L} \tag{5.35}$$

In this case the signal source for the input circuit of the transistor is matched to the transistor input admittance. An important requirement is to minimise the noise figure (F) of the transistor as an amplifier as may be seen from equation (5.27). For minimum noise figure there is an optimum source impedance which differs from the matched condition. However, in the case of a bipolar transistor a match is not very far from optimum. If an attempt is made to maintain practically the undamped Q by making $R_{in}/t^2 \gg R_L$ then the noise figure of the equivalent bipolar amplifier will suffer severely.

The overall noise performance of the oscillator is proportional to F/Q^2 (5.27), so that an improvement of $2:1$ in damped Q is not worthwhile if the price is an increase of $4:1$ or more in F.

For an FET there is likely to be a more beneficial tradeoff as the optimum source impedance from a noise figure point of view is much lower than the input impedance and hence the high input impedance, when transformed across the tank circuit, will produce relatively little damping. This has the advantage that, unlike a nearly matched bipolar transistor, the close to unity overall loop gain does not imply unity gain in the FET itself, which has to make up for the mismatch loss. This reduces the noise figure degradation which occurs in an active device with unity gain.

A further factor producing degradation in Q is the coupling necessary to provide a final power output from the oscillator. For simplicity assume that an FET is used and it is arranged that the Q degradation due to the FET input loading is negligible. Then the loaded $Q(Q_L)$ is related to the unloaded $Q(Q_u)$ by:

$$C = \frac{P(Q_u - Q_L)}{Q_u}$$

$$\therefore \quad Q_u \frac{C}{P} = Q_u - Q_L$$

$$\therefore \quad Q_L = Q_u \left(1 - \frac{C}{P}\right) \tag{5.36}$$

For a good noise performance it would seem to be desirable to aim for $Q_L/Q_u \simeq 1$, resulting in a very small value of C. The performance may then be limited by noise in any amplifier which follows the oscillator. If bipolar transistors are used and $t^2 = R_{in}/R_L$ then half the total power $(P/2)$ is required for the input drive to the transistor. The remaining power $(P/2)$ is available for output circuit losses and final power output C. For many purposes $C = P/4$ is not too far from optimum in these circumstances. In this case $Q_L = 3Q_u/4$. This assumes that the loading due to the transistor output impedance can be made negligible.

5.6 The Choice of Parameters

To achieve the best possible phase noise performance for a simple signal source it is necessary to make a correct choice of a number of parameters. These are:

(a) The resonant circuit—this has already been discussed.
(b) The impedance transformations—this has already been discussed.
(c) The type of active device—in most cases the choice is initially between a bipolar transistor and an FET. Having made this choice it is necessary to choose a suitable unit for the frequency and power level involved. As already pointed out an FET is often much to be preferred as it will permit a higher loaded resonant circuit Q and will usually have a better flicker noise performance. It will also allow a bigger C/P ratio and will thus have to generate a smaller RF power (P). It should be noted that equation (5.27) gives no information as to the offset frequency f_m, below which flicker noise begins to make a noticeable contribution. This will be determined by the choice of transistor, the power level and the oscillator linearity. If the required power level is such as to be a major determinant of the achievable noise figure, it may be worth considering an oscillator/amplifier combination. If our primary requirement is good phase noise performance at fairly small offset frequencies, the linearity of the amplifier is less important than that of the oscillator, as also is its noise figure.
(d) The frequency—on examining equation (5.26) it will be noted that, for a

given noise figure, power generated, resonator Q and offset frequency f_m, $N_{op}/P \propto f_0^2$. Thus, except in so far as the circuit Qs differ in the two cases, a low frequency oscillator will have a superior phase noise performance to that of a higher frequency oscillator. Thus in many instances it may be preferable to choose a lower frequency oscillator and follow it by a frequency multiplier to give the required output frequency.

5.7 Examples

In this Section we shall calculate the probable performance of various types of oscillator using realistic values for the parameters involved. The first example chosen is one where experimental results are also available for comparison.

5.7.1 120 MHz Crystal Oscillator
An engineered design of a crystal oscillator (see 5 Figure 6) had the following design characteristics:

> Frequency—120 MHz
> Transistor—2N918
> Crystal—Operated in series resonance:
>   5th Overtone AT cut
>   Unloaded $Q = 2 \times 10^4$
> $f_0 = 120$ MHz
> $r_s = 50$ ohms
>   Estimated loaded $Q = 4 \times 10^3$

120 MHz Crystal Oscillator　　**5 Figure 6**

Noise figure—estimated at 20 dB. This estimate was the result of measurements on a 2N918 amplifier with similar mean DC operating and impedance conditions at 120 MHz. This measurement may have been somewhat in error as the DC current conditions may not have been fully equivalent to the pulsed current under operating conditions.

> Power Generated $P = 15$ mW
>
> Power Output $C = 5$ mW

From equation (5.27)

$$\frac{N_{op}}{C} = \frac{FkT}{C} \frac{1}{8Q^2} \left(\frac{f_0}{f_m}\right)^2 /\text{Hz}$$

$$\therefore \quad \frac{N_{op}}{C} = \frac{FkT}{40} \frac{1}{16 \times 10^6} \, 1{\cdot}44 \times 10^{16} \frac{1}{f_m^2}$$

(Reckoning kT in mW per Hz)

$$\text{at } f_m = 1 \text{ kHz}$$

$$\frac{N_{op}}{C} = (20 - 174 - 16 + 89{\cdot}5 - 60) \text{ dB/Hz}$$

$$\therefore \quad \frac{N_{op}}{C} = -140{\cdot}5 \text{ dB/Hz} \tag{5.37}$$

The phase noise performances of two sample oscillators were measured at X band and at a number of offset frequencies, after frequency multiplication by 72 times. The measured results, scaled back to 120 MHz, are compared with the theoretical predictions in 5 Figure 7.

Considering that the oscillator was not really linear as assumed in the theoretical analysis, and considering also that the experimental parameters and tolerances were not known to a high order of accuracy, the agreement between calculated and measured performance is considered to be quite good. The poorer experimental performance above 20 kHz was due to the effects of white noise added by the amplifier which was included as part of the module design, and into which the oscillator output was fed at a level of 5 mW.

Due to the limitations of the noise measuring equipment used (Reference 12), no measurements were taken at offset frequencies of less than 1 kHz.

5.7.2 5 MHz High Grade Oscillator

For some applications, such as station master oscillators for satellite ground stations carrying low data rate traffic, the 'phase noise' performance at extremely low offset frequencies is important. For such an application the SMO frequency is usually specified as 5 MHz. At this frequency third overtone

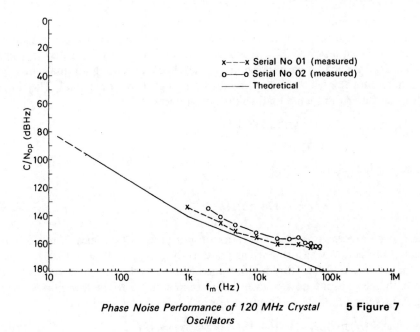

Phase Noise Performance of 120 MHz Crystal **5 Figure 7**
Oscillators

.AT cut crystals can be manufactured with values of Q in the region of 5×10^5 (Reference 13).

At this relatively low carrier frequency a very good noise figure may be achieved, if suitable transistors are used at a low standing current. Such a low standing current is also essential to obtain good $1/f$ transistor noise performance. Excellent linearity is achievable by the use of a separate gain control loop. With a suitable transistor and these circuit techniques it would seem to be possible to obtain a noise figure as low as 1 dB at 100 Hz offset frequency which worsens as $1/f$ below this frequency and is constant above (but see Section 5.9.3.1). A possible power dissipation in the crystal is of the order of 2 mW. The output power C into the following amplifier might then be 0.5 mW $= -3$ dBm. Assume the working $Q = 1 \times 10^5$.

Using equation (5.27) the phase noise performance at an offset frequency of 100 Hz, is given by:

$$\frac{N_{op}}{C} = \frac{FkT}{C}\frac{1}{8Q^2}\left(\frac{f_0}{f_m}\right)^2$$

$$\left(\frac{N_{op}}{C}\right)_{100} = (+1 - 174 + 3 - 9 - 100 + 134 - 40)\ \text{dB/Hz}$$

$$\left(\frac{N_{op}}{C}\right)_{100} = -185\ \text{dB/Hz} \tag{5.38}$$

This performance is in fact too good: it shows up two facts: the danger of carelessly estimating F when the loop gain is unity (see Section 5.9.3.1) and a limitation of equation (5.27) if used uncritically for a very high performance oscillator. In such a case noise in a matched resistive load may not be negligible and this will not be filtered by the high Q of the crystal. The noise power density from such a load will be given by:

$$N_{o \, (load)} = -174 \text{ dBm/Hz (kT)}$$

$$N_{op \, (load)} = -177 \text{ dBm/Hz}$$

Now $C = 0.5 \text{ mW} = -3 \text{ dBm}$

$$\therefore \quad \frac{N_{op}}{C} \text{ (load)} = -174 \text{ dB/Hz} \tag{5.39}$$

Thus, even a perfect oscillator with an output power of 0.5 mW is unlikely to reach a phase noise performance as good as $(N_{op}/C)_{100} = -174$ dB/Hz. As far as the author is aware this performance has not yet been achieved even at rather higher output power levels, but commercial 5 MHz low phase noise oscillators are available with output power levels of a few milliwatts

$$\text{and } (N_{op}/C)_{100} = -164 \text{ dB/Hz (Reference 14)} \tag{5.40}$$

5.7.3 A 600 MHz Voltage Controlled Oscillator

The design of a 600 MHz VCO with a relatively wide voltage controlled tuning range of $\pm 10\%$ is considered. Crystals are not available at such a high carrier frequency and in any case would not permit the tuning range. Thus, the tuning element would have to be a distributed circuit of some kind, such as a trough line or strip line. It may be shown that for an unsophisticated circuit, where the tuning range requirement is allowed to dictate the coupling between the resonant circuit and the transistor, the working Q will be low (see Appendix IV). This is not an optimum design but is given as an example of the sort of performance to be expected when design simplicity is important. In this case we assume $Q = 10$ (Appendix IV).

The required power output is assumed to be 30 mW. The total RF power generated will be in the region of 120 mW. A low level, low noise transistor is therefore unlikely to be suitable. A small UHF power transistor (eg 2N3866) is likely to have a noise figure of about 20 dB.

$$\frac{N_{op}}{C} = \frac{FkT}{C} \frac{1}{8Q^2} \left(\frac{f_0}{f_m}\right)^2$$

$$\therefore \quad \left(\frac{N_{op}}{C}\right)_{10 \text{ kHz}} = (+20 - 174 - 15 - 29 + 176 - 80) \text{ dB/Hz}$$

$$= -102 \text{ dB/Hz} \tag{5.41}$$

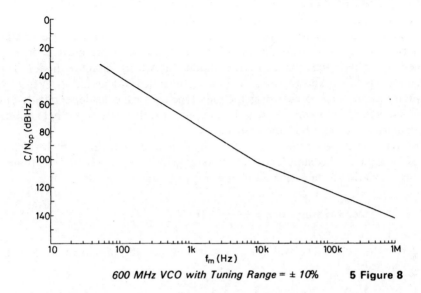

600 MHz VCO with Tuning Range = ± 10%   **5 Figure 8**

This performance is not outstanding and could probably be improved by choice of a more suitable transistor. The Flicker noise in such a transistor is not known but it is likely to be fairly high. It is assumed that it becomes appreciable below 10 kHz. Using this frequency as the transition between $1/f^3$ and $1/f^2$ noise, the resultant overall performance is plotted in 5 Figure 8.

5.8 Integration of Phase Noise of an Oscillator

By following the methods outlined in Section 3.7.2 it is easy to integrate the total phase noise over an offset frequency band f_1 to f_2 as long as f_1 is not zero, for then the integral does not converge. The integral concerned is (equation 3.40)

$$\bar{\phi}^2 = \int_0^b \left(\frac{2N_{op}}{C}\right)_f df \; \text{rads}^2 \tag{5.42}$$

The reason for the non-convergence in the case of a typical oscillator is that for small values of f_m, $(N_{op}/C)_f$ is proportional to $1/f^3$. This gives a singularity at $f = 0$.

5.9 Other Factors which Affect Oscillator Phase Noise

Equation (5.27) gives the best attainable phase noise performance of an oscillator with a single resonator. The performance *may* be improved by the use of more than one resonator and it may very easily be degraded, in some cases to a very large degree, by poor design.

5.9.1 Disturbances on the Tuning Voltage of a VCO

For many purposes it is necessary that an oscillator shall be electrically tunable over a specified frequency band. Such oscillators are most frequently tuned by voltage sensitive capacitive diodes known as varactor diodes. The oscillator is then known as a Voltage Controlled Oscillator or VCO. A varactor diode tuned oscillator is the only type we shall consider explicitly but the following discussion is readily adapted to the case of other tuning mechanisms such as YIG tuned oscillators.

An important parameter of a VCO is the total frequency range over which it must be capable of tuning. Let the tuning range be $(f_2 - f_1)$ Hz. Let the tuning voltage variation for a frequency change $(f_2 - f_1)$ Hz be V volts.

$$\therefore \quad \text{mean tuning slope} = \frac{f_2 - f_1}{V} \text{ Hz/V} \tag{5.43}$$

The varactor tuning law will not be linear. Let the ratio of maximum slope to mean slope be L.

$$\therefore \quad \left(\frac{df}{dv}\right)_{max} = \frac{L(f_2 - f_1)}{V} \text{ Hz/V} = K_0'(\text{Hz/V}) \tag{5.44}$$

Let the unwanted ripple (or noise voltage/$\sqrt{\text{Hz}}$) on the varactor tuning voltage at a frequency f_m be V_m.

Now (equation 4.3)

$$\phi_0 = \frac{\delta f_0}{f_m}$$

$$\therefore \quad \phi_{0\,max} = \frac{df}{dV} \frac{V_m}{f_m} \text{ radians rms} \tag{5.45}$$

If $\phi_{0\,max} < 0.1$ radian

$$\left(\frac{2N_{op}}{C}\right)^{1/2} = \phi_{0\,max}$$

$\therefore$ maximum equivalent phase power density to carrier ratio is given by:

$$\left(\frac{2N_{op}}{C}\right)^{1/2}_{f_m} = \frac{L(f_2 - f_1)}{V} \frac{V_m}{f_m} \tag{5.46}$$

$\therefore$ maximum allowable equivalent ripple or noise voltage in 1 Hz bandwidth at f_m for a specified maximum resultant N_{op}/C is:

$$V_m = \left(\frac{2N_{op}}{C}\right)^{1/2}_{f_m} \frac{V f_m}{L(f_2 - f_1)} = \left(\frac{2N_{op}}{C}\right)^{1/2} \frac{f_m}{K_0'}. \tag{5.47}$$

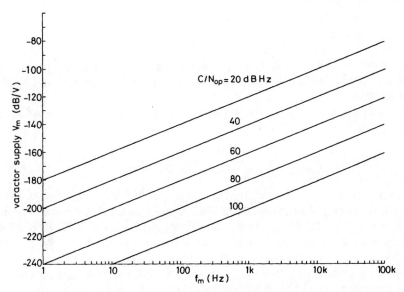

Permissible rms Voltage (ripple and noise/√Hz) **5 Figure 9**
on Varactor Supply of VCO with Tuning Sensitivity of 150 MHz/V

In dB with respect to 1 Volt

$$V_m = 20 \log_{10}\left[\left(\frac{2N_{op}}{C}\right)^{1/2}_{f_m} \frac{Vf_m}{L(f_2 - f_1)}\right] \text{dBV}$$

$$\therefore \quad V_m = \left[+3 + 10 \log_{10}\left(\frac{N_{op}}{C}\right)_{f_m} - 20 \log_{10}K'_0 \right.$$

$$\left. + 20 \log_{10} f_m \right] \text{dBV} \tag{5.48}$$

For a VCO with a wide tuning range and V a few tens of volts at most, this is an extremely severe requirement. For example an X-band oscillator with a tuning range of 1000 MHz and a tuning voltage range of 20 volts would have a *mean* tuning slope of 50 MHz/V. A graph of the maximum permissible ripple voltage for a maximum tuning slope of 150 MHz/V is given as 5 Figure 9. It will be seen that at $f_m = 50$ Hz and a required resultant

$$\left(\frac{C}{N_{op}}\right) \text{ of } 100 \text{ dB Hz, } V_m = -226 \cdot 5 \text{ dBV} \tag{5.49}$$

(For a discussion of the approximation involved in assuming the bandwidth of a hum line to be 1 Hz, see Section 5.9.4).

Thus in this case the maximum allowable rms ripple voltage on the varactor tuning supply is about 5 picovolts!! For the case where this supply was derived directly from a voltage stabiliser, a very special stabiliser design would be required, both from the point of view of ripple suppression and in respect of the noise voltage in a 1 Hz bandwidth at 50 Hz.

It would also be necessary to prevent electrostatic and magnetic pick up, and to control all earth loop coupling in an endeavour to meet such a requirement.

The graph of 5 Figure 9 may readily be adapted for other values of tuning slope by adding

$$20 \log_{10} \left[\frac{150}{\text{Actual Tuning Sensitivity (MHz/V)}} \right] \text{ to } V_m \text{ (dB)} \qquad (5.50)$$

In many cases the varactor tuning supply is fed from the output of an integrated circuit operational amplifier. A high quality low noise amplifier such as the TDA 1034 has a Power Supply Ripple Rejection (PSRR) of 86 dB. This greatly simplifies the required smoothing on the power supply but does *not* relax the requirements of any *unbalanced* voltage pick up due to poor shielding or earth loops. In addition, if any DC offset voltage is required to set the centre frequency of the VCO then the full ripple requirements must be met on this supply.

5.9.2 Disturbances on Other Power Supply Rails
The effects peculiar to voltage controlled oscillators have been considered in Section 5.9.1 above. We shall now consider other ways in which noise or ripple on power supply rails produces degradation in the phase noise performance of any oscillator. In general these effects are less easy to calculate theoretically as they are dependent on circuit design parameters which are not usually well known. Nevertheless the sensitivity of an oscillator frequency to power supply variation may readily be measured using a frequency counter. This provides a value of df/dV for equation (5.45). Equations (5.47) and (5.48) may then be used, writing:

$$20 \log \frac{dV}{df} \text{ in place of } 20 \log \left[\frac{V}{L(f_2 - f_1)} \right]$$

The graph of 5 Figure 9 may also be used as previously described. Whilst it is hard to calculate df/dV from first principles it is helpful to consider the causes qualitatively to assist in design modifications if the initial design is excessively sensitive to power supply ripple. Even so any low phase noise oscillator requires a very low ripple, low noise power supply.

5.9.2.1 Due to Non-Linearities
If an oscillator is highly non-linear due, for example, to voltage limiting during each RF cycle, it will produce harmonics to which the resonator will appear reactive. A change in supply voltage will alter the limiting conditions, the

harmonic content, the reactive loading and the output frequency. The only cure is to alter the operating conditions to reduce the degree of non-linear operation.

5.9.2.2 Due to Transistor Capacitance Changes

Transistors, either bipolar or field effect, are subject to changes in junction capacitance with voltage applied across the junctions. The larger the proportion of the total frequency determining capacitance which is due to transistor junction capacitances the more sensitive the output frequency to the bias conditions. In a VHF, and even more an SHF, oscillator it may sometimes be difficult to provide sufficient fixed (and stable) capacitive loading to reduce this effect to negligible proportions, but every effort should be made to do this, otherwise the phase noise performance may turn out to be appreciably worse than that predicted by equation (5.27) unless the power supply is very good indeed (microvolts of ripple at most).

5.9.3 Disturbances at the Output

5.9.3.1 Added Noise

If a low power output low noise oscillator feeds a resistive load, the thermal noise of the load will not be filtered by the oscillator resonator. The available thermal noise power N_0 from a load at room temperature is -174 dBm. Thus N_{op} for the thermal load $= -177$ dBm. If the oscillator has a power output of -10 dBm, (N_{op}/C) cannot be better than -167 dB/Hz at any offset frequency, however large. If an amplifier follows, this will contribute noise greater than thermal and the performance will be further degraded; this case is considered much more fully in Section 6.3 below.

It has not, perhaps, been sufficiently emphasised that, in many oscillators, output circuit noise may be a significant factor. Under steady state conditions the oscillator loop gain is unity and often the gain of the active device is close to unity; hence collector noise and collector circuit noise will have more effect on the noise figure F than would be true if the gain were higher. F in equation (5.27) is the noise figure of the transistor as an amplifier *under the same operating conditions* as pertain to the oscillator. These operating conditions include near unity gain. Hence a *guess* of the likely noise figure, knowing the type of transistor, may be very optimistic, as the intuitively expected figure will be that pertaining to a high gain configuration.

5.9.3.2 Mismatch Effects

Variations in the VSWR of an external load, into which an oscillator feeds, will tend to affect the phase noise performance. As will be apparent from 5 Figure 4, variations in G_0, the load admittance, will vary the admittance of the resonator and hence its Q. After transformation by $t^2:1$ this will also result in a different source admittance, for the transistor input circuit, from the optimum value, thus probably resulting in degradation in the noise figure F.

Thus in equation (5.27), rewritten for convenience:

$$\frac{N_{op}}{C} = \frac{FkT}{C} \frac{1}{8Q^2} \left(\frac{f_0}{f_m}\right)^2$$

F and Q may both be expected to change; Q may be either increased or decreased, but F is likely to increase if the design were originally optimised for a nominal load. In addition the ratio of C (the power output) to P (the total power generated) is likely to change.

Thus for optimum phase noise performance with loads with a large VSWR tolerance, it is normally necessary to isolate the oscillator from the load, using ferrite isolators or an amplifier.

5.9.4 Vibration

Vibration of a resonator will almost invariably modulate its resonant frequency and thus the oscillator, of which the resonator forms a part, will produce phase noise sidebands at an offset frequency equal to the vibration frequency. Thus for good performance under vibration the physical design of resonator or the choice of crystal (Reference 9) is very important.

A typical value of vibration sensitivity for a rugged crystal suitable for a low noise oscillator (Reference 15) is:

$$\Delta f = \frac{f_0}{10^9} X \tag{5.51}$$

where Xg is the vibration excitation: g being the acceleration due to gravity, f_0 is the resonant frequency and Δf is the peak frequency shift.

With care and rugged layout a crystal oscillator may be designed so that the crystal is the major contributor to vibration induced frequency modulation.

Now

$$\phi = \frac{\Delta f}{\sqrt{2}} \cdot \frac{1}{f_m} \qquad \text{[see (4.2) and (4.3)]}$$

and

$$\frac{\phi^2}{2} = \frac{N_{op}}{C} \qquad \text{(see 3.42)}$$

$$\therefore \quad \left(\frac{N_{op}}{C}\right)_{f_m} = \frac{\Delta f^2}{4f_m^2} \tag{5.52}$$

From (5.51) and (5.52)

$$\left(\frac{N_{op}}{C}\right)_{f_m} = \frac{f_0^2}{10^{18}} \frac{X^2}{4f_m^2} \tag{5.53}$$

Strictly this should be written:

$$\left(\frac{N_p}{C}\right)_{f_m} = \frac{f_0^2}{10^{18}} \frac{X^2}{4f_m^2}$$

(5.54)

as the duration of the vibration, and hence its bandwidth, has not been specified. Equation (5.53) is thus only really valid for a duration of 1 second. However the duration is often not known and the value of Δf given by equation (5.51) is only a typical value, so that in practice equation (5.53) is a useful guide to the order of magnitude.

As an example consider a 5 MHz crystal oscillator sinusoidally vibrated at 1 g, first at 10 Hz and then at 50 Hz.

$$\left(\frac{N_{op}}{C}\right)_{10} = \frac{(5 \times 10^6)^2}{10^{18}} \frac{1}{400} = 6 \cdot 25 \times 10^{-8}$$

$$= -72 \text{ dB/Hz}$$

(5.55)

$$\left(\frac{N_{op}}{C}\right)_{50} = \frac{(5 \times 10^6)^2}{10^{18}} \frac{1}{4 \times 2500} = 2 \cdot 5 \times 10^{-9}$$

$$= -86 \text{ dB/Hz}$$

(5.56)

It should be noted that the phase noise performance is degraded as $1/f_m^2$ or 20 dB per decade fall in vibration frequency.

As will be seen in Chapter 6 the phase noise power of an oscillator is multiplied by n^2 if the frequency of the oscillator is multiplied by n. Assume that the frequency of the 5 MHz oscillator is multiplied by 200 to give an output at 1000 MHz. The results given in (5.55) and (5.56) will be degraded by $20 \log_{10} 200 = 46$ dB and become:

$$\left(\frac{N_{op}}{C}\right)_{10} = -26 \text{ dB/Hz}$$

(5.57)

$$\left(\frac{N_{op}}{C}\right)_{50} = -40 \text{ dB/Hz}$$

(5.58)

The effects of vibration at 1 g have thus become almost catastrophic at a 1 GHz output frequency and low vibration frequencies. After a further multiplication to a 10 GHz output frequency the results would be truly catastrophic, in fact at 10 Hz vibration frequency the linear approximation (Section 2.5) would no longer be valid because the resultant phase jitter greatly exceeds $0 \cdot 1$ radians. Even at $f_m = 100$ Hz

$$\left(\frac{N_{op}}{C}\right)_{100}$$

would be degraded from -46 dB/Hz at 1 GHz to -26 dB/Hz at 10 GHz.

5.9.5 Manual Tuning Mechanisms

For many applications an oscillator must be tuned over a frequency band by altering one or more of the physical dimensions of its resonator. This may be the sole method of tuning: in other cases, a varactor diode may be used to achieve voltage tuning over small frequency ranges, supplemented by mechanical tuning when large frequency shifts are demanded. For UHF and SHF oscillators the mechanical tuning will usually be achieved by moving a short-circuiting plunger in a coaxial cavity or a waveguide resonator.

At any frequency above about 100 MHz it is normally considered good practice to adopt a non-contacting tuning mechanism, such as a split stator capacitor in the case of capacitance tuning and some type of choked plunger, or high capacitance insulated plunger, for resonator tuning. In some cases it is possible to vary the length of the inner conductor of a coaxial resonator by attaching bellows to the open (high potential) end. The control of these bellows may be achieved by movement of a rod inside the hollow centre conductor. Such practices should be followed even when the oscillator phase noise performance is non-critical.

For a low phase noise oscillator *it is always unacceptable* to use a tuning plunger which achieves its short circuiting effect by means of physical contact, such as the use of spring fingers. Contact changes of the plunger will produce step changes (Δf) in frequency. As an example consider an oscillator using a $\lambda/4$ coaxial resonator at an output frequency of 3 GHz. The electrical length (l) of the resonator will be given by:

$$l = 2 \cdot 5 \text{ cms} \tag{5.59}$$

$$f_0 = \frac{c}{4l} \tag{5.60}$$

For a small change in length Δl (cms).

$$\Delta f = -\frac{c}{4l^2} \Delta l \tag{5.61}$$

$$\therefore \quad \Delta f = -f_0 \frac{\Delta l}{l} \tag{5.62}$$

$$\therefore \quad \Delta f = -3 \times 10^9 \frac{\Delta l}{2 \cdot 5} \text{ Hz} \tag{5.63}$$

Thus, very approximately:

$$\Delta f = -1 \times 10^9 \, \Delta l \text{ Hz} \tag{5.64}$$

and a 1 kHz frequency shift will result from a change in length of 1×10^{-6} cms. A sudden change in length due to microscopic changes in contact between the spring fingers and the surface with which they make contact, will thus produce a step change in frequency, corresponding to a ramp function in phase. Even when the tuning mechanism is nominally stationary and not subject to vibration or temperature changes, the effects will still occur due to microscopic local stress changes. Such frequency jumps will usually become even more frequent and of larger magnitude under the conditions of stress due to temperature changes or vibration.

The use of a signal source incorporating such a poorly designed oscillator will seriously degrade the performance of almost any communication system in which it is used. An analysis of the resulting degradation of the bit error rate of a digital communication system incorporating such an oscillator is carried out in Section 10.5.6.

5.9.6 Dual Resonators

Some oscillators, particularly those using microwave thermionic devices such as disc seal triodes and klystrons may incorporate two, or even more, high Q resonators.

Whilst no attempt will be made here to carry out an analysis of this case, it would be expected that the noise power density both amplitude and phase would fall more rapidly with increase in offset frequency than in the case of a single cavity oscillator. The phase noise power density might be expected to fall by 40 dB per decade with increasing offset frequency over part of the offset frequency range.

The phase noise performance of a very low noise dual cavity X-band klystron oscillator developed by Ferranti Ltd in the 1950's (Reference 16) is plotted in 6 Figure 3 curve (*e*). To the author's knowledge the performance of this oscillator at 100 kHz offset frequency has not been bettered by any X-band source.

5.9.7 Subsequent Filtering

The phase noise characteristics at the output of an oscillator may be improved by passing the carrier through a narrow band, bandpass filter which will pass the carrier and close-in sidebands whilst rejecting sidebands at the higher offset frequencies. Except for the case when the basic oscillator has poor frequency stability and might therefore drift outside the filter passband, the lower limit of the filter bandwidth will be constrained only by the achievable ratio of f_0/Q_f, where f_0 is the oscillator frequency and Q_f is the filter Q.

As an example consider a 1 MHz oscillator followed by a single resonator with a Q of 100. The sidebands would be suppressed 20 dB for offset frequencies greater than

$$\frac{10f_0}{2Q} = \frac{10^7}{200} = 50 \text{ kHz.}$$

In some applications particularly where improvement at small offset frequencies is required a crystal filter may be provided.

5.10 A Simple Misunderstanding

Consider a simple oscillator operating at 100 MHz with a loaded Q of only 100. The 3 dB half bandwidth would appear to be

$$\frac{f_0}{2Q} = \frac{10^8}{2 \times 10^2} = 500 \text{ kHz.}$$

It might seem that the 3 dB bandwidth would extend out to 500 kHz and therefore that equation (5.27) would only be applicable for offset frequencies greater than 500 kHz.

This is *not* so because the noise filtering is due to the feedback Q (Q') which is given by:

$$Q' = Q \frac{2}{\pi} \frac{P}{FkTB} \qquad \text{(see 5.16)}$$

Even for a total power generated as low as 0 dBm and a poor noise figure (F) of 30 dB, $P/FkTB$ is a very large number. It may easily be evaluated by working in dB.

The bandwidth $B = 1$ MHz $= 60$ dB and $kT = -174$ dBm

$$\therefore \quad \frac{P}{FkTB} = (0 - 30 + 174 - 60) \text{ dB}$$

$$= 84 \text{ dB} = 10^{8.4}$$

$$\therefore \quad Q' = 100 \times \frac{2}{\pi} \times 10^{8.4}$$

$$\therefore \quad Q' = 1.6 \times 10^{10}$$

The 3 dB half bandwidth for noise filtering purposes is thus

$$\frac{f_0}{2Q'} = \frac{10^8}{3.2 \times 10^{10}} = 0.003 \text{ Hz}$$

CHAPTER 6

FREQUENCY MULTIPLIER CHAINS

6.1 The Need for Frequency Multiplication

It is apparent from equation (5.27) that, when all other parameters are fixed,

$$\frac{N_{op}}{C} \propto \frac{f_0^2}{Q^2} \tag{6.1}$$

At frequencies up to the order of 120–150 MHz it is possible to use a quartz crystal as resonator and thus achieve simultaneously a high Q and good frequency stability and accuracy. At still higher frequences the phase noise performance of a simple oscillator will be degraded both due to the higher value of f_0 and simultaneously the lower Q of a suitable resonant circuit which can no longer be a quartz crystal. A further penalty is that the frequency accuracy and stability will be degraded. Thus it is apparent that a considerable improvement in performance might be achieved by using a crystal oscillator followed by a frequency multiplier to give the required output frequency. This would appear to be a beneficial tradeoff even if the multiplication process produced a phase noise degradation proportional to the square of the multiplication ratio. This degradation would be exactly offset by the f_0^2 term in the numerator of the RHS of equation (6.1), leaving the improvement in Q as a bonus, to say nothing of the improved frequency stability.

A further bonus is often available in that the noise figure of the transistor oscillator may be considerably better if the oscillator is operated at a sub-multiple of the required output frequency.

6.2 The Effect of Frequency Multiplication

6.2.1 An Amplitude Modulated Wave
Consider an amplitude modulated wave given by:

$$e = \sqrt{2C}\,[1 + M'(t)]\,\sin w_1 t$$

The frequency is $w_1/2\pi$. If the frequency is multiplied by

$$n\left(n = \frac{w_2}{w_1}\right):$$

$$e = \sqrt{2C}\,[1 + M'(t)] \sin nw_1 t$$

$$\therefore \quad e = \sqrt{2C}\,[1 + M'(t)] \sin w_2 t \quad \text{(referred to unity gain)}$$

It should be noted that the modulation index is unchanged by *ideal* frequency multiplication and that hence the AM noise sidebands will also be unchanged. A real frequency multiplier, being a non-linear device, may multiply the AM sidebands or, in so far as it limits, it may suppress all AM. In practice the AM performance is a function of the details of the actual design adopted.

6.2.2 A Phase Modulated Wave

Rewriting equation (2.4) with w_1 in place of w, the equation for a phase modulated signal is given by:

$$e_1 = \sqrt{2C} \sin (w_1 t + \theta \sin pt) \tag{6.2}$$

The phase of this signal is:

$$\Phi_1 = (w_1 t + \theta \sin pt)$$

The frequency is given by:

$$\frac{1}{2\pi}\frac{d\Phi_1}{dt} = f_1(t) = \frac{1}{2\pi}(w_1 + p\theta \cos pt) \tag{6.3}$$

Let us multiply the frequency by n to give a new frequency $f_2(t)$ and new signal e_2.

$$\therefore \quad f_2(t) = nf_1(t) = \frac{1}{2\pi}(nw_1 + np\theta \cos pt)$$

This may be written:

$$f_2(t) = \frac{1}{2\pi}(w_2 + np\theta \cos pt) \tag{6.4}$$

The phase of this signal is given by:

$$\Phi_2 = \int (w_2 + np\theta \cos pt)dt$$

$$\therefore \quad \Phi_2 = (w_2 t + n\theta \sin pt) \quad \text{(neglecting the irrelevant}$$
$$\text{integration constant)} \tag{6.5}$$

$$\therefore \quad e_2 = \sqrt{2C} \sin (w_2 t + n\theta \sin pt) \quad \text{(referred to unity gain)} \tag{6.6}$$

Equation (6.6) represents the signal after frequency multiplication and is similar to equations (6.2) and (2.4) for a phase modulated signal, apart from the change in carrier frequency and the value of the phase modulation index which is $n\theta$ instead of θ. It will be noted further that the modulating frequency $(p/2\pi)$, which is also the sideband offset frequency (f_m), is unchanged.

Thus, we may conclude that as far as phase modulation is concerned, the effect of frequency multiplication by a factor n, is to multiply the modulation index by n.

The effect of frequency multiplication may also be pictured physically as follows. Consider the 'instantaneous frequency' of a frequency modulated wave. Frequency multiplication by a factor n will multiply the 'instantaneous frequency' by n without altering the rate at which the carrier is being modulated. Thus the carrier frequency f_0 will become nf_0 and the peak positive frequency $(f_0 + \Delta f)$ will become $(nf_0 + n\Delta f)$. The effective frequency deviation thus becomes $n\Delta f$. Now the phase modulation index $(M$ or $\theta) = \Delta f/f_m$ (equation 2.13). Hence the new modulation index is:

$$\theta' = \frac{n\Delta f}{f_m} = n\theta$$

All the analyses of Chapters 2 and 3 may now be considered to apply to a signal which has been subject to a frequency multiplication n, by substituting $n\theta$ for θ, $n\phi_0$ for ϕ_0, $(n\phi)^2$ for ϕ^2 etc.

This equivalence is subject to one important restriction which is that when $n\theta$ is substituted for θ the inequality of equation (2.9) and $\tan(n\theta) = n\theta$ shall still be valid approximations. As long as this is so, and it is almost always so for a carrier unintentionally modulated by noise, the carrier peak amplitude $\simeq \sqrt{2C}$ and only the first order sidebands are significant, each having a peak amplitude of $\sqrt{2C}\,n\theta/2$. As a guide it is suggested that the validity of the approximations should be considered whenever $n\theta$ exceeds 0·1 radians or the DSB signal to noise ratio falls below 23 dB.

6.2.3 Effect of Frequency Multiplication on Phase Noise

Let $(N_{op}/C)_1$ and $(N_{op}/C)_2$ be the phase noise density to carrier ratios and $\overline{\phi_{o1}^2}$ and $\overline{\phi_{o2}^2}$ the corresponding mean square phase jitter densities prior to, and after frequency multiplication respectively. From equation (3.43)

$$\left(\frac{N_{op}}{C}\right)_1 = \frac{\overline{\phi_{o1}^2}}{2}$$

After frequency multiplication

$$\phi_{o2} = n\phi_{o1}$$

and

$$\left(\frac{N_{op}}{C}\right)_2 = \frac{\overline{\phi_{o2}^2}}{2}$$

$$\therefore \quad \left(\frac{N_{op}}{C}\right)_2 = \frac{\overline{(n\phi_{o1})^2}}{2} = n^2\left(\frac{N_{op}}{C}\right)_1 \tag{6.7}$$

Thus, the effect of frequency multiplication by a factor n is to multiply the phase noise power to carrier ratio by n^2.

6.2.4 Effect of Frequency Multiplication on Thermal Noise

For the case of thermal noise superposed on a carrier there are two cases to consider as follows:

(a) Single sideband noise
(b) Double sideband noise

6.2.4.1 Superposed SSB Thermal Noise

Let thermal noise of noise density N_0 and frequency $(f_0 + \delta f)$ be superposed upon a clean carrier of power level C and frequency f_0 as shown in 6. Figure 1.
   The modulation indices are:

$$M' = \sqrt{\frac{N_0}{C}} \tag{6.8.1}$$

$$\theta = \sqrt{\frac{N_0}{C}} \tag{6.8.2}$$

The equations for the two real and two potential sidebands are given in equations (3.16) which are repeated here for convenience.

$$\text{Upper PM sideband} = +\sqrt{\frac{N_0}{2}} \sin(w + p)t \tag{6.9.1}$$

$$\text{Upper AM Sideband} = +\sqrt{\frac{N_0}{2}} \sin(w + p)t \tag{6.9.2}$$

$$\text{Lower PM Sideband} = -\sqrt{\frac{N_0}{2}} \sin(w - p)t \tag{6.9.3}$$

Phasor Diagram   6 Figure 1

$$\text{Lower AM Sideband} = + \sqrt{\frac{N_0}{2}} \sin (w - p)t \tag{6.9.4}$$

After frequency multiplication by n, the carrier frequency will be multiplied by n to become $nw/2\pi = nf_0$, the PM modulation index will be multiplied by n and the AM modulation index will be unchanged. Thus after frequency multiplication the sidebands will become:

Upper PM sideband

$$\text{UPM}_2 = + n \sqrt{\frac{N_0}{2}} \sin (nw + p)t \tag{6.10.1}$$

Upper AM sideband

$$\text{UAM}_2 = + \sqrt{\frac{N_0}{2}} \sin (nw + p)t \tag{6.10.2}$$

Lower PM sideband

$$\text{LPM}_2 = - n \sqrt{\frac{N_0}{2}} \sin (nw - p)t \tag{6.10.3}$$

Lower AM sideband

$$\text{LAM}_2 = + \sqrt{\frac{N_0}{2}} \sin (nw - p)t \tag{6.10.4}$$

It will be noted that the lower sidebands no longer completely cancel. The combined lower sideband is:

$$\text{LS}_2 = \left(-n \sqrt{\frac{N_0}{2}} + \sqrt{\frac{N_0}{2}} \right) [\sin (nw - p)t]$$

$$\therefore \quad \text{LS}_2 = - \sqrt{\frac{N_0}{2}} (n - 1)[\sin (nw - p)t] \tag{6.11}$$

The power in the combined lower sideband is:

$$\text{PLS}_2 = \frac{N_0}{4} (n - 1)^2 \tag{6.12}$$

and the power in each of the upper sidebands is:

$$\text{Power UPM}_2 = n^2 \frac{N_0}{4} \tag{6.13}$$

$$\text{Power UAM}_2 = \frac{N_0}{4} \tag{6.14}$$

If $n \gg 1$ then the result of frequency multiplication is to turn low level single sideband noise into predominately DSB phase noise. When the approximation involved in writing $(n - 1) = n$ is justified, then the resultant PM sidebands, due to superposed SSB noise followed by a frequency multiplication of n times, each have a power level of $n^2 N_0/4$ and the SSB phase noise power density to carrier ratio is given by:

$$\frac{N_{op}}{C} = n^2 \frac{N_0}{4C} \qquad (6.15)$$

An interesting feature emerges in this case of high order frequency multiplication of noise superposed on a carrier. To the degree of accuracy involved in the approximation that the AM sidebands are negligible compared with the PM sidebands, the upper and lower resultant sidebands are conformable.

6.2.4.2 Superposed DSB White Thermal Noise

Consider the case where DSB white noise of noise power density N_0 is superposed on a 'clean' carrier of power level C and frequency $f_0 = w/2\pi$. The total power in the two 1 Hz wide sidebands at $(f_0 \pm \delta f) = 2N_0$. Of this $N_0/2$ appears in each of the upper and lower phase noise sidebands and the power of each of the AM noise sidebands is also $N_0/2$. After frequency multiplication the AM sidebands will be unchanged and the initial PM modulation index will be multiplied by n to become $n\phi$.

Let N_{op1} and N_{op2} be the phase noise power densities before and after frequency multiplication respectively and N_{o1} be the noise density prior to frequency multiplication. Assume unity power gain for the carrier,

i.e. $C_2 = C_1 = C$

From equation (6.7)

$$N_{op2} = n^2 N_{op1}$$

Now

$$N_{op1} = \frac{N_{o1}}{2} \quad \text{(half PM)}$$

$$N_{op2} = n^2 \frac{N_{o1}}{2} \qquad (6.16)$$

$$\text{and} \quad \left(\frac{N_{op}}{C}\right)_2 = n^2 \left(\frac{N_o}{2C}\right)_1 \qquad (6.17)$$

If n is a large number so that, after frequency multiplication, the AM contribution to the noise sidebands is negligible, a measurement of the noise density to sideband carrier ratio would in fact, be largely a measurement of N_{op2}/C. That is the spectrum analyser or other measuring equipment would indicate the 'phase noise' density rather than the noise density as the total noise is practically all phase noise.

Thus, it should be remembered that frequency multiplication by a factor n, multiplies phase noise power by n^2 and, to a fair approximation pure noise power by $n^2/2$. (c.f. equations 6.7 and 6.17).

6.3 Added Amplifier Noise

In many cases the power output (C) of a low phase noise oscillator may be relatively low and in fact insufficient to meet the specified output level for the signal source. This may be true for a fundamental frequency source. It is even more likely to be true if the oscillator frequency is a sub-multiple of the required frequency and the necessary frequency multiplication is achieved by a varactor, or step recovery diode, multiplier, which will inevitably attenuate the signal. Thus, an amplifier will be required. In some instances there may be a choice as to where the amplifier is located in the circuit. The two possibilities are:

(*a*) Prior to frequency multiplication.
(*b*) After frequency multiplication.

We shall consider in turn:

(1) A fundamental frequency source consisting of an oscillator operating at the required output frequency, followed by a fundamental frequency amplifier.
(2) An oscillator/frequency multiplier combination with circuit arrangements either:

(*a*) Amplification prior to frequency multiplication.
(*b*) Amplification after frequency multiplication.

6.3.1 A Fundamental Frequency Source
Let the noise figures of the oscillator and amplifier be F_0 and F_A respectively. Assume the narrow band noise figure to be constant over the frequency range $f_0 \pm \delta f$ where δf is large compared with the offset frequency range of interest. The amplifier available noise power density (N_{oA}) referred to the amplifier *input* will be given by:

$$N_{oA} = F_A kT \qquad (6.18)$$

Let the total noise power density be N_{ot}, that of the oscillator N_{oo} and that of the amplifier N_{oA}.

Let the carrier power at the input to the amplifier be C.

$$\therefore \quad \frac{N_{ot}}{C} = \frac{N_{oo}}{C} + \frac{N_{oA}}{C} \tag{6.19}$$

Substituting the values of N_{oo}/C and N_{oA}/C given by equations (5.27) and (6.18):

$$\frac{N_{ot}}{C} = \frac{F_o kT}{C} \frac{1}{4Q^2} \left(\frac{f_0}{f_m}\right)^2 + \frac{F_A kT}{C} \tag{6.20}$$

Both the noise density and the carrier power will be subject to the same amplifier power gain: at the amplifier output their ratio will be unchanged. Let C_x be the output power of the source and N_{ox} the output noise density. Then:

$$\frac{N_{ox}}{C_x} = \frac{F_0 kT}{C} \frac{1}{4Q^2} \left(\frac{f_0}{f_m}\right)^2 + \frac{F_A kT}{C} \tag{6.21}$$

Now, in the absence of frequency multiplication or limiting,

$$\frac{N_0}{C} = \frac{2N_{op}}{C} \quad \text{and} \quad \frac{N_{ox}}{C_x} = \frac{2N_{opx}}{C_x} \tag{6.22}$$

where N_{opx} is the phase noise density at the signal source output.

$$\therefore \quad \frac{N_{opx}}{C_x} = \frac{F_0 kT}{2C} \frac{1}{4Q^2} \left(\frac{f_0}{f_m}\right)^2 + \frac{F_A kT}{2C} \tag{6.23}$$

Consider the case where F_0 and F_A are similar in value: this is often so because the higher power level of the amplifier frequently offsets the difficulty in achieving a good oscillator noise figure.

For small values of f_m, the oscillator phase noise, which is the first term of (6.23), will be greater than the amplifier contribution (second term). When

$$\frac{1}{4Q^2} \left(\frac{f_0}{f_m}\right)^2 = 1$$

or

$$f_m = \frac{f_0}{2Q} \tag{6.24}$$

the oscillator and amplifier contributions to the combined phase noise density will be equal and the oscillator performance will be degraded 3 dB by the addition of the amplifier.

When $f_m \gg f_0/2Q$ the (white) amplifier phase noise will dominate the overall performance and therefore N_{ox}/C_x and N_{opx}/C_x will remain at a constant value with increasing frequency, up to an offset frequency where the selectivity of the amplifier begins to attenuate the noise sidebands. For many oscillator amplifier combinations, where the oscillator loaded Q is high compared with the Q of any resonant circuits in the amplifier, the source phase noise performance consists of three regions. For small offset frequencies, up to f_c:

$$\left(\frac{N_{op}}{C}\right)_{f_m} \propto \frac{1}{f_m^3} \tag{6.25}$$

For an intermediate range of offset frequencies:

$$\left(\frac{N_{op}}{C}\right)_{f_m} \propto \frac{1}{f_m^2} \tag{6.26}$$

For a still higher range of offset frequencies:

$$\left(\frac{N_{op}}{C}\right)_{f_m} \quad \text{is independent of } f_m \tag{6.27}$$

6.3.2 A Source with Amplification Prior to Frequency Multiplication

In the case where an oscillator feeds directly into an amplifier, the phase noise density to carrier ratio at the amplifier output is given by equation (6.23), which is repeated here for convenience.

$$\frac{N_{opx}}{C_x} = \frac{F_0 kT}{2C} \frac{1}{4Q^2} \left(\frac{f_0}{f_m}\right)^2 + \frac{F_A kT}{2C}$$

If the amplifier output is fed into a frequency multiplier, which might consist of a varactor diode frequency multiplier chain or a step recovery diode (SRD) multiplier, then the frequency of the final output may be multiplied by a factor n.

Frequency multiplication by a factor n will multiply the phase noise power to carrier ratio by n^2. Let the noise density to carrier ratio at the output be N_{op2}/C_2. Thus, after frequency multiplication:

$$\frac{N_{op2}}{C_2} = n^2 \left[\frac{F_0 kT}{2C} \frac{1}{4Q^2} \left(\frac{f_0}{f_m}\right)^2 + \frac{F_A kT}{2C}\right] \tag{6.28}$$

Inserting the amplifier prior to frequency multiplication thus has the disadvantage that the phase noise contribution of the amplifier is multiplied by n^2. It has a number of offsetting advantages compared with the arrangement where the initial amplification follows the frequency multiplier. These are:

(a) To obtain sufficient power output from the oscillator to drive most high order frequency multiplier chains involves a serious sacrifice in oscillator performance. The noise figure, the Flicker Noise performance and the Q may all be seriously degraded.

(b) The noise figure of the amplifier (F_A) will be good, as the amplifier is centred on the relatively low oscillator frequency.

It should be realised that the foregoing discussion refers to the first power amplifier only. As soon as the carrier power has been amplified to a sufficiently high level, added noise at normal levels will have little effect on the noise/carrier ratio. Frequency multipliers, other than those using transistors, will attenuate the carrier, perhaps by as much as $1/n$, and it is necessary to ensure that, during the multiplication process, the carrier power level is not allowed to fall below the value where subsequent superposed noise may be significant. In some cases it may be necessary to provide an amplifier between two frequency multiplying stages.

6.3.3 A Source with Amplification After Frequency Multiplication

If the final required output frequency is not too high, and hence the frequency multiplication ratio fairly small, it may be possible to provide sufficient power output from the basic oscillator, to drive the frequency multiplier directly. Even so this is only likely to be true if the frequency stability requirements are modest enough to permit a fairly high power level in the master resonant circuit. If this is so, and it is arranged that the amplifier follows the frequency multiplier, the amplifier phase noise contribution will not suffer multiplication by n^2. For the case where the phase noise at large offset frequencies (where amplifier noise often dominates) is of great importance, this arrangement may give a superior performance. This is particularly true when the output frequency is still low enough that the noise figure of the amplifier has not been seriously degraded compared with that which could have been obtained if the amplifier had been operated at the oscillator frequency rather than the output frequency.

A comparison of the likely performance for the two cases of an amplifier before or after the frequency multiplier may readily be made if the carrier attenuation due to the frequency multiplication is known. For the purpose of illustration let it be assumed that a carrier power C at the input to the multiplier becomes a power C/n at the multiplier output. This is roughly typical of many multipliers. Let F_{A1} be the noise figure of an amplifier at the input (oscillator) frequency. Let F_{A2} be the noise figure of an amplifier at the output frequency.

If an input amplifier were used the phase noise contribution of this amplifier, referred to the output, would be:

$$\frac{N_{op2}}{C_2} = n^2 \frac{F_{A1}kT}{2C} \tag{6.29}$$

For an amplifier, inserted after the multiplier, the phase noise contribution would be:

$$\frac{N_{op2}}{C_2} = F_{A2}kT \div \frac{2C}{n} = \frac{nF_{A2}kT}{2C} \tag{6.30}$$

An improved performance would be obtained in the second case only if:

$$\frac{nF_{A2}\,kT}{2C} < \frac{n^2 F_{A1}\,kT}{2C}$$

i.e.   when $F_{A2} < nF_{A1}$ (6.31)

6.4 Example—An X-band Crystal Oscillator/Multiplier Source

We will consider the engineered equipment, the crystal oscillator of which was described in Section 5.7.1. This oscillator feeds into an amplifier and thence into a multi stage varactor frequency multiplier with an overall multiplication ratio of 72. The first stage of the amplifier has the following characteristics:

Transistor used   = 2N3866
Input Power (C) = 5 mW = 7 dBm
Noise Figure (F_A) = 20 dB
Power Gain   = 16 dB
Power Output   = 200 mW

From equation (6.23) the amplifier contribution to the phase noise density to carrier ratio, before frequency multiplication, is:

$$\frac{N_{opA}}{C} = \frac{F_A\,kT}{2C}$$

$$\therefore \quad \frac{N_{opA}}{C} = (+20 - 174 - 3 - 7) \text{ dB/Hz}$$

$$= -164 \text{ dB/Hz}$$

The calculated phase noise performance of the oscillator described in Section 5.7.1 above is shown in 5 Figure 7. Frequency multiplication by n degrades the phase noise/carrier ratio by n^2.

$$n^2 = (72)^2 = 37 \cdot 1 \text{ dB}$$

The calculated overall phase noise performance at the output frequency of 8640 MHz, together with the measured performance of two units are shown in 6 Table 1 and 6 Figure 2. The table also shows the individual contributions of the oscillator and the amplifier both at 120 MHz (columns 2 and 4 respectively) and at the X Band output frequency (columns 3 and 5 respectively). In column 6 the phase noise powers from columns 3 and 5 are added to give the overall noise density to carrier ratio N_{op}/C in dB. The results of measurements on two different units, made to the same design are given in columns 7 and 8.

The agreement between predicted and measured values is considered to be reasonable. It will be seen, by comparing columns 3 and 5 of 6 Table 1, that

6 Table 1   X-Band Source

Comparison of Predicted and Measured Performance $\left(\dfrac{N_{op}}{C}\right)$ dB/Hz

		Theory				Measurement	
fm kHz	Osc PM 120 MHz	Osc PM X Band	Amp PM 120 MHz	Amp PM X Band	PM Total X Band	Unit 001 PM X Band	Unit 002 PM X Band
	-dBc	-dBc	-dBc	-dBc	-dBc	-dBc	-dBc
1	140·5	103·4	164	126·9	103·4	96·3	
2	146·5	109·4	164	126·9	109·3		97·7
3	150	112·9	164	126·9	112·7	107·6	103·5
5	154·5	117·4	164	126·9	116·9	114	109
10	160·5	123·4	164	126·9	121·8	118	115
20	166·5	129·4	164	126·9	125	122·6	118·6
30	170	132·9	164	126·9	125·9		119·5
40	172·5	135·4	164	126·9	126·3	123·2	118
50	174·5	137·4	164	126·9	126·5		122·5
60	176	138·9	164	126·9	126·6	125·2	123·3
70	177·4	140·3	164	126·9	126·7		124·7
80	178·5	141·4	164	126·9	126·7	126	124·7
1	2	3	4	5	6	7	8

the 'flat' amplifier noise begins to dominate in the region of 20 kHz offset frequency.

The accuracy of the measurements was estimated to be approximately $\pm0\cdot5$ dB. These measurements were carried out using X-band noise measuring equipment described in Reference 12.

It is perhaps worthwhile pointing out that the agreement between theory and practice implies that the varactor frequency multipliers did not themselves contribute significantly to the phase noise of the source in any way. The satisfactory results recorded were only achieved after carrying out modifications to the initial varactor multiplier designs to eliminate parasitic oscillation. (Reference 17).

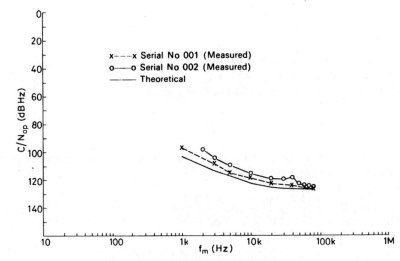

X-Band Source (8640 MHz). 120 MHz Crystal Oscillator × 72   **6 Figure 2**
Theoretical and measured performance

6.5 Limitations of Simple Oscillator and Oscillator/Multiplier Sources

Although a simple oscillator may be tuned over a relatively broad band, oscillator/multiplier sources usually provide only a single frequency output or, at the best, output over a relatively narrow tuning range. This is due to the necessity, in any but the crudest sources, to provide adequate filtering between and after the frequency multiplier stages, to prevent the passage of unwanted harmonics of the basic oscillator. In addition, if system requirements impose a tight tolerance on the output frequency or frequencies, and the frequency drift is to be maintained at a very low value, then the fundamental oscillator must be a crystal oscillator (or atomic standard) with a very limited tuning range. Any attempt to increase the tuning range of a crystal oscillator substantially will inevitably cause degradation of its frequency stability. Thus, if a multiplicity of accurately selectable output frequencies is required, this reason alone will necessitate the adoption of a frequency synthesiser as a signal source.

Even for the case where only a single output frequency is required, simple oscillator or oscillator/multiplier chains have further limitations with respect to phase noise performance. These limitations become increasingly apparent as the required output frequency increases. In considering these limitations it is therefore convenient to compare the phase noise performance of various types of signal sources with a final output at X-band.

A comparison of the two extreme approaches to the problem of providing a low phase noise X-band source, is illuminating. These are, on the one hand the use of a high quality crystal oscillator followed by an amplifier and frequency multipliers; and on the other hand the use of a relatively high power X-band oscillator.

Let the crystal oscillator frequency be f_1.

Let the X-band output frequency be f_2.

Then the frequency multiplication ratio n is given by:

$$n = \frac{f_2}{f_1}$$

The phase noise performance of the crystal oscillator/multiplier source is given by equation (6.28) which, with f_1 substituted for f_0, is:

$$\frac{N_{op2}}{C_2} = \frac{F_0 kT}{8C}\left(\frac{nf_1}{Qf_m}\right)^2 + n^2 \frac{F_A kT}{2C} \tag{6.32}$$

$$\therefore \quad \frac{N_{op2}}{C_2} = \frac{F_0 kT}{8C}\left(\frac{f_2}{Qf_m}\right)^2 + n^2 \frac{F_A kT}{2C} \tag{6.33}$$

For a source consisting of an X band oscillator, using equation (5.27), replacing N_{op} by N_{opx}, C by C_2, f_0 by f_2, and F_0 and Q by F_x and Q_x (to denote values for an X-band oscillator) gives:

$$\frac{N_{opx}}{C_2} = \frac{F_x kT}{8C_2}\left(\frac{f_2}{Q_x f_m}\right)^2 \tag{6.34}$$

This is identical in form to the first term on the right hand side (RHS) of (6.33) and differs only in the values to be substituted for F, C and Q. Dividing (6.34) by the first term on the RHS of (6.33):

$$\frac{F_x kT}{8C_2}\left(\frac{f_2}{Q_x f_m}\right)^2 \frac{8C}{F_0 kT}\left(\frac{Qf_m}{f_2}\right)^2$$

$$= \frac{F_x}{F_0}\frac{C}{C_2}\left(\frac{Q}{Q_x}\right)^2 = (\text{say}) \; L \tag{6.35}$$

L is the ratio of the X-band oscillator phase noise density/carrier ratio to that of the crystal oscillator referred to the output frequency but without the amplifier contribution.

$(F_x/F_0)(C/C_2)$ is likely to be greater than unity due to the very much poorer noise figure of an X-band high power oscillator, even if this is somewhat offset by C_2 being considerably greater than C. However the Q of the crystal oscillator is likely to be much greater than that of the X-band oscillator so that:

$$\left(\frac{Q}{Q_x}\right)^2 \gg 1 \quad \text{and} \quad L \gg 1$$

From (6.34) and (6.35)

$$\frac{N_{opx}}{C_2} = \frac{F_0 kT}{8C} \left(\frac{f_2}{Qf_m}\right)^2 \times L$$

$$\frac{N_{opx}}{C_2} = \frac{F_0 kT}{8C} \left(\frac{f_2}{Qf_m}\right)^2 + (L-1)\frac{F_0 kT}{8C} \left(\frac{f_2}{Qf_m}\right)^2$$

Thus at the offset frequency f'_m for which:

$$\frac{n^2 F_A kT}{2C} = (L-1)\frac{F_0 kT}{8C} \left(\frac{f_2}{Qf'_m}\right)^2 \tag{6.36}$$

the phase noise density to carrier ratio of the crystal oscillator multiplier chain and the X-band oscillator will be equal.

Therefore for equality of performance:

$$\left(f'_m \frac{Q}{f_2}\right)^2 = (L-1)\frac{F_0 kT}{8C} \frac{2C}{n^2 F_A kT}$$

$$\therefore \quad f'^2_m = \frac{(L-1)}{n^2} \frac{F_0}{F_A} \left(\frac{f_2}{2Q}\right)^2$$

But

$$\left(\frac{f_2}{n}\right)^2 = f_1^2$$

$$\therefore \quad f'^2_m = (L-1)\frac{F_0}{F_A} \left(\frac{f_1}{2Q}\right)^2 \tag{6.37}$$

$$f'^2_m = L\frac{F_0}{F_A} \left(\frac{f_1}{2Q}\right)^2 - \frac{F_0}{F_A} \left(\frac{f_1}{2Q}\right)^2$$

Substituting for L from (6.35)

$$f'^2_m = \frac{F_x}{F_0} \frac{F_0}{F_A} \frac{C}{C_2} \left(\frac{Q}{Q_x}\right)^2 \left(\frac{f_1}{2Q}\right)^2 - \frac{F_0}{F_A} \left(\frac{f_1}{2Q}\right)^2$$

$$f'^2_m = \frac{F_x}{F_A} \frac{C}{C_2} \left(\frac{f_1}{2Q_x}\right)^2 - \frac{F_0}{F_A} \left(\frac{f_1}{2Q}\right)^2 \tag{6.38}$$

To the extent that L is a reasonable approximation for $(L-1)$

$$f'^2_m = \frac{F_x}{F_A} \frac{C}{C_2} \left(\frac{f_1}{2Q_x}\right)^2 \tag{6.39}$$

For values of f_m below f'_m the crystal oscillator/multiplier chain will have a superior noise performance to the X-band oscillator. For larger offset frequencies the X-band oscillator will achieve a better phase noise performance. Thus, for a simple signal source with an output at a microwave frequency, it is in general impossible to obtain simultaneously good phase noise performance over the whole offset frequency range. These limitations may be overcome by the incorporation of phase lock loops into the signal source, which leads naturally to the adoption of frequency synthesiser techniques.

The use of phase lock loops in signal sources is discussed in Chapter 7 and synthesisers are discussed in Chapter 8. Before doing so, however, the performance achieved by some simple signal sources will be outlined in Section 6.6 and a brief discussion of spurious outputs from oscillator/multiplier chains will be given in Section 6.7.

6.6 The Performance of some X-Band Sources

In this section the phase noise performance, of a number of simple signal sources with actual or potential outputs at X-band, is compared. The output frequencies are not identical but are close enough to make the comparison meaningful.

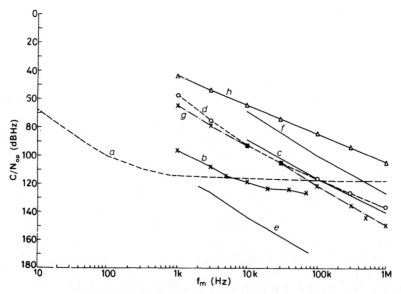

Phase Noise Performance of some X-Band Sources   **6 Figure 3**

The sources considered are as follows:

(*a*) An Austron 1120S 5 MHz crystal oscillator followed by a X 1600 frequency multiplier, giving an output at 8 GHz. (Reference 14).

(*b*) A 120 MHz crystal oscillator followed by a X72 frequency multiplier, giving an output at 8640 MHz. (See Section 6.4).

(*c*) A Frequency West mechanically tuned bipolar transistor cavity oscillator at 1.2 GHz followed by a X6 frequency multiplier to give an output at 7.2 GHz (Reference 18).

(*d*) A 2 K25 reflex klystron (Reference 20) at 9.4 GHz.

(*e*) A Ferranti low noise two cavity klystron oscillator (Reference 16).

(*f*) A GaAs FET X band oscillator (Reference 19) at 10.14 GHz.

(*g*) An X band Gunn oscillator (Reference 20) at 11.1 GHz.

(*h*) An X band Impatt oscillator (Reference 20).

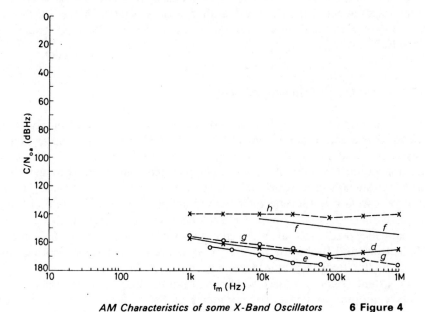

AM Characteristics of some X-Band Oscillators   **6 Figure 4**

Klystron oscillator noise theory, which is closely analogous to the treatment given in Chapter 5, is outlined in Reference 21.

The phase noise performance of each of the above signal sources is shown in 6 Figure 3 and the AM noise performance of some of these sources is shown in 6 Figure 4. With the exception of sources (*a*) and (*c*) all these values are measured values. The measurements on signal source (*b*) were carried out at RRE Malvern during October 1966 by Mr N. R. Court on signal sources supplied by the author. The measured values for the other sources have been abstracted from the quoted references and, if necessary, converted by calculation to the same form N_{op}/C, when initially expressed as a frequency deviation. In the case of signal source (*a*), the Austron data for their 5 MHz reference oscillator (Austron 1120S) has been scaled to X-band, assuming it to be followed by an amplifier with a noise figure of 3 dB and frequency multiplication by 1600 times. Such an amplifier noise figure would be difficult to achieve even at 5 MHz unless the first amplifier stage gain (and hence its power level) were limited so that its noise figure could be maintained. In practice, it would also be difficult, but not impossible, adequately to control the spurious output level of unwanted harmonics of the 5 MHz oscillator.

The superb performance of signal source (*e*) in respect of both AM noise and phase noise is due to a large extent to the use of two high Q cavities. The use of two cavities was briefly discussed in Section 5.9.5. Other factors contributing to this performance are the use of optimum electron gun and cavity gap designs. This klystron was specially developed by Ferranti for an RRE CW radar requirement with which the author was associated in the 1950's and early 1960's. For details of some of this work see Section 10.1 and References 22 and 23.

6.7 Spurious Outputs from Signal Sources

In this chapter phase jitter in signal sources has been attributed purely to the effects of statistical noise contributions (thermal noise, shot noise etc) in oscillators and amplifiers modified by the effects of any frequency multiplication. However, there are other possible contributions to the phase jitter of a signal source which are due to various non-statistical causes. In the case of oscillators, the effects of power supply ripple and vibration were discussed in Section 5.9. Another effect which may be present is the existence of parasitic oscillation at a specific unwanted frequency or frequencies. Magnetrons and Backward Wave Oscillators are particularly susceptible, and even varactor frequency multipliers may manifest this effect. To a greater or lesser extent, signal sources which incorporate frequency multipliers will produce outputs at unwanted harmonics of the basic oscillator frequency. Such effects may also be apparent in the outputs of frequency synthesisers unless special precautions are taken.

The question of spurious outputs and the way in which their effects may be integrated with the phase noise to determine the overall phase jitter in any given offset frequency band are treated in Section 8.5 below.

CHAPTER 7

THE USE OF PHASE LOCK LOOPS

7.1 Introduction

It is apparent from 6 Figure 3 that an SHF signal source consisting of a simple crystal oscillator multiplier chain cannot simultaneously achieve good phase noise performance over the whole offset frequency range. Comparison of curves (a) and (b) shows that the 5 MHz SMO with its superb performance at low offset frequencies is limited at high offset frequencies by amplifier noise. Basically this is due to the low level of power output (C) which is inherent in this type of oscillator.

A better overall performance would be obtained if it were possible to take the performance shown by curve (a) up to an offset frequency of about 5 kHz and that of curve (b) at higher offset frequencies. In principle this would be possible by using source (b) to provide the final output, but to lock the phase of this source to that of the frequency multiplied 5 MHz source using a phase lock loop (PLL) with a natural frequency of about 5 kHz.

However this would necessitate modifying the 120 MHz crystal oscillator of source (b) to make it capable of voltage tuning. The price of doing this, for a narrow tuning range, would probably be 10–12 dB degradation of phase noise performance and the phase noise performance at large offset frequencies (say 1 MHz), although good, would not meet a very stringent requirement such as:

$$(C/N_{op})_{1 \text{ MHz}} = 130 \text{ dB Hz}.$$

Consider the performance of a very good X-band cavity voltage controlled oscillator (VCO) with a narrow tuning range and the following parameters:

$F = 20$ dB
$Q = 500$
$C = +10$ dBm
$f_0 = 8640$ MHz

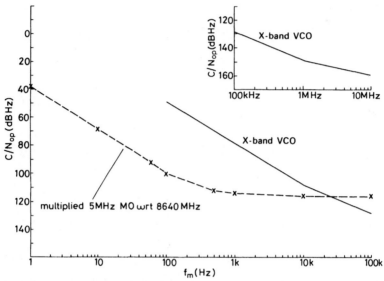

Illustrating a very good Composite X-Band Source   **7 Figure 1**

Using equation (5.27)

$$\left(\frac{N_{op}}{C}\right)_{10\text{ kHz}} = -108 \text{ dB/Hz}.$$

The calculated performance of such an oscillator, assuming f_c to be at about 10 kHz, is plotted on 7 Figure 1.

The performance of the 5 MHz SMO, after multiplication to 8640 MHz [curve (*a*) of 6 Figure 3] is also plotted on 7 Figure 1. The phase noise curves of the *X*-band VCO and the multiplied 5 MHz MO cross at about 25 kHz. If the *X*-band VCO were phase locked to the multiplied MO using a phase lock loop (PLL) with a natural frequency of about 25 kHz then the composite source would have an overall performance the graph of which would follow that of the multiplied 5 MHz MO up to 25 kHz and that of the *X*-band VCO for all higher offset frequencies.

Thus the overall phase noise performance of signal sources over an offset frequency range from the lowest to the highest offset frequencies may be greatly improved by the use of phase lock loops which will now be considered in more detail.

7.2 Phase Lock Loop Configuration

In the simplest case, when frequency multiplication or division is not involved, a PLL consists of a signal source, which may be regarded as the reference, to which it is required to phase lock a voltage controlled oscillator (VCO). The

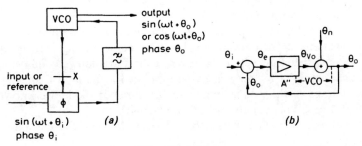

PLL—Block Diagram **7 Figure 2**

reference and VCO signals are fed into the two inputs of a phase detector. Only the baseband output of the phase detector is used; outputs at carrier frequency or harmonics thereof being rejected by a following low pass filter, which also serves to shape the PLL response. The amplitude of the baseband output signal from the phase detector is proportional to the 'phase error' between the two signals. This error signal is fed to the VCO thus achieving phase error correction by closing a negative feedback loop. A simple block diagram of the PLL is given in 7 Figure 2(*a*) and the mathematical model in 7 Figure 2(*b*).

Once locked, a high gain loop will only have a small mean residual error. The mean output of the phase detector will be small for a mean error which is either very close to zero or to $\pi/2$ radians, depending on the type of phase detector used. Any mean change in the VCO phase [the output phase (θ_0)] will produce a correction voltage at the phase detector output which will change the VCO phase in such a way as to reduce the mean error either towards zero or towards a fixed value of $\pi/2$ radians. As we are usually only concerned with the output phase noise that is the AC (modulation frequency) component of the output phase, a fixed phase shift of $\pi/2$ radians, rather than zero, between the input and the VCO will not usually be significant. Thus we shall *define* θ_0 with respect to the *mean* output phase; in effect neglecting any fixed phase shift of $\pi/2$ radians which may exist between input and output.

In those special cases when the absolute phase of the VCO must be equal to the input phase θ_i, a fixed phase shifter providing $\pi/2$ radians of phase shift may be inserted at the output of the PLL if it is required. An example of such an application would be where the VCO was used ultimately to feed a coherent AM detector which requires a local oscillator which is in phase with the carrier component of the input signal.

It is illuminating initially to consider the fundamental characteristics of each of the main loop components and to combine them step by step to obtain an understanding of the overall loop performance. This probably gives a better physical insight than commencing with a formal analysis using the Laplace Transformation. This physical insight is helpful when we come to consider the effect of component imperfections.

7.2.1   Voltage Controlled Oscillator

Consider the VCO, which is probably a varactor tuned oscillator, to have a tuning sensitivity of K_0' Hz per volt of varactor tuning voltage V_0. Its sensitivity may therefore also be expressed as $2\pi K_0'$ radians/second per volt.

The mean value of K_0' may easily be estimated if the total frequency range and the voltage of the power supply to the varactor diode are known. [See Section 5.9.1 especially equations (5.43) and (5.44)].

In accordance with the notation of Reference 24 the VCO constant will be written as

$$K_0 \frac{\text{radians}}{\text{second}} \text{ per volt} \qquad (\text{i.e. } K_0 = 2\pi K_0').$$

Let the VCO output voltage at carrier frequency w be:

$$V(t) = V \sin [wt + \theta_n(t) + \theta_v(t)] \tag{7.1}$$

where $\theta_n(t)$ is the phase jitter due to VCO noise and $\theta_v(t)$ is the phase jitter due to the tuning voltage $V_v(t)$.

Consider the phase jitter in a 1 Hz baseband bandwidth at an angular frequency p. The phase jitter component $\theta_{vo}(t)$, due to the control voltage $V_0(t)$, at an angular frequency p will only be due to the control voltage component at angular frequency p: (this assumes linearity, which is valid for small correction voltages at angular frequency p).

We shall therefore consider the control voltage to be given by:

$$V_0(t) = V_0 \cos pt \tag{7.2}$$

The oscillator phase change due to this voltage will be given by:

$$\frac{d\theta_{vo}}{dt} = K_0 V_0 \cos pt \text{ rads per second} \tag{7.3}$$

$$\therefore \quad \theta_{vo} = \int K_0 V_0 \cos pt \, dt \tag{7.4}$$

In the notation of the Laplace Transformation

$$\theta_{vo}(s) = \frac{K_0}{s} V_0 \tag{7.5}$$

It will be noted that, as the VCO tuning sensitivity (K_0) is in radians per second per volt, the VCO acts as an integrator (7.4) in any phase control loop. It differs from a transistor or other negative feedback amplifier (including a simple automatic frequency control system) in which only the introduction of specific CR networks (or the effect of stray capacitances) produces integration. In a position control servo, the motor may act as a perfect double integrator.

7.2.2 *Phase Detector*

The phase detector has two RF inputs, nominally at the same frequency, and produces an output voltage which ideally is proportional to their phase difference. In the case of a PLL, such as that shown in 7 Figure 2, the two inputs are the Reference Oscillator (or input signal) and the VCO. The phase detector gain constant is denoted by K_d and is expressed in volts output per radian of phase difference between the two input signals.

Phase detectors may take any one of a number of forms which include:

(*a*) Analogue phase detectors, which are basically mixers with an IF centred at zero frequency. An analysis of the performance is given in Section 2.8 above for small phase errors. A practical limitation of this type of phase detector is that the output is proportional to the sine of the phase difference rather than to the phase difference. But for small phase errors

$$\theta_e \simeq \sin \theta_e \quad \text{(where } \theta_e \text{ is the phase error).}$$

(*b*) A 'digital' phase detector using an 'exclusive or gate'. The output voltage rises linearly as the error (θ_e) goes from 0 to π radians. The maximum input frequency will be limited by the type of logic circuit used.

(*c*) A 'digital' phase detector using IC flip flops. This is an edge triggered detector. The output voltage rises linearly as the phase error goes from 0 to 2π radians. Once again the maximum input frequency is limited by the type of integrated circuit used. A more detailed description of digital phase detectors follows.

(*b*) *The 'Exclusive Or Gate' Phase Detector*

In the TTL integrated circuit range, 7486 contains quadruple 2 input exclusive or gates. This is used as an example. The connections and truth table are shown in 7 Figure 3 (Reference 25). The mode of operation is illustrated in 7 Figure 4.

Truth Table TTL IC 7486

Inputs		Output Y
A	*B*	
L	L	L
L	H	H
H	L	H
H	H	L

7 Figure 3(a)

The greater the phase difference between the two input waveforms up to π radians the wider will be the output pulses. Suitable filtering may therefore extract a DC output proportional to the phase difference between the two

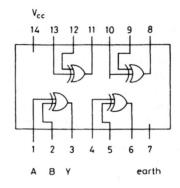

TTL IC 7486   **7 Figure 3(b)**

input waveforms. This form of phase detector suffers from the disadvantage that its application is limited to the case where the input waveforms are symmetrical. From 7 Figure 4 it is apparent that with a difference of π radians between the two inputs the output voltage will be DC at the high logic level of the particular logic family used. When the phase difference is zero the output level will be equal to the low logic level. As the phase difference increases from zero the pulse width of the output will increase and the *mean* value of the DC output voltage will increase linearly from the 'low' to 'high' logic levels. Assuming a minimum difference in logic levels of 2 volts, the phase detector gain constant K_d will be $2 \cdot 0 / \pi$ volts per radian $= 0 \cdot 64$ volts per radian.

Another difficulty with this type of phase detector us due to its characteristic in the region of 2π (i.e. 0) radians phase error. As may be seen from 7 Figure 4

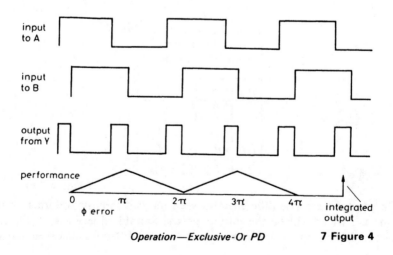

Operation—Exclusive-Or PD   **7 Figure 4**

the integrated correction voltage has the same sense regardless of whether the phase error is slightly greater than or slightly less than 2π. One sense of error will be reduced, and the other increased, by the operation of the PLL.

For satisfactory operation it is therefore necessary to work at a phase difference of $\pi/2$ radians between the VCO and reference waveforms; then the incremental correction voltage will change sign whenever the error with respect to $\pi/2$ radians changes sign. The AC phase error will then be satisfactorily controlled by the PLL. This may be achieved by applying a low noise DC offset voltage, equal to minus half the peak output of the phase detector, to the integrating amplifier which follows the phase detector.

(c) Edge Triggered Phase Detectors

Integrated circuit flip flops, of either *R-S* or *J-K* types, may be used as edge triggered phase detectors in a variety of ways. Such detectors are not restricted to operation with symmetrical input waveforms but, for optimum performance, particularly with regard to unwanted noise output, they should normally be fed with differentiated or square wave inputs so that the locations of the waveform edges in the time domain are precise. Only one specific example of such phase detectors will be discussed. Once the principles are understood others may easily be devised after consulting IC manufacturers' data sheets.

A phase detector using a 74S113 integrated circuit, together with its pin connections and Truth Table, are shown in 7 Figure 5 (Reference 25). As connected, J is permanently low and K is permanently high for both flip flops.

Input data (J low) is transferred to the output on the falling edge of a clock pulse. A low input to Pr sets Q to H independently of the clock. This is effected by feeding the $\bar{Q}$ outputs of both flip flops into one Nand Gate of a 7400 IC. When $\bar{Q}_1$ and $\bar{Q}_2$ are both high Pr goes low on both flip flops.

Inputs				Outputs	
Pr	C_k	J	K	Q	$\bar{Q}$
L	X	X	X	H	L
H	$\downarrow$	L	L	Q_0	$\bar{Q}_0$
H	$\downarrow$	H	L	H	L
H	$\downarrow$	L	H	L	H
H	$\downarrow$	H	H	Toggle	
H	H	X	X	Q_0	$\bar{Q}_0$

74S113 Truth Table   **7 Figure 5(a)**

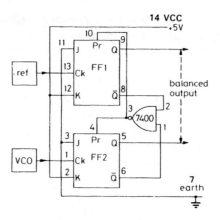

A 74S113 Phase Detector with Balanced   **7 Figure 5(b)**
Output

The waveforms for several different values of phase error are shown in 7 Figure 6; whilst 7 Figure 7 shows the overall input/output characteristic when used in conjunction with a balanced active filter amplifier. It is apparent that the balanced design has eliminated possible difficulties in the region of zero phase error: the correction voltage changes sign when the sign of the phase error changes.

The gain constant of this type of phase detector is equal to the difference between high and low logic voltages divided by 2π.

Thus typically $K_d = 2.0/2\pi = 0.32$ volts per radian.

(d) Other Types of Phase Detector
Other types of integrated circuits are available which have been designed specifically as phase detectors. An example is the MC4044 for which $K_d = 0.12$ V/radian.

7.2.3 Loop Filter (and loop amplifier)

7.2.3.1 Passive Filter
It is necessary to provide a low pass filter after the phase detector, to reject carrier frequency components and high frequency noise. In the simplest case this filter might consists of a single resistor and capacitor of time constant $RC = \tau_1$ (see 7 Figure 8).

In this case

$$\frac{V_{out}}{V_{in}} = \frac{1}{1 + jwCR}$$

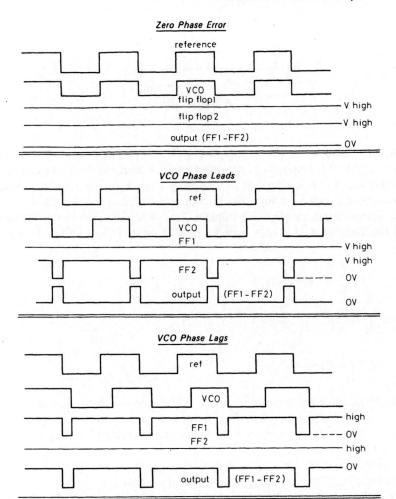

Phase Detector using 74S113(TTL)   **7 Figure 6**

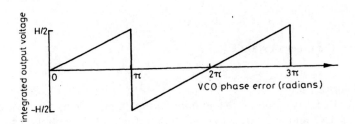

Output Characteristic 74S113 PD   **7 Figure 7**

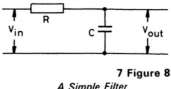

7 Figure 8
A Simple Filter

We have used w for the baseband angular frequency output of the phase detector, which corresponds to the offset angular frequency p. To be consistent with the notation in the rest of the book we should have used p. In this case we have used w as the formulae will then be more familiar to the reader.

For convenience we shall use Laplace Transforms but, for those unfamiliar with this technique, it is only necessary to know that the Laplace Transform of:

$$\frac{d\theta_0(t)}{dt} \quad \text{is} \quad s\theta_0(s)$$

$$\text{of} \quad \frac{d^2\theta_0(t)}{dt^2} \quad \text{is} \quad s^2\theta_0(s)$$

$$\text{of} \quad \int \theta_0(t) \, dt \quad \text{is} \quad \frac{\theta_0(s)}{s}$$

in which the same notation θ_0 is used for either the waveform $\theta_0(t)$ in the time domain or its transform $\theta_0(s)$ in the s domain, according to the context.

As far as steady state sinusoidal signals are concerned s may always be replaced by (jw).

In short we shall use $F(s)$ as a generalised immittance function and as a shorthand way of manipulating differential equations.

We shall write the *Transfer Function*

$$\frac{V_{out}}{V_{in}} \quad \text{as} \quad F_1(s)$$

$$\therefore \quad F_1(s) = \frac{1}{1 + sCR} = \frac{1}{1 + s\tau} \tag{7.6}$$

where τ is the time constant CR.

7.2.3.2 *Active Filter*
In some PLLs, to obtain the required loop gain, it may be necessary to provide a DC amplifier after the phase detector. Let the voltage gain of this amplifier be A. It is usual, in this case to use the amplifier and RC network as

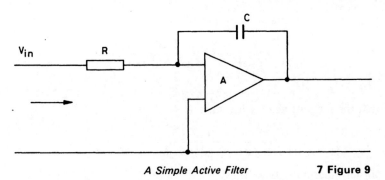

A Simple Active Filter   **7 Figure 9**

a Miller Integrator with an effective time constant of $(1 + |A|)RC$. For $|A| \gg 1$ the effective time constant is given approximately by:

$$\tau_1 = |A|RC \quad\quad\quad (7.7)$$

For this case of an active filter the transfer function:

$$F_2(s) = \frac{A}{1 + CRAs} = \frac{A}{1 + A\tau s} \quad\quad (7.8)$$

An active filter is shown in 7 Figure 9.

7.2.3.3 Lag—Lead Active Filter
In order to optimise the loop performance (as is also the case for a Position Control Servo), it is necessary to add a phase lead network with a time constant $\tau_2 = R_2 C$.

This addition converts the PLL into a loop in which the loop gain and the damping factor may be separately controlled.

Consider the active lag-lead filter shown in 7 Figure 10.

Initially we shall write Z_1 and Z_2 as $Z_1(s)$ and $Z_2(s)$ as though they were both functions of (s).

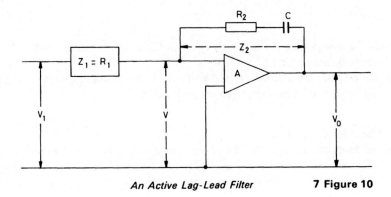

An Active Lag-Lead Filter   **7 Figure 10**

By inspection the signal at the input to the amplifier is given by:

$$V = \frac{V_1 Z_2(s)}{Z_1(s) + Z_2(s)} + V_0 \frac{Z_1(s)}{Z_1(s) + Z_2(s)}$$

But $V_0 = AV$

(note that the magnitude of A is in fact $-A$).

$$\therefore \quad \frac{V_0}{A} - \frac{V_0 Z_1(s)}{Z_1(s) + Z_2(s)} = \frac{V_1 Z_2(s)}{Z_1(s) + Z_2(s)}$$

The transfer function $F_3(s) = V_0/V_1$

$$\therefore \quad F_3(s) = \frac{A Z_2(s)}{Z_2(s) + (1 - A)Z_1(s)} \tag{7.9}$$

Now $Z_1(s) = R_1$

and $Z_2(s) = R_2 + \dfrac{1}{sC}$

$$\therefore \quad F_3(s) = \frac{A(R_2 + 1/sC)}{R_2 + 1/sC + R_1(1 - A)}$$

$$\therefore \quad F_3(s) = \left(\frac{A(sR_2 C + 1)}{sC}\right)\left(\frac{sC}{sR_2 C + 1 + sR_1 C(1 - A)}\right)$$

Putting $R_1 C = \tau_1$ and $R_2 C = \tau_2$

$$\therefore \quad F_3(s) = \frac{A(s\tau_2 + 1)}{s\tau_2 + 1 + s\tau_1(1 - A)} \tag{7.10}$$

This is identical to the equation for $F_2(s)$ on p. 9 of Reference 24. For large values of A writing $A' = -A$ and assuming

$$|s\tau_1(1 + A')| \gg |s\tau_2 + 1|$$

$$F_3(s) \simeq -\frac{A'(s\tau_2 + 1)}{s\tau_1(1 + A')} \simeq -\frac{s\tau_2 + 1}{s\tau_1} \tag{7.11}$$

It should be noted that the active filter amplifier gain (A) cancels in (7.11) and hence does not appear in (7.12), (7.14) or (7.21).

Equation (7.8) may also be derived from (7.11) for the case where $\tau_2 = 0$ (lag filter only) over the frequency range where $|A\tau s| \gg 1$.

7.2.4 The Open Loop Gain

Open the loop at the input to the phase detector (point X of 7 Figure 2) and inject a carrier at the correct frequency (ω) but with a phase error (θ in) referred to the correct reference phase with respect to the input. The phase of the output signal (θ out) from the VCO is then measured.

$$\frac{\theta \text{ out}}{\theta \text{ in}} = -(\text{open loop gain}) = -A''\beta$$

Where A'' is the forward gain and β is the feedback factor.

Reference to 7 Figure 2 makes it apparent that $\beta = $ unity. That is the phase jitter of the VCO is fed directly to the phase detector.

We may readily calculate the open loop gain: it is minus the product of the individual gains or transfer functions.

$$\therefore \quad A''\beta = \left[K_d F(s) \frac{K_0}{s}\right] = K_d \left(\frac{s\tau_2 + 1}{s\tau_1}\right)\frac{K_0}{s} \tag{7.12}$$

7.2.5 The Closed Loop Gain

From amplifier negative feedback theory the closed loop gain H of a feedback amplifier is:

$$H = \frac{A''}{1 + A''\beta} \tag{7.13}$$

where A'' is the forward gain and β is the feedback factor. In our case $\beta = 1$.

$\therefore$ the closed loop gain is given by:

$$H(s) = \frac{K_d F(s) K_0/s}{1 + K_d F(s) K_0/s}$$

$$\therefore \quad H(s) = \frac{K_0 K_d F(s)}{s + K_0 K_d F(s)} = \frac{\theta_0}{\theta_i} \tag{7.14}$$

The closed loop gain function $H(s)$ is in fact the transfer function between the input and output of the PLL regarded as a network. The input is the required phase (that of the reference input) and the output is the phase of the VCO (shifted by a constant value of $\pi/2$ if necessary). This is illustrated in 7 Figure 11. The input phase θ_i and the output phase θ_0, should not be confused with (θ in) and (θ out) which referred to the open loop performance.

7.2.6 Loop Natural Frequency and Damping Factor

From (7.14)

$$[s + K_0 K_d F(s)]\theta_0 = [K_0 K_d F(s)]\theta_i$$

θ_i can be any arbitrary time function including zero. If we put $\theta_i = 0$ we shall get the natural unforced transient given by:

$$[s + K_0 K_d F(s)]\theta_0 = 0$$

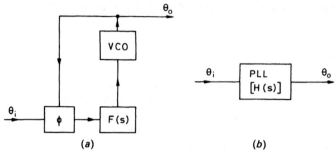

The PLL as a Network   **7 Figure 11**

Substituting for $F(s)$ the value given by (7.11) we get

$$\left[s + K_0 K_d \frac{(s\tau_2 + 1)}{s\tau_1} \right] \theta_0 = 0$$

(The change of sign of the open loop transfer function is required for the stability of the phase lock loop). In practice the output of the phase detector will be connected to give the right polarity to achieve negative feedback.

$$\therefore \quad \left[s^2 + K_0 K_d \frac{\tau_2}{\tau_1} s + \frac{K_0 K_d}{\tau_1} \right] \theta_0 = 0 \tag{7.15}$$

The physical meaning will be more apparent if we write this as a differential equation:

$$\frac{d^2\theta_0}{dt^2} + K_0 K_d \frac{\tau_2}{\tau_1} \frac{d\theta_0}{dt} + \frac{K_0 K_d}{\tau_1} \theta_0 = 0 \tag{7.16}$$

Consider first the case where $\tau_2 = 0$. For this case

$$\frac{d^2\theta_0}{dt^2} = -\frac{K_0 K_d}{\tau_1} \theta_0 \tag{7.17}$$

The acceleration of θ_0 is equal to a constant times itself with a negative sign: the angular acceleration is proportional to the deflection but in the opposite direction. This is the equation of Simple Harmonic Motion (that of a simple pendulum).

Assume then $\theta_0 = A \sin wt$ \hfill (7.18)

$$\therefore \quad \frac{d^2\theta_0}{dt^2} = -w^2 A \sin wt \tag{7.19}$$

w is the natural angular frequency of oscillation of the loop. This will be denoted w_n. From (7.17) and (7.18)

$$\frac{d^2\theta_0}{dt^2} = -\frac{K_0 K_d}{\tau_1} A \sin w_n t \tag{7.20}$$

From (7.19) and (7.20)

$$w_n^2 = \frac{K_0 K_d}{\tau_1} \tag{7.21}$$

and

$$f_n = \frac{1}{2\pi} \left(\frac{K_0 K_d}{\tau_1} \right)^{1/2} \tag{7.22}$$

w_n is the natural angular frequency of the PLL and f_n is its natural frequency.

Consider now the case when τ_2 is not zero, that is revert to equation (7.16) substituting

$$w_n^2 \quad \text{for} \quad \frac{K_0 K_d}{\tau_1}$$

$$\therefore \quad \frac{d^2\theta_0}{dt^2} + w_n^2 \tau_2 \frac{d\theta_0}{dt} + w_n^2 \theta_0 = 0 \tag{7.23}$$

This is a linear differential equation with constant coefficients. We should expect an exponential solution.

$$\text{Try} \quad \theta_0 = Ae^{mt} \tag{7.24}$$

$$\therefore \quad \frac{d\theta_0}{dt} = mAe^{mt}$$

and

$$\frac{d^2\theta_0}{dt^2} = m^2 Ae^{mt}$$

$$\therefore \quad m^2(Ae^{mt}) + w_n^2 \tau_2 m(Ae^{mt}) + w_n^2(Ae^{mt}) = 0 \tag{7.25}$$

Dividing through by Ae^{mt} gives the auxiliary equation for m:

$$m^2 + w_n^2 \tau_2 m + w_n^2 = 0 \tag{7.26}$$

Solving the quadratic equation for m

$$m = \frac{-w_n^2 \tau_2 \pm \sqrt{(w_n^2 \tau_2)^2 - 4w_n^2}}{2}$$

$$m = -w_n^2 \frac{\tau_2}{2} \pm \left[\left(w_n^2 \frac{\tau_2}{2} \right)^2 - w_n^2 \right]^{1/2}$$

$$\therefore \quad m = -w_n^2 \frac{\tau_2}{2} \pm \left[w_n^4 \frac{\tau_2^2}{4} - w_n^2 \right]^{1/2} \tag{7.27}$$

The bracketed term will be zero, giving two equal real negative roots when:

$$w_n^4 \frac{\tau_2^2}{4} = w_n^2$$

i.e. when

$$\frac{w_n \tau_2}{2} = 1 \tag{7.28}$$

In this case (7.24) with unequal ms, becomes in the limit for equal ms

$$\theta_0 = (A + Bt) \exp\left(- w_n^2 \frac{\tau_2}{2}\right) \tag{7.29}$$

After a change in steady state demand the response will approach the new demanded level exponentially.

The bracketed term will be imaginary when in (7.27)

$$\frac{w_n \tau_2}{2} < 1 \tag{7.30}$$

In this case there will be two complex roots which are complex conjugates. The response will be a damped sinusoidal oscillation: damped, because of the negative real part of the exponentials.

When $w_n \tau_2/2 > 1$ the bracketed term will be real and there will be two unequal negative real roots, both corresponding to exponentially decaying responses.

The case $[w_n \tau_2/2] = 1$ represents the transition between non-oscillating and oscillating responses. It represents the case of critical damping which gives the fastest response possible without any over-shoot. Thus $w_n \tau_2/2$, termed the damping factor, is unity for critical damping. The damping factor is usually designated ζ. Thus

$$\zeta = \frac{w_n \tau_2}{2} \tag{7.31}$$

But (7.21)

$$w_n = \left(\frac{K_0 K_d}{\tau_1}\right)^{1/2}$$

$$\therefore \quad \zeta = \frac{\tau_2}{2}\left(\frac{K_0 K_d}{\tau_1}\right)^{1/2} \tag{7.32}$$

It is apparent from (7.32) that if $\tau_2 = 0$, the damping factor is zero and we have an undamped oscillation at angular frequency w_n. This is in fact the case which led to equation (7.17).

Thus, in 7 Figure 10, the inclusion of R_2 is essential: if it is omitted the PLL will have an undamped oscillatory response; $(\tau_2 = R_2 C)$.

It is hardly necessary to remind the reader that changes in the loop gain, due to any cause, will also change the loop natural frequency and damping factor (7.22 and 7.32).

7.3 Phase Lock Loop Characteristics

7.3.1 Frequency Transmission Characteristics

The transfer function is the transmission characteristic of the loop for noise on the input phase reference.

Rewriting equation (7.14) for convenience

$$H(s) = \frac{K_0 K_d F(s)}{s + K_0 K_d F(s)}$$

For an active filter using a high gain amplifier (7.11).

$$F(s) = \frac{s\tau_2 + 1}{s\tau_1}$$

For this case

$$H(s) = \frac{K_0 K_d(s\tau_2/\tau_1) + (K_0 K_d/\tau_1)}{s[s + (K_0 K_d/s)(s\tau_2/\tau_1) + (K_0 K_d/s\tau_1)]}$$

$$\therefore \quad H(s) = \frac{sw_n^2 \tau_2 + w_n^2}{s^2 + w_n^2 \tau_2 s + w_n^2}$$

$$\therefore \quad H(s) = \frac{2\zeta w_n s + w_n^2}{s^2 + 2\zeta w_n s + w_n^2} \tag{7.33}$$

For sinusoidal phase inputs the transfer function of the PLL is such that low frequencies pass through unattenuated and frequencies well above f_n are attenuated. However it differs from a simple low pass filter in that f_n is not the 3 dB point, which is in fact dependent both on f_n and the damping factor. The response, normalised to w_n, for various values of damping factor, is given in Figure 2.3 of Reference 24. Frequently the damping factor is chosen to be 0·707. This gives a good response with only reasonable overshoot. (Approx 2 dB).

In this case the response falls by about 6 dB per octave, (20 dB per decade) increase in frequency beyond $1·5 f_n$.

7.3.2 Residual Phase Error

If we wanted to use a PLL to lock a noisy VCO to a low noise reference, our primary concern would be the degree to which the VCO noise was reduced. We would normally be less concerned about what proportion of the (low level) noise of the reference was transmitted through the loop to the output (Section 7.3.1 above).

Referring to 7 Figure 2(b) it may be seen that our main concern is the degree by which θ_n is reduced by the operation of the loop. To calculate this it is necessary first to calculate the output phase (θ_0) in terms of the input phase (θ_i) and the VCO phase jitter θ_n.

From inspection of 7 Figure 2(b):

$$\theta_0 = \theta_{vo} + \theta_n$$

$$\theta_{vo} = A''\theta_e$$

$$\therefore \quad \frac{\theta_{vo}}{A''} = \theta_e = \theta_i - \theta_0$$

$$\therefore \quad \frac{\theta_{vo}}{A''} = \theta_i - \theta_{vo} - \theta_n$$

$$\therefore \quad \theta_{vo}\left(\frac{1}{A''} + 1\right) = \theta_i - \theta_n$$

$$\therefore \quad \theta_{vo} = \frac{A''(\theta_i - \theta_n)}{1 + A''}$$

But (7.13) putting $\beta = 1$

$$H = \frac{A''}{1 + A''}$$

$$\therefore \quad \theta_{vo} = H(\theta_i - \theta_n)$$

$$\therefore \quad \theta_0 = H(\theta_i - \theta_n) + \theta_n$$

$$\therefore \quad \theta_0 = H\theta_i + \theta_n(1 - H)$$

As all the terms are functions of (s) this may be written more fully as:

$$\theta_0(s) = H(s)\theta_i(s) + \theta_n(s)[1 - H(s)] \tag{7.34}$$

This is an important equation. It is also useful to find the error (θ_e)

$$\theta_e = \theta_i - \theta_0$$

$$\therefore \quad \theta_e = \theta_i - H\theta_i - \theta_n(1 - H)$$

$$\therefore \quad \theta_e(s) = \theta_i(s)[1 - H(s)] - \theta_n(s)[1 - H(s)] \tag{7.35}$$

Thus the error in responding to (θ_i) is of the same form as the error in reducing θ_n. Hence we may speak unambiguously of the error response of the phase lock loop $|1 - H(s)|$. This should be intuitively obvious.

For the two cases when, firstly, $\theta_n = 0$ and, secondly, $\theta_i = 0$.

$$|\theta_e(s)| = |\theta_i(s)| \, |1 - H(s)| \tag{7.35.1}$$

$$|\theta_e(s)| = |\theta_n(s)| \, |1 - H(s)| = |\theta_0(s)| \tag{7.35.2}$$

In the present context of a noisy VCO being locked to a low noise reference oscillator, our interest is in equation (7.35.2).

It is now necessary to evaluate equation (7.35.2). But (7.14)

$$H(s) = \frac{K_0 K_d F(s)}{s + K_0 K_d F(s)}$$

$$\therefore \quad \frac{\theta_0(s)}{\theta_n(s)} = \frac{s + K_0 K_d F(s) - K_0 K_d F(s)}{s + K_0 K_d F(s)}$$

$$\therefore \quad \frac{\theta_0(s)}{\theta_n(s)} = \frac{s}{s + K_0 K_d F(s)} \tag{7.36}$$

Assuming a high gain active filter (equation 7.11)

$$F(s) = \frac{s\tau_2 + 1}{s\tau_1}$$

$$\therefore \quad \frac{\theta_0(s)}{\theta_n(s)} = \frac{s}{s + K_0 K_d(s\tau_2/s\tau_1) + (K_0 K_d/s\tau_1)}$$

$$\therefore \quad \frac{\theta_0(s)}{\theta_n(s)} = \frac{s^2}{s^2 + K_0 K_d(\tau_2/\tau_1)s + (K_0 K_d/\tau_1)}$$

But (7.21)

$$\frac{K_0 K_d}{\tau_1} = w_n^2$$

and (7.32)

$$\zeta = \frac{\tau_2}{2}\left(\frac{K_0 K_d}{\tau_1}\right)^{1/2}$$

$$\therefore \quad \frac{\theta_0(s)}{\theta_n(s)} = \frac{s^2}{s^2 + 2\zeta w_n s + w_n^2} \tag{7.37}$$

We may find the steady state value of $[\theta_0(s)/\theta_n(s)]$ as a function of offset frequency by writing jw for s. Normally in this book we have used w for the carrier angular frequency and p for the angular offset frequency. However little confusion should be caused here by the further use of the more familiar w in place of p.

$$\therefore \quad \frac{\theta_0(jw)}{\theta_n(jw)} = \frac{w^2}{-w^2 + 2jw\zeta w_n + w_n^2} \tag{7.38}$$

The modulus of the denominator is given by:

$$|w_n^2 - w^2 + 2jw\zeta w_n| = [(w_n^2 - w^2)^2 + 4w^2\zeta^2 w_n^2]^{1/2}$$
$$= [w_n^4 - 2w_n^2 w^2 + w^4 + 4w^2\zeta^2 w_n^2]^{1/2}$$

Put $\zeta = 0.707$ (i.e. $2\zeta^2 = 1$)

$$\therefore \quad \left|\frac{\theta_0(jw)}{\theta_n(jw)}\right|_{\zeta=0.707} = \frac{w^2}{(w_n^4 + w^4)^{1/2}} \qquad (7.39)$$

Let $w = w_n \chi$

Consider the cases where $\chi < 1$

   i.e. $w < w_n$

Except when w is relatively close to w_n

$$w_n^4 \gg w^4$$

For these cases:

$$\left|\frac{\theta_0(jw)}{\theta_n(jw)}\right|_{\zeta=0.707} = \frac{w_n^2 \chi^2}{w_n^2} \qquad (7.40)$$

$$= \chi^2 = \left(\frac{w}{w_n}\right)^2 = \left(\frac{f}{f_n}\right)^2$$

Thus from shortly below f_n the modulus of the reduction factor in the phase jitter falls as f^2 as frequency falls. Now θ_0 represents a phase error with respect to the demanded phase which is zero ($\theta_i = 0$).

In the case with which we are primarily concerned (phase noise) it represents peak output phase jitter with respect to that of the Reference input.

If θ_0 is the peak phase jitter in this sense then ϕ_0 (ϕ is rms) $= \theta_0/\sqrt{2}$. Now

$$\overline{\phi_0^2} = \frac{2N_{op}}{C} \quad (3.42)$$

The power ratio is equal to the square of the phase jitter. Alternatively, θ_0/θ_n is the ratio of the two modulation indices and hence is analogous to a *voltage* ratio.

A fall in θ by f^2/f_n^2 below f_n, would represent a fall in modulation power by f^4/f_n^4. This is more conveniently expressed as a fall in sideband power by 12 dB per octave, or 40 dB per decade, fall in frequency below f_n.

When $\chi = 1$: $w = w_n$

$$\left|\frac{\theta_0(jw)}{\theta_n(jw)}\right|_{\zeta=0.707} = \frac{w^2}{\sqrt{2}\,w^2} = \frac{1}{\sqrt{2}} \qquad (7.41)$$

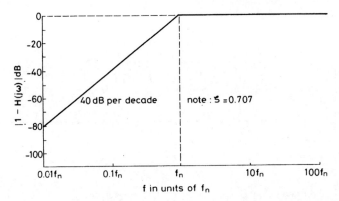

PLL Error $|\theta_e/\theta_i| = |1 - H(jw)|$ **7 Figure 12**

When $\chi > 1 : w > w_n$ and (7.39)

$$\left|\frac{\theta_0(jw)}{\theta_n(jw)}\right|_{\zeta=0\cdot707} = 1 \tag{7.42}$$

and the error is equal to the VCO jitter: the loop does not correct for the VCO jitter at all.

The error response of a high gain second order loop calculated as described above is plotted in 7 Figure 12.

If we have a relatively noisy VCO, with constant phase noise density over a large offset frequency range

$$\overline{\phi_0^2} = 2\frac{N_{op}}{C} = \text{a constant } K$$

Then

$$\phi_0 = \sqrt{\frac{2N_{op}}{C}} = \sqrt{K}$$

and the phase jitter density is also constant over this offset frequency range.

Assume that we phase lock this VCO to a low phase noise reference oscillator with a flat phase noise characteristic 60 dB down with respect to the VCO noise. See 7 Figure 13. The phase jitter density and hence the phase noise density ($N_{op}/C = \overline{\phi_0^2}/2$) at the output will follow the phase noise curve of the VCO at frequencies appreciably above f_n. Below f_n the output phase noise will fall by 40 dB per decade until the reference level is reached. At the cost of an error only in the near vicinity of f_n, which never exceeds 3 dB, we may approximate the performance of the PLL by two straight line segments intersecting at f_n.

The resultant performance of the phase locked VCO is shown by curve (c) of 7 Figure 13.

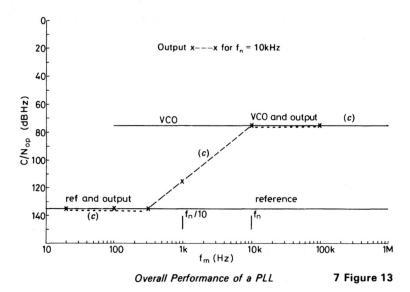

Overall Performance of a PLL   **7 Figure 13**

If it is required to know the phase noise density performance of a controlled VCO at a specific offset frequency in known relation to f_n without the trouble of drawing a graph, then this may readily be determined as follows for a clean reference.

Let the uncontrolled VCO phase noise density/carrier ratio, at an offset frequency f, be $(N_{op}/C)_f$.

Let the output phase noise density to carrier ratio, at an offset frequency f, be $(N'_{op}/C)_f$. For

$$\frac{f}{f_n} > 1 : \left(\frac{N'_{op}}{C}\right)_f = \left(\frac{N_{op}}{C}\right)_f \tag{7.43.1}$$

For

$$\frac{f}{f_n} < 1 : \left(\frac{N'_{op}}{C}\right)_f = \left(\frac{N_{op}}{C}\right)_f \left(\frac{f}{f_n}\right)^4 \tag{7.43.2}$$

7.3.3 Mathematical Convergence of $1/f^3$ Noise Operated on by a PLL

Many oscillators have a phase noise characteristic which follows a $1/f^3$ power law at small offset frequencies. If an attempt is made to find the value of the integrated phase jitter down to zero frequency the integral does not converge. (See Section 5.8).

$$\overline{\phi^2} = \int_0^b \left(\frac{2N_{op}}{C}\right)_f df \ \text{rads}^2 \tag{7.44}$$

$$\left(\frac{2N_{op}}{C}\right)_f = 2\left(\frac{N_{op}}{C}\right)_{fc}\left(\frac{fc}{f}\right)^3 \tag{7.45}$$

fc is the frequency below which the $1/f^3$ law applies.

$$\therefore \quad \overline{\phi^2} = \int_0^b 2\left(\frac{N_{op}}{C}\right)_{fc}\left(\frac{fc}{f}\right)^3 df \text{ rad}^2 \tag{7.46}$$

If a VCO with such a characteristic is controlled by a PLL which provides a reduction in phase jitter by $(f/f_n)^4$ below f_n, then the integral becomes convergent.

Let the integrated phase jitter after PLL correction be ϕ'. Then:

$$\overline{\phi'^2} = \int_0^b 2\left(\frac{N_{op}}{C}\right)_{fc}\left(\frac{fc}{f}\right)^3\left(\frac{f}{f_n}\right)^4 df \text{ rads}^2 \tag{7.47}$$

$$\therefore \quad \overline{\phi'^2} = \int_0^b 2\left(\frac{N_{op}}{C}\right)_{fc}\left(\frac{fc^3}{f_n^4}\right)fdf \text{ rads}^2. \tag{7.48}$$

This integral is convergent because it becomes equal to zero at $f = 0$ and has a finite value at all other offset frequencies.

7.4 The Need for Frequency Division

In designing a UHF or SHF signal source to give a good phase noise performance over a large offset frequency range it is necessary to lock the phase of a UHF or SHF oscillator to that of a much lower frequency oscillator with a good close to carrier phase noise performance, as discussed in Section 7.1. The obvious way to do this is to multiply the frequency of the low frequency source to that of the output VCO and then to operate the phase detector of the PLL at the UHF or SHF output frequency.

This approach has three disadvantages as follows:

(*a*) It is difficult to provide a controlled variable multiplication ratio if a range of output frequencies is required.
(*b*) The phase detector would have to operate at the output frequency. This would normally preclude the use of highly efficient digital phase detectors.
(*c*) The phase detector might well be required to operate over a large frequency range.

An alternative approach is to divide the frequency of the output VCO by a factor sufficient to reduce it to that of the reference oscillator. The phase detector would then operate at the (low) reference oscillator frequency. This approach has the advantage that programmable digital frequency dividers are available. A change in the division ratio will then necessitate a change in output frequency which (over the pull in range of the PLL) (Reference 24) will happen automatically as the loop pulls in.

A simple example of a PLL with frequency division within the loop is shown in 7 Figure 14.

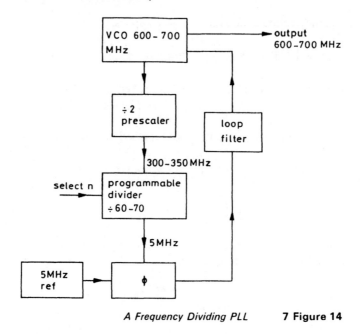

A Frequency Dividing PLL   **7 Figure 14**

7.5 Programmable Dividers

7.5.1 Principles of Operation

The object of this discussion is not to describe a particular integrated circuit but to illustrate how a programmable divider *might* be built up from elementary flip flops and gates. This treatment will, it is hoped, dispel the air of mystery which could be felt by an engineer who is not very familiar with digital electronics. Competent digital engineers should skip this sub-section.

Consider a master/slave *J-K* flip flop connected as shown in 7 Figure 15. *J* and *K* are connected permanently to logic level 1, as also is the preset terminal. The flip flop is cleared, that is $Q = 0$ and $\bar{Q} = 1$ whenever Clr = 0 and Pr = 1. Thus with the connections shown we may clear the flip flop at any time merely by putting Clr to logic 0.

Putting $J = K = 1$ converts the *J-K* flip flop to a *T* type flip flop which will change state on the *falling* edge of every input pulse to the clock terminal (as long as Clr is at logic 1) giving $Q_{(n+1)} = \bar{Q}_{(n)}$.

Now consider four of these flip flops cascaded as shown in 7 Figure 16. The four flip flops are labelled $F_0 - F_3$ inclusive and the corresponding outputs are $Q_0 - Q_3$.

The input waveform and the waveforms at $Q_0 - Q_3$ are shown in 7 Figure 17, starting at a time immediately after the flip flops have been cleared ($Q_0 = Q_1 = Q_2 = Q_3 = 0$). The periods relating to a complete cycle of the input

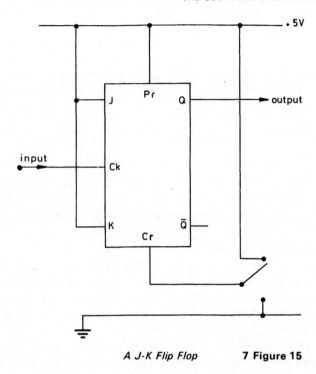

A J-K Flip Flop   **7 Figure 15**

waveform have also been numbered. As each flip flop triggers on the *falling* edge of its input waveform, which for flip flop F_n is the output waveform of flip flop $F_n - 1$, it is apparent that each flip flop divides the frequency of its input waveform by 2. At the end of cycle 15 all flip flops have returned to zero and the whole process may be repeated.

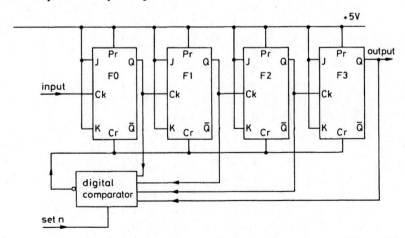

A Programmable Frequency Divider   **7 Figure 16**

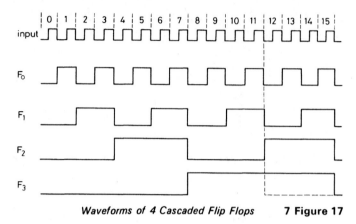

Waveforms of 4 Cascaded Flip Flops   **7 Figure 17**

The logic levels at the outputs of $F_0 - F_3$ for each cycle of the input waveform are also shown in 7 Table 1. This table was derived directly by inspection of the waveform diagram 7 Figure 17. It is interesting to note that the binary number, reading F_3, F_2, F_1, F_0 in that order, corresponds to the input waveform cycle number in decimal form. For example during cycle 13 we have 1101 which is just 13 in binary form.

If we feed $Q_3 \rightarrow Q_0$ into the appropriate 4 inputs of a 4 digit comparator which has been programmed to contain (say) the decimal number 12 (1100)

7 Table 1 *Logic States for 4 Cascaded Flip Flops*

n	F_3	F_2	F_1	F_0
0	0	0	0	0
1	0	0	0	1
2	0	0	1	0
3	0	0	1	1
4	0	1	0	0
5	0	1	0	1
6	0	1	1	0
7	0	1	1	1
8	1	0	0	0
9	1	0	0	1
10	1	0	1	0
11	1	0	1	1
12	1	1	0	0
13	1	1	0	1
14	1	1	1	0
15	1	1	1	1

7 Table 2   Some Prescalers

Type No.	Max. Input Frequency	Division Ratio
SP 8619	1·5 GHz	÷ 4
SP 8634	700 MHz	÷ 10
SP 8643	350 MHz	÷ 10/11

then the output of the comparator will go to 1 at the beginning of period 12. This output may be applied through an inverter to the 'clear' inputs of all four flip flops which will all return to zero and the cycle of operations will restart. This condition is shown as a dashed line on 7 Figure 17. The frequency of the output from F_3 is now one twelfth of the input frequency. Changing the number, preset into the digital comparator will thus change the overall division ratio to that of the preset number, as long as this does not exceed 16 for a 4 stage divider.

7.5.2 Commercial Programmable Frequency Dividers
Complete frequency dividers which include all the flip flops, the digital comparator and all necessary gates, are available as single integrated circuit packages.

For example in the TTL range 74192 is a four stage programmable digital frequency divider. This is a positive edge triggered divider which will work up to a maximum input frequency of 25 MHz.

In many PLL designs it may be necessary to use a divider input frequency very much higher than 25 MHz. In this case it is necessary to use dividers with ECL input stages and, for the highest frequencies to precede the programmable divider by a 'prescaler', which is a high speed divider with either a fixed division ratio, or a ratio with a very limited programmable variation such as 5/6 or 10/11.

There are many commercially available prescaler IC's. Some examples are given in 7 Table 2.

7.6 The Effects of Frequency Division on Correction Ratio and Phase Lock Loop Parameters

Consider the frequency dividing PLL shown in 7 Figure 14. Assume the programmable divider to be set to give a division ratio of 60, which corresponds to an overall division of 120 and an output frequency of 600 MHz. As the VCO output frequency is divided by 120 by the time it reaches the input to the phase detector, the VCO phase jitter deviation will also be divided by this ratio. This is the inverse of the case covered fully in Section 6.2.3. above. In this case the VCO phase noise at the input to the phase detector, will be reduced by $(20 \log_{10} 120)$ dB = 41·6 dB.

For purposes of illustration assume that the uncontrolled VCO phase noise density at 600 MHz is exactly 80 dB worse than that of the 5 MHz reference over the full offset frequency range from zero to f_n (the natural frequency of the PLL). At the phase detector (PD) input the VCO phase noise density will be only 38·4 dB (80–41·6) worse than the reference. Even an idealised perfect PLL (not realisable in practice) which produced complete correction of any error below f_n could in this case only produce 38·4 dB correction. If this had been achieved, the instantaneous phase of the VCO input to the PD would become identical to that of the 5 MHz reference. Hence the output of the phase detector would be zero and no further correction of VCO phase jitter would be possible. Thus the effect of frequency division by n in a PLL is to lessen the phase jitter correction ratio by n and the phase noise reduction ratio by n^2.

Having, by way of introduction, considered the effect of frequency division in a PLL rather intuitively, it is relatively simple to incorporate the necessary modifications to the mathematical analysis of the PLL carried out in Sections 7.2 and 7.3 above.

As the phase jitter from the VCO is divided by n at the input to the phase detector, the effective phase detector gain constant becomes K_d/n. Thus the PLL open loop gain (7.12) becomes:

$$A''\beta = \frac{K_d}{n}F(s)\frac{K_0}{s} \tag{7.49}$$

The Transfer Function (7.14) becomes:

$$H(s) = \frac{K_0\dfrac{K_d}{n}F(s)}{s + K_0\dfrac{K_d}{n}F(s)} \tag{7.50}$$

Equation (7.21) becomes:

$$w_n^2 = \frac{K_0 K_d}{\tau_1 n} \tag{7.51}$$

and

$$f_n = \frac{1}{2\pi}\left(\frac{K_0 K_d}{\tau_1 n}\right)^{1/2} \tag{7.52}$$

Equation (7.32) becomes:

$$\zeta = \frac{\tau_2}{2}\left(\frac{K_0 K_d}{\tau n}\right)^{1/2} \tag{7.53}$$

These equations are in fact quite general, in that they may be used for any value of 'n' including unity.

The 40 dB per decade reduction below f_n suffers a restriction at the low

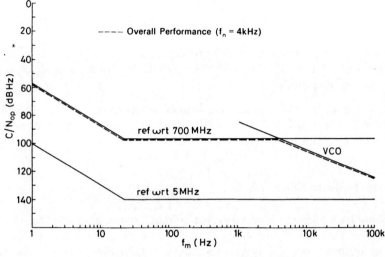

Performance of a Frequency Dividing PLL **7 Figure 18**

frequency end in that the VCO noise can never be reduced below a figure represented by the reference noise $+20 \log_{10} n$. The simplest way of taking account of this is to calculate the natural frequency and damping factor from (7.52) and (7.53), to add $20 \log_{10} n$ to the phase noise characteristics of the reference oscillator and then analyse the performance as though it were a fundamental loop operating at the output frequency. That is all noise sources would be referred to the output frequency.

A graphical analysis of the phase noise performance of the loop of 7 Figure 14 is shown in 7 Figure 18, using realistic figures for the 5 MHz reference and the UHF VCO. The reference oscillator noise performance is shown at 5 MHz and also with respect to 700 MHz, which is the maximum output frequency where n, and thus the phase noise degradation, is at its greatest. The effective degradation of the 5 MHz oscillator with respect to 700 MHz is $20 \log_{10} (700/5) = 42{\cdot}9$ dB.

The resultant overall performance, calculated by using the straight line approximation to the PLL performance, (40 dB per decade correction below f_n) is also shown. The value of f_n was chosen roughly to coincide with the offset frequency where the uncorrected VCO and the reference oscillator curves intersect.

Having determined the overall performance and the required PLL natural frequency in this way, the detailed design of the PLL is readily decided as follows:

1. Measure or calculate, the VCO gain characteristic K_0(rads/sec)/V.
2. Select a phase detector suitable for operation at 5 MHz. Measure or calculate the phase detector gain constant K_d.

3. Select a suitable operational amplifier for the active filter.
4. Calculate τ_1 from (7.52); all other parameters in this equation now being known.
5. Assume a damping factor of 0·707 and calculate τ_2 from (7.53).
6. Select a suitable value of C in 7 Figure 10. Calculate R_1 and R_2 from the known values of τ_1 and τ_2.

The foregoing method of design is adequate except in one respect: the effects of any noise generated within the phase detector, the loop amplifier or even the dividers have not been considered.

7.7 Phase Detector Noise Floors

In a phase lock loop the phase detector voltage output, after modification by the transfer function of the loop filter, is used to control the VCO. Any spurious output from the phase detector will therefore produce unwanted VCO phase jitter. In our earlier treatment of phase detectors (see Section 7.2.2) they have been considered to be ideal devices (at least for small errors) in the sense that the output voltage is directly proportional to the phase difference between the two input signals. Real phase detectors, particularly digital phase detectors, approximate very closely to this ideal apart from the fact that, in common with all electronic devices, there is some contamination of the output signal by thermal noise associated with the resistive components (or due to shot noise in the active devices) of the phase detector.

In addition the phase detector is normally followed by an active filter, the amplifier of which will also generate random noise. Both these contributions will be considered.

The thermal noise voltages will normally be very low compared with the wanted voltage output from the phase detector. In such cases the noise contribution of the PD can be neglected. However in other cases, where the input signals to the phase detector are both very pure, or alternatively have almost identical phase jitter characteristics, the required correction voltage from the phase detector will be very small. In these cases the noise output from the phase detector/amplifier combination will form the ultimate limit on the achievable phase noise reduction. Where this is the case we refer to the "noise floor" of the phase detector.

7.7.1 Analogue Phase Detectors
For convenience the abbreviation PD will be used in referring to a phase detector.

Consider a mixer, used as an analogue PD as shown in 7 Figure 19.

If the output signal to noise ratio is to be as good as possible, it is necessary that the input signal level should be as large as possible. To achieve adequate

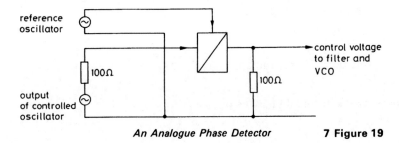

An Analogue Phase Detector **7 Figure 19**

signal linearity it is also necessary that the 'local oscillator' level (regarding the PD as a mixer) should be large compared with the 'signal' level. For a balanced diode bridge semiconductor mixer, the maximum 'local oscillator' level will be about + 10 dBm and the maximum input signal level about − 10 dBm. To achieve the best output S/N ratio we should therefore drive at these levels.

When driven at these levels, a good mixer will give a conversion loss of about 6 dB, thus giving an output signal of − 16 dBm at the difference frequency. A good low noise mixer will also have a Noise Temperature Ratio (NTR), see Appendix II, of about 1·3 : 1 for output frequencies greater than a few MHz. For low output frequencies the NTR will increase, probably as 1/f. Choose a mixer which has a low NTR specified at the lowest possible output frequency, for this corresponds to the offset frequency in which we are interested.

We shall assume an NTR of 1·3 and *initially* neglect the rise in value at low offset frequencies.

As a DC coupled output is required, it will be necessary to have a low resistance output load to prevent a backward bias voltage developing which would produce a degradation of the noise performance.

7.7.1.1 Phase Detector Without Active Filter Amplifier

(a) Signal Output
When used as a mixer, with input signal and local oscillator at two different frequencies, an output signal of − 16 dBm at the difference frequency will develop a voltage of 0·05 volts rms across the 100 ohm output load. The peak voltage of the output sinusoid will be 0·0707 volts. For two signals of the same frequency this same peak voltage will occur for a phase error of $\pi/2$ radians. Let the output voltage of the PD be V_d and the phase error be θ_e. The output voltage of an analogue PD is proportional to the sine of the phase error.

$$\therefore \quad V_d = 0·0707 \sin \theta_e. \tag{7.54}$$

$$\therefore \quad \frac{dV_d}{d\theta_e} = 0·0707 \cos \theta_e. \tag{7.55}$$

For small values of phase error $\cos \theta_e \simeq 1$

$\therefore$ for small errors:

$$\frac{dV_d}{d\theta_e} = 0{\cdot}0707V \text{ per radian.} \tag{7.56}$$

For a small sinusoidal phase error

$$\theta_e(t) = \theta_e \sin pt$$

$$\frac{dV_d}{d\theta_e} = 0{\cdot}0707 \text{ volts peak per peak radian error.}$$

Let ϕ_e be the rms phase error

$$\therefore \quad \frac{dV_d}{d\phi_e} = 0{\cdot}0707 \text{ volts rms (across 100 ohm) per rms radian error.} \tag{7.57}$$

As we shall be comparing V_d, the signal output voltage, with the noise output of the PD at the same point in the circuit (thus at the same impedance level) we may work in voltage dB.

$$\therefore \quad V_d = -23 \text{ dBV per rms radian} \tag{7.58}$$

(b) *Noise Output*

The noise temperature of a mixer and resistive load, referred to the output, is given by

$$T_m \text{ out} = T_0(NTR + 1) \quad \text{where} \quad T_0 \text{ is } 290°\text{K.} \tag{7.59}$$

[See Appendix II equation (A.52)]

$\therefore$ the available noise power density at the mixer/load output is

$$N_0 = 2{\cdot}3 \ kT \text{ watts/Hz} \tag{7.60}$$

$$\therefore \quad N_0 = 2{\cdot}3 \times 1{\cdot}38 \times 10^{-23} \times 290 \text{ W/Hz}$$

$$= 9{\cdot}2 \times 10^{-21} \text{ W/Hz}$$

This represents a matched rms noise voltage density (from 100 Ω) of $\sqrt{9{\cdot}2 \times 10^{-21} \times 100}$ V/$\sqrt{\text{Hz}}$

$\therefore$ o/c noise voltage (V_{nm}):

$$V_{nm} = 2 \times 9{\cdot}59 \times 10^{-10} = 1{\cdot}918 \times 10^{-9} \text{ V/}\sqrt{\text{Hz}}$$

$$= -174{\cdot}3 \text{ dBV/}\sqrt{\text{Hz}} \tag{7.61}$$

(c) Noise Floor
But the output sensitivity of the PD is -23 dBV per rms radian input error.

$\therefore$ the equivalent PD output phase jitter density due to PD noise is:

$$\phi_{oe} = -174\cdot3 + 23 \text{ dB rads/}\sqrt{\text{Hz}}$$

$$\therefore \quad \overline{\phi_0^2} = -151\cdot3 \text{ dB rads}^2/\text{Hz} \tag{7.62}$$

$$\text{But } \overline{\phi_0^2} = 2\,\frac{N_{op}}{C} \tag{7.63}$$

$$\therefore \quad \frac{N_{op}}{C} \text{ threshold} = -154.3 \text{ dB/Hz.} \tag{7.64}$$

This is a somewhat optimistic figure as it assumes a mixer NTR of only $1\cdot3$, and that the VCO control circuit input impedance does not contribute noise. For this simple case, without a following active filter amplifier, a noise floor of -145 dBc/Hz would be about the best that could be achieved in practice. (Note: dBc denotes dB wrt the carrier).

7.7.1.2 An Analogue Phase Detector and Active Filter
Bearing in mind the necessity to operate at low impedance levels, to obtain optimum noise performance, a circuit for a PD/active amplifier combination is given in 7 Figure 20. A low noise integrated circuit amplifier TDA 1034 has been selected to provide realistic values for the calculations.

The noise characteristics of the TDA 1034 (Reference 26) are given below:

Frequency	30 Hz	1 kHz	Unit
Noise voltage	7	4	Nanovolts/$\sqrt{\text{Hz}}$
Noise current	$2\cdot5$	$0\cdot6$	pA/$\sqrt{\text{Hz}}$

R_1 has been chosen to be 1 k ohm. Its value should be as low as possible subject to the restriction that $\tau_1 = R_1 C$. τ_1 is fixed by the required loop natural frequency (see Equation 7.52) and C cannot be so large as to demand the use of an electrolytic capacitor which would be excessively noisy. In some circuit configurations, but not in others, R_3 may be required: its noise contribution has not been included.

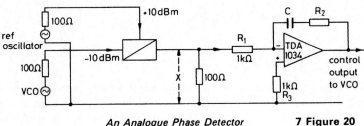

An Analogue Phase Detector **7 Figure 20**

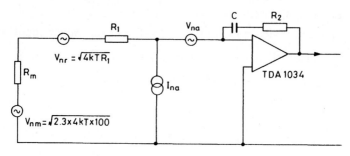

Circuit Used to Calculate Mixer/Amplifier   **7 Figure 21**
Noise

To calculate the effective noise density at the amplifier input we use the equivalent circuit shown in 7 Figure 21.

$$V_{nm} = \frac{1\cdot918}{10^9} \text{ V}/\sqrt{\text{Hz}} \text{ (see 7.61)} \tag{7.65}$$

$$V_{nr} = \sqrt{4 \times 1\cdot38 \times 10^{-23} \times 290 \times 10^3} = 4 \text{ nanovolts}/\sqrt{\text{Hz}} \tag{7.66}$$

The input impedance of the amplifier will be very high compared with $(R_1 + R_m) = 1\cdot1$ k ohm.

∴ Voltage at input to amplifier due to the amplifier current generator is as follows:

at   30 Hz   1 kHz

$$V_{nI} = \frac{2\cdot5 \times 1\cdot1 \times 10^3}{10^{12}} \qquad \frac{0\cdot6 \times 1\cdot1 \times 10^3}{10^{12}} \tag{7.67}$$

$$= 2\cdot75 \qquad 0\cdot66 \text{ nanovolts}/\sqrt{\text{Hz}}$$

$$V_{na} = 7 \qquad 4 \text{ nanovolts}/\sqrt{\text{Hz}} \tag{7.68}$$

The total noise voltage density at the amplifier input (V_{on}) will be given by:

$$V_{on} = (V_{nm}^2 + V_{nr}^2 + V_{nI}^2 + V_{na}^2)^{1/2} \text{ V}/\sqrt{\text{Hz}} \tag{7.69}$$

$$\therefore \quad V_{on} = \left(\frac{1\cdot918^2 + 4^2 + 2\cdot75^2 + 7^2}{10^{18}}\right)^{1/2} \text{ V}/\sqrt{\text{Hz}} \text{ at 30 Hz}$$

$$= 8\cdot73 \text{ nanovolts}/\sqrt{\text{Hz}} \text{ at 30 Hz} \tag{7.70}$$

and

$$V_{on} = 6\cdot0 \text{ nanovolts}/\sqrt{\text{Hz}} \text{ at 1 kHz} \tag{7.71}$$

The contribution due to the NTR of the mixer is only part of the first term in the expression for V_{on} see (7.59 and 7.61). Thus a *modest* increase in NTR for low offset frequencies will have little effect on the value of V_{on} even at 1 kHz.

Consider the noise floor at 30 Hz and 1 kHz.

Offset frequency	30 Hz	1 kHz	*Units*
V_{on}	8·73	6·0	nanovolts/$\sqrt{\text{Hz}}$
V_{on}	−161	−164·4	dBV/$\sqrt{\text{Hz}}$
Sensitivity [see (7.58)]	− 23	− 23	dBV/rms radian/$\sqrt{\text{Hz}}$
ϕ_e	−138	−141·4	dB rms rads/$\sqrt{\text{Hz}}$
$\therefore \ \overline{\phi_0^2}$	−138	−141·4	dB rads^2/Hz
$\therefore \ \dfrac{N_{op}}{C} =$	−141	−144·4	dB/Hz　　　(7.72)

Thus the noise floor of such a PD/active filter amplifier combination is −141 dBc at 30 Hz and −144 dBc at 1 kHz. It should be noted that if, for any reason, it is necessary to use a value of R several times 1 k ohm or a more noisy amplifier, then the noise floor of the PD/amplifier combination may be considerably worse than the figures calculated above. A mixer with a very poor NTR at low frequencies would also degrade the noise floor well below these values.

7.7.2 Digital—using 'Exclusive Or' Gate PD

The operation of such a PD has been described in Section 7.2.2(b) where it is shown that such a PD has a gain constant given by:

$$K_d = 0·64 \text{ volts per radian.}$$

This is a conservative value as it is based on the minimum voltage differential between 'high' and 'low' outputs given in the specification for 'Series 74' TTL. The output impedance of a 'Series 74' gate is less than 100 ohms (Reference 27) which drives a 1 k ohm resistor and a TDA 1034 integrated circuit (see 7 Figure 22).

As a good approximation (assuming the power supply to the 7486 IC to be relatively noise free) the noise voltage across X due to the PD with its maximum source impedance of 100 ohms will not be larger than that for the analogue PD of 7 Figure 19. The noise due to the amplifier will be identical to

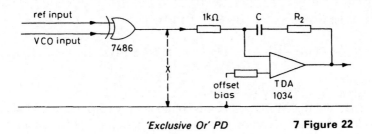

'Exclusive Or' PD　　　　　　**7 Figure 22**

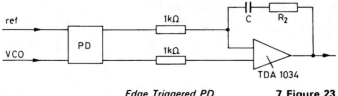

Edge Triggered PD   **7 Figure 23**

that of 7 Figure 21 and the total will be closely equal to that given by Equations (7.70) and (7.71).

However the signal output voltage is greater in the digital case than in the analogue case by $0.64/0.0707$ (see 7.57) $= 19$ dB.

Thus the signal to noise ratio and hence the noise floor will be 19 dB better than for the analogue PD/amplifier combination. Comparison with equation (7.72) gives the following values:

Offset frequency   30 Hz   1 kHz

Noise Floor $\left(\dfrac{N_{op}}{C}\right) =$ -160 dB/Hz -163 dB/Hz   (7.73)

7.7.3 Digital using Edge Triggered PD

The operation of such a PD is described in Section 7.2.2(c) where it is shown that such a PD has a minimum gain constant of approximately:

$$K_d = 0.32 \text{ volts/radian}$$

Assume it to be connected in circuit as shown in 7 Figure 23.

Compared with the noise contributions of the 1 k ohm resistor and the TDA 1034 IC, the noise output of the PD will be relatively negligible and therefore little error will occur by assuming the effective noise voltage at the input to the active filter amplifier to be identical to that in the case of an analogue PD (equations 7.70 and 7.71).

The signal output voltage is greater than in the analogue case in the ratio $0.32/0.0707$ (see 7.57) that is by 13.1 dB.

Thus the signal to noise ratio and hence the noise floor will be 13.1 dB better than for the analogue PD/active filter combination. Comparison with equation (7.72) gives the following values:

Offset frequency   30 Hz   1 kHz

Noise Floor $\dfrac{N_{op}}{C} =$ -154.1 -157.5 dB/Hz   (7.74)

7.7.4 The use of a Saturating Amplifier

The output of a digital PD will consist of nominally constant amplitude pulses the width of which depends on the phase error. Amplifier noise will be superposed both on the pulse peaks and the (for small phase errors) much longer periods between pulses. If the amplifier is designed so that it is normally biassed to cut off and then driven into saturation by the pulses, amplifier noise will be largely suppressed except during the rise time and the fall time of the pulses. By this means a considerable reduction in noise floor may be achieved at the price of not being able to use this amplifier as an active filter. An active filter might be provided as a separate unit following some amplification (and noise floor improvement) by using a high and low level saturated amplifier immediately after a digital PD.

If this approach were adopted, the noise floor (referred to the PD frequency) would tend to improve by 3 dB per octave reduction in PD frequency. That this is so is apparent from the following considerations. For a given phase error, the pulse length will be inversely proportional to frequency: constant rise and fall times, due to IC limitations, will represent a smaller proportion of the total time. Thus, for a given design, the time during which noise makes a contribution will represent a proportionally decreasing percentage of the cycle time.

Assuming first that the PD operating frequency is fixed by other considerations, the noise floor improvement achievable by the use of a saturating amplifier is fairly limited in practice for two reasons as follows:

(*a*) Any noise contributions from the *output* circuit of the saturated amplifier and from the following active filter amplifier are only effectively reduced by the saturating amplifier voltage amplification of the *peak* pulse output of the PD. The peak pulse output of the PD has a minimum guaranteed value of 2·4 volts and in some units may be in the region of 4 volts. Thus to get even a 6 dB improvement in noise floor means that the output stage of the saturating amplifier must provide an 8 volt output pulse. A 20 dB improvement would require a 40 volt pulse output. This would dictate the power supply voltage required by the unit and probably necessitate the use of discrete transistors rather than IC's for the saturating amplifier and active filter amplifier.

(*b*) The saturated amplifier will contribute directly to the loop gain. It may not be a problem to allow this to increase, although, for a given PLL natural frequency and damping factor, τ_1 and τ_2 will have to be scaled appropriately. If, for any reason, the increase in loop gain cannot be tolerated, it would be necessary to reduce the VCO sensitivity $K_0[(\text{rads/sec})/\text{V}]$. This is not always easy if a large VCO tuning range is required. A considerable increase in VCO complexity, such as the provision of two tuning varactors, one as part of the PLL and the other for programmed tuning, may result. For UHF and SHF VCO's, added complexity usually implies other penalties in VCO performance.

If a low PD operating frequency is tentatively chosen so that a saturating amplifier shall give the maximum possible noise floor improvement then two factors must be borne in mind.

(*a*) A very low PD operating frequency (carrier frequency) is not compatible with a high value of PLL natural frequency (f_n) which may be required for other reasons.

(b) The use of a low PD operating frequency *usually* results in a larger frequency division in the PLL. This in itself degrades the effective noise floor (when referred to the VCO frequency) by $20 \log_{10} n$ where n is the division ratio. (See Section 7.7.5). This degradation of 6 dB per octave more than offsets the 3 dB per octave improvement in the PD noise floor (with respect to the PD operating frequency) which results from an increased value of n.

7.7.5 The Effect of Frequency Division

Consider first a normal PLL without frequency division. An oscillator phase jitter of $\phi_{0\text{VCO}}$ radians rms relative to the reference signal will produce a PD output (V_{do}) of $K_d \phi_0$ volts rms.

i.e. $\quad V_{do} = K_d \phi_{0\text{VCO}}$ $\qquad\qquad$ (7.75)

This is also the voltage at the input to the active filter. An rms noise voltage $V_{on}/\sqrt{\text{Hz}}$ at the input to the active filter is equal to that which would be produced by a VCO rms phase jitter of V_{on}/K_d.

Let the VCO phase jitter which is, in this sense, equivalent to the PD/amplifier noise, be denoted $\phi_{0\,\text{PD}}$.

$\therefore \quad \phi_{0\,\text{PD}} = \dfrac{V_{on}}{K_d}$ $\qquad\qquad$ (7.76)

Thus the ratio of equivalent jitter associated with PD noise to that due to the VCO itself is:

$$\frac{\phi_{0\,\text{PD}}}{\phi_{0\,\text{VCO}}} = \frac{V_{on}}{K_d} \frac{K_d}{V_{do}} = \frac{V_{on}}{V_{do}} \qquad\qquad (7.77)$$

Consider now a PLL where the VCO frequency is divided by a factor n before feeding the PD. The VCO phase jitter $\phi_{0\,\text{VCO}}$ will be divided by n at the input to the PD. The PD output voltage will be given by:

$$V_{do} = K_d \frac{\phi_{0\,\text{VCO}}}{n} \qquad\qquad (7.78)$$

Thus, in a frequency dividing PLL, the PD gain constant becomes (K_d/n) volts per radian, whilst the noise output due to the PD/amplifier will be unchanged. Hence, compared with a fundamental frequency PLL, the signal/PD noise voltage ratio will be multiplied by $1/n$. Hence the PD noise floor will deteriorate by $20 \log_{10} n$.

In a PLL with a large division ratio the PD noise floor may easily determine the loop performance. As an example consider a good PD/amplifier combination with a fundamental noise floor of -160 dBc/Hz. Assume the

VCO frequency to be 100 MHz and that of the PD 10 kHz, necessitating a frequency division of 10,000. The effective noise floor will become $-160 + 20 \log 10^4 = -80$ dBc/Hz.

7.8 Frequency Conversion Within a PLL

Frequency division within a PLL serves two purposes. Firstly it locks a high frequency VCO to a lower frequency reference. Secondly, by using programmable dividers, it provides a range of VCO output frequencies.

The first purpose might equally well be served by a conventional mixing process, operating the PD at the intermediate frequency. An example of such an arrangement is shown in 7 Figure 24 and a similar approach is adequate for any PLL from which only a single output frequency is required or alternatively where the local oscillator used to drive the mixer, is itself programmable over the appropriate frequency ranges.

As there is no frequency division within this loop, the PD noise floor will not be degraded and will be that of a 50 MHz (analogue) PD of about -145 dBc/Hz. The reference signal to the PD is at a relatively low frequency (the IF) and thus may readily be designed to give a good phase noise performance. The other input to the PD at IF is obtained by beating the local oscillator with the VCO. As the instantaneous frequency (and hence the 'phase') of the IF is the difference between that of the VCO and the local oscillator, phase noise on the local oscillator will be directly added to that of the VCO at intermediate frequency.

Thus the signal into Port 2 of the PD will have phase noise characteristics which are the sum of those of the VCO and the local oscillator.

The PLL action will be to control the VCO phase so as to reduce the combined IF phase jitter towards that of the Reference. An 'ideal PLL' would

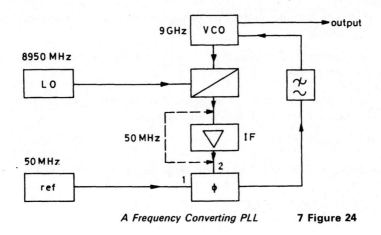

A Frequency Converting PLL   **7 Figure 24**

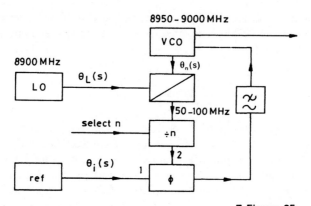

7 Figure 25
A PLL Incorporating Mixing and Frequency Division

achieve this at all offset frequencies below f_n. A real loop would give 40 dB per decade reduction below f_n. A superficial consideration might then lead one to suppose that the VCO phase noise, and hence the output phase noise, had been reduced in this way towards that of the Reference oscillator. In fact the sum of the phase noise powers of the VCO and local oscillator will have been reduced in this way. Even assuming a 'clean' Reference signal, the residual VCO phase noise after PLL correction must always be at least as great as that of the local oscillator used for frequency conversion.

Consider now a PLL which incorporates both a mixer and frequency division as shown in 7 Figure 25.

Let: $\theta_0(s)$ be the phase jitter of the PLL output under *closed loop* conditions.

$\theta_n(s)$ represent the phase jitter of the uncorrected VCO alone (i.e. under *open loop* conditions).

$\theta_L(s)$ represent the phase jitter of the uncorrected LO alone (open loop conditions).

$\theta_i(s)$ represent the phase jitter of the reference alone (open loop conditions).

Open the loop at the input to the active filter amplifier. Under these conditions the mixer output at IF will be given by:

$$\theta_{\text{IF}} = \theta_n(s) + \theta_L(s)$$

The input to port 2 of the PD (θ_{PD}) after frequency division by n, will be:

$$\theta_{\text{PD}}(s) = \frac{1}{n} [\theta_n(s) + \theta_L(s)]$$

In a similar loop but without the mixer and LO, the input to port 2 of the PD $[\theta'_{PD}(s)]$ would be:

$$\theta'_{PD}(s) = \frac{\theta_n(s)}{n}$$

Thus the frequency division ratio 'n' operates on LO noise (θ_L) in the same way as it operates on VCO noise (θ_n).

Under *open loop* conditions the error signal at the PD $[\theta_E(s)]$ (not to be confused with θ_e the closed loop error signal) will be:

$$\theta_E(s) = \theta_i(s) - \frac{1}{n}[\theta_n(s) + \theta_L(s)]$$

On *closing* the loop the PLL action will be to control the phase of the VCO so as to reduce the error signal at the PD, that is to make

$$\frac{1}{n}[\theta_n(s) + \theta_L(s)] \quad \text{tend towards } \theta_i(s).$$

As explained in Section 7.6 the effect of the frequency division ration (n) can be allowed for by multiplying the phase jitter of the reference by n to give $n\theta_i(s)$ when referred to the output.

Hence $\theta_n(s)$ and $\theta_L(s)$, that is VCO phase jitter and LO phase jitter, suffer no direct degradation due to the division ratio 'n'. In a graphical analysis of PLL performance the phase noise density of the LO should be added to n^2 times that of the reference. That is the action of the loop is to pull the phase noise density of the VCO down towards that of $[n^2\overline{\phi^2_{0\,REF}} + \overline{\phi^2_{0L}}]$.

Thus for the arrangement of 7 Figure 25 we may conclude that the phase noise of the local oscillator is not increased by 20 log n dB as is effectively the case for the phase noise of the reference oscillator.

It should be noted that the incorporation of a mixer, and possibly an IF amplifier, into a PLL affects the loop gain. Let A_m and A_{IF} be the mixer and IF amplifier gain constants. Then equations (7.52) and (7.53) become:

$$f_n = \frac{1}{2\pi}\left(\frac{K_0 K_d A_m A_{IF}}{\tau_1 n}\right)^{1/2} \tag{7.79}$$

$$\zeta = \frac{\tau_2}{2}\left(\frac{K_0 K_d A_m A_{IF}}{\tau_1 n}\right)^{1/2} \tag{7.80}$$

CHAPTER 8

FREQUENCY SYNTHESISERS

If a signal source is required to provide a large number of easily selected, accurately controlled and stable output frequencies with good phase noise performance, then the use of a synthesiser can hardly be avoided.

8.1 Types of Synthesisers

Synthesisers may be divided into three basic classes:

(*a*) Non coherent synthesisers which use more than one basic reference crystal or crystal oscillator.
(*b*) Coherent synthesisers which use the method of direct synthesis.
(*c*) Coherent indirect synthesisers which rely primarily on phase lock loops.

Coherent synthesisers (types *b* and *c*) derive all their output frequencies from a single high quality crystal oscillator.

Non coherent synthesisers were the first in the field but they become very expensive if a high order of frequency stability and good spectral purity are required. This is because a number of very high quality, and hence expensive crystal oscillators is required. Synthesisers of this type will not be further discussed.

8.1.1 Direct Synthesis
If the output of the master oscillator is fed into harmonic generators and also into decade frequency dividers a large number of frequencies is potentially available. These may be combined by addition and/or subtraction in mixers to produce yet further output frequencies all of which have a high order of frequency stability, as they have been derived from the master oscillator.

The two major limitations with synthesisers of this type are that it is difficult if not impossible to obtain a really good phase noise performance and about equally difficult to obtain adequate suppression of spurious outputs if a really good overall phase jitter performance is specified and the output frequency is in the SHF range. For output frequencies up to and including the VHF band such synthesisers can be made with an adequate performance for nearly all applications. Details of the performance which may be achieved by good design are given below (Reference 28).

Output frequency range up to 160 MHz.

$$\frac{N_{op}}{C} \text{ at } f_m = 1 \text{ Hz} \qquad -85 \text{ dB/Hz}$$

10 Hz	-95 dB/Hz
100 Hz	-105 dB/Hz
1 kHZ	-115 dB/Hz
10 kHz	-125 dB/Hz
100 kHz	-130 dB/Hz
1 MHz	-135 dB/Hz

Spurious outputs < -70 dBc.

If, by adding a $\times 60$ multiplier, the maximum output frequency were increased to 9600 MHz, then, unless very special steps were taken, both the phase noise and the spurious signal output level would be degraded by 35·6 dB. Such a signal source would no longer be suitable for critical applications. The simplest way of improving the phase noise performance and reducing spurious outputs (both at large offset frequencies) would be by the provision of a VCO at 1920 MHz maximum frequency (or 9·6 GHz) and to phase lock this VCO to the synthesiser. The resultant overall equipment can no longer be regarded as a direct synthesiser.

Direct synthesisers will not therefore be further discussed.

8.1.2 Synthesis using Phase Lock Loops

Indirect synthesisers may take a large variety of forms and vary greatly in complexity depending upon the specification which they have to meet. The only thing they all have in common is the use of one or more PLLs. In the rest of this chapter we shall be concerned exclusively with synthesisers based on the use of PLLs.

8.2 Factors affecting Choice of Configuration

Requirements imposed by the system in which the synthesiser is to be used (see Chapter 10) would result in specifying at least the following parameters:

(*a*) Output frequency range.
(*b*) The minimum size of selectable frequency increment.
(*c*) Frequency accuracy and stability.

(*d*) Phase noise characteristics versus offset frequency or maximum integrated phase jitter over specific offset frequency bands.

(*e*) The choice and characteristics of the master oscillator if this were separate from the synthesiser itself.

(*f*) Maximum level of discrete spurious outputs referenced to the carrier output level.

(*g*) Power output level and its tolerance, over the frequency range, the range of environmental conditions and with time.

(*h*) The worst voltage standing wave ratio of the load which the synthesiser feeds.

(*i*) The vibration environment.

(*j*) The temperature range and probably other environmental conditions.

(*k*) The characteristics of the power supplies which feed the synthesiser.

(*l*) Constraints on physical dimensions.

(*m*) Ease of operation and display of the selected output frequency.

(*n*) The maximum resetting time after selecting a new frequency.

(*o*) The reliability, probably expressed as a 'Mean Time Between Failures' (MTBF).

Whilst all these factors may affect the choice of synthesiser configuration, the most important ones, in any event the ones most relevant to the subject matter of this book, are (*a*) to (*f*).

For a simple approach, using frequency multipliers and frequency dividing PLLs only, the higher the output frequency, the more difficult it will be to achieve a good phase noise performance. To illustrate this consider that a low phase noise 5 MHz master oscillator has been specified and that an output frequency range of 9–10 GHz is required, output frequencies being selectable in 5 MHz steps.

Neglecting some of the practical constraints, the simplest approach conceptually would be to use a 9–10 GHz VCO with a frequency dividing PLL and to operate the phase detector at 5 MHz. This would imply a programmable frequency divider with a division ratio n of 1800–2000. Quite apart from the fact that dividers are not at present available for operation at this high input frequency, the phase noise performance of the loop components would be affected in the following way when referred to the output frequency:

(*a*) Phase noise performance of master oscillator
    —degraded by $20 \log_{10} n$ (see Section 7.6).
(*b*) Phase detector noise floor
    —degraded by $20 \log_{10} n$ (see Section 7.7.5).

Thus for the -maximum division ratio of 2000 the MO phase noise performance and the PD noise floor will each be degraded by 66 dB. A block diagram of such a synthesiser is given in 8 Figure 1.

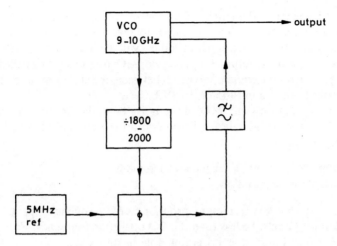

An Over-Simplified X-Band Synthesiser   **8 Figure 1**

For a PD noise floor of -140 dBc (at 5 MHz) the effective noise floor would become -74 dBc and even for a perfect Master Oscillator the performance within the loop bandwidth could never be better than -74 dBc and outside the loop bandwidth it would be determined by the phase noise performance of the X-band VCO.

Consider now what would be required if a minimum frequency increment of 100 kHz had been specified. It would be necessary to operate the PD at 100 kHz resulting in the configuration shown in 8 Figure 2.

The phase noise contribution of the 5 MHz Reference Oscillator will be improved by $20 \log_{10} 50$ by the initial frequency division and then degraded by $20 \log_{10} 100{,}000$ (with respect to 10 GHz) by the frequency division within the loop. This gives a total master oscillator degradation of 66 dB. The fact

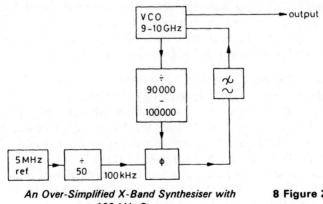

An Over-Simplified X-Band Synthesiser with   **8 Figure 2**
100 kHz Steps

that this is the same as the circuit of 8 Figure 1 is not a coincidence. For any system, other than ones using mixers to achieve frequency translations, the master oscillator phase noise performance will be degraded by $20 \log_{10} n$, where n is the ratio of output frequency to MO frequency, regardless of the partition of the total required division between initial division of the MO frequency and the division ratio of the PLL.

Thus, considering only the MO phase noise degradation, the choice of PD operating frequency is immaterial. From two other points of view as follows it may be very important:

(a) Minimum programmable frequency increment.
(b) PD noise floor degradation.

It is obvious that the minimum programmable frequency step in a *single* phase lock loop, cannot be less than the PD operating frequency. For example, if the divider in 8 Figure 2 is programmable in unit steps from $n = 90,000$ to $n = 100,000$, there will be 10,000 different output frequencies available, between 9000 and 10,000 MHz (i.e. in steps of 100 kHz).

The PD noise floor degradation is $20 \log_{10} n$ dB. In the case of 8 Figure 2 at 10 GHz output frequency, this is 100 dB, which is liable to be very serious. Even assuming a very good phase detector with a basic noise floor of -160 dBc/Hz at 100 kHz, when referred to the output frequency this becomes -60 dBc/Hz. Thus, however good the MO, even a perfect PLL cannot achieve a phase noise performance better than $N_{op}/C = -60$ dBc/Hz from the lowest conceivable offset frequencies up to f_n. At offset frequencies above f_n the phase noise performance will be determined by that of the uncontrolled VCO.

If still smaller frequency increments were specified, and the simple single PLL approach were retained, then the PD noise floor degradation could become catastrophic. It would then become necessary to adopt a more complex configuration using more than one phase lock loop, probably incorporating mixers in some of the loops to effect frequency transformations.

In addition to PD noise floor problems associated with a small demanded frequency increment, there are two other factors which may necessitate a multi-loop approach. These factors are still operative even for the less demanding case where only large frequency increments are required:

(a) Frequency Divider limitations. For SHF synthesisers the maximum frequency limitations of digital dividers (say 1·5 GHz) will prevent the use of an SHF VCO in a single loop system. This difficulty may of course be circumvented by using (say) an L band VCO in a PLL followed by a fixed ratio frequency multiplier, whilst remembering that the frequency increment provided by the PLL will also be multiplied by this ratio.
(b) The phase noise performance, at offset frequencies greater than f_n, will be determined by the uncontrolled phase noise performance of the VCO. (If the option mentioned in (a) above is adopted then the phase noise performance

will be that of the L band VCO degraded by $20 \log_{10} n$, where n is the frequency multiplication ratio of the output multiplier).

We shall consider (*b*) in more detail as it is of considerable importance. In this discussion the effects of the PD noise floor will be neglected as its effects have already been treated.

A master oscillator (reference) will usually be selected with a very good phase noise performance close to the carrier. To achieve this it is necessary to operate the basic oscillator at a low power level (see Chapter 5). To achieve an adequate final output power it is usually necessary to cascade it with a low noise amplifier. Although the noise figure of this amplifier will usually be very good it will add broad band noise to the output of the basic crystal oscillator. Even if such an amplifier is not necessary, the noise density contributed by a resistive load at the input to the synthesiser cannot be less than -174 dBm at room temperature. As an illustration, if the carrier level is 1 mW (0 dBm), $N_0/C = -174$ dB/Hz and $N_{op}/C = -177$ dB/Hz.

This will have a constant value for offset frequencies greater than those for which noise from the oscillator itself is dominant. This represents an absolute theoretical limit on the far from carrier phase noise performance for a room temperature MO, the output level of which never falls below 0 dBm in the chain from the output of the basic oscillator to the input to the PD in the PLL. In practice no master oscillator is as good as this and we shall assume for our illustration a good practical 5 MHz MO which has $N_{op}/C = -140$ dB/Hz for all offset frequencies greater than 20 Hz.

If we use this to feed a PLL such as that in 8 Figure 2, or in fact any PLL which translates directly to 10 GHz, then the MO performance will be degraded by $20 \log_{10} 10,000/5 = 66$ dB to become -74 dBc/Hz at all offset frequencies greater than 20 Hz. Even a perfect PLL can do no better than reduce the VCO phase noise to -74 dBc/Hz at all offset frequencies up to f_n. At offset frequencies greater than a few hundred Hz at most, this is not a particularly good performance and would be unacceptable for many critical applications.

The difficulty cannot be evaded by reducing the f_n of the loop to a very low figure (e.g. 30 Hz) for then the VCO phase noise characteristic will determine the performance over most of the offset frequency range. A typical performance for a fairly good X-band VCO (or its equivalent at 1·25 GHz after frequency multiplication by 8) is 77 dBc/Hz at 10 kHz, 52 dBc/Hz at 1 kHz and 22 dBc/Hz at 100 Hz.

For critical applications this obviously gives an unacceptable performance for the intermediate offset frequency range even if f_n is carefully chosen.

This difficulty, together with the noise floor problem, forces the adoption of a multiple PLL approach to SHF synthesiser design and infact even for UHF synthesisers.

There are so many approaches to the design of multi-loop synthesisers that it is difficult (or perhaps the author is insufficiently farseeing) to produce general design principles. It is therefore useful to consider specific examples of designs which have been produced.

8.3 Analysis of some Examples

8.3.1 A UHF Synthesiser

(a) *Specification*
Assume that a synthesiser is required to meet the following specification.
(i) Frequency range—570–690 MHz.
(ii) Step size − 1 kHz.
(iii) Frequency accuracy and stability: no degradation of MO performance.
(iv) Integrated phase jitter.

Offset frequency range	Phase jitter (millirads rms)
50–200 Hz	5
200 Hz–200 kHz	8
200 kHz–2 MHz	5

(v) The phase jitter is inclusive of the contributions of the 5 MHz reference MO which, wrt 5 MHz, are as follows:

Offset frequency	$N_{op}/C \ dB/Hz$
50	− 125
100	− 129
200	− 132
1 kHz	− 136
10 kHz	− 140

(vi) Discrete spurious outputs not to exceed:

Over the offset frequency range	dBc
±20 kHz	− 55
±20 kHz to ±30 kHz	− 60
±30 kHz to ±40 kHz	− 65
> ±40 kHz	− 70

The discrete spurious signals and the phase noise, when integrated together over the appropriate bands, shall result in an overall phase jitter not exceeding that given by (iv) above.

Discrete spurious signals at the output of the reference oscillator at 5 MHz will be no greater than − 120 dBc.

(b) Overall Description of Synthesiser

A synthesiser to meet this specification, and other requirements, will now be described (courtesy Microwave Associates) and the phase noise performance analysed using techniques which were outlined in Chapter 7.

A simplified block diagram, laid out in such a way as to highlight those factors which influence the phase noise performance, is given in 8 Figure 3. It will be noted that three phase lock loops are used.

To obtain an accurate assessment of the performance of the synthesiser contributions and clearly to separate them from those of the (5 MHz) Master Oscillator; measurements were carried out, both during design and production, using a 5 MHz MO with better phase noise characteristics than those of the MO specified for the final system.

As our present purpose is to compare phase noise performance calculated using the theory developed in previous chapters, with measured results on an engineered synthesiser we shall use in the calculations a commercial 5 MHz reference oscillator with a known phase noise performance somewhat better than the laboratory 5 MHz reference used for the measurements. This will reveal the full capabilities of this design of synthesiser whilst still enabling theory and measurement to be compared over most of the offset frequency range.

The phase noise characteristics of the MO used for the calculations are as follows:

f_m	$\dfrac{N_{op}}{C}$
1 Hz	-100 dB/Hz
10 Hz	-130 dB/Hz
22 Hz	-140 dB/Hz
> 22 Hz	-140 dB/Hz

(c) Loop 1

In loop 1 a 400–480 MHz VCO is divided by an integer in the range 40,000 to 47996 to feed a 10 kHz PD.

Frequency division is achieved by a fixed $\div 4$ prescalar followed by a programmable divider, programmable in unit steps from 10,000 to 11,999. A digital phase detector is used with a basic noise floor approaching -160 dBc. The 5 MHz external master oscillator is divided by 500 to provide the 10 kHz reference for Loop 1.

The MO phase noise characteristics with respect to 5 MHz (shown on 8 Figure 4) will be improved by ($20 \log_{10} 500$) dB ($= 54$ dB) and degraded by ($20 \log_{10} 47996$) dB ($= 93 \cdot 6$ dB) when referred to the VCO maximum frequency of 480 MHz. Thus, wrt 480 MHz, the 5 MHz MO phase noise will be degraded by 39·6 dB. The result is plotted on 8 Figure 4. The PD noise floor at 10 kHz, of -160 dBc/Hz, will be degraded by 93·6 dB by the loop

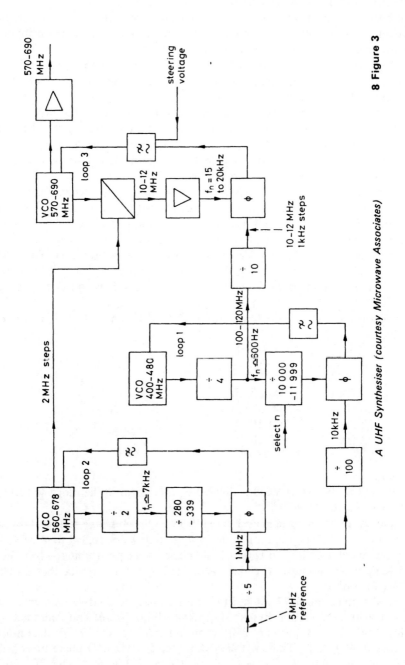

A UHF Synthesiser (courtesy Microwave Associates)

8 Figure 3

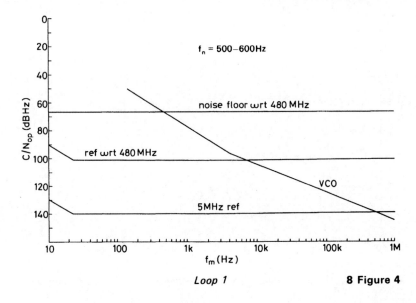

Loop 1   **8 Figure 4**

division ratio (wrt 480 MHz). This is also plotted on 8 Figure 4, as is the measured VCO characteristic which is in good agreement with a calculation using equation (5.27) with:

$$F = 20 \text{ dB}$$

$$C = +15 \text{ dBm}$$

$$Q = 10$$

$$f_0 = 480 \text{ MHz}$$

$$f_m = 10 \text{ kHz}$$

The overall performance of Loop 1 ($f_n = 600$ Hz) using the straight line approximation to PLL performance (see 7 Figure 12) is plotted on 8 Figure 5 wrt 480 MHz and also wrt 12 MHz (32 dB improvement) which is the maximum input frequency to Loop 3 from Loop 1. The input to Loop 3 from Loop 1 goes into the reference arm of the Loop 3 PD and has a frequency range of 10–12 MHz in 1 kHz steps as required.

(d) Loop 2

Loop 2 locks a 560–678 MHz VCO to the 5 MHz reference using a 1 MHz PD. The VCO passes through a divide by 2 prescaler and a programmable divider with $n = 280 - 339$. This permits selection of the output frequency in 2 MHz steps.

With a noise floor at 1 MHz of about -155 dBc/Hz and a maximum division ratio of 678 ($20 \log_{10} 678 = 56 \cdot 6$ dB) the effective noise floor is

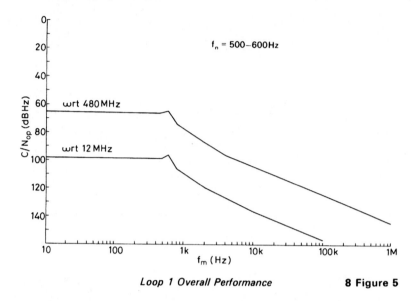

Loop 1 Overall Performance   **8 Figure 5**

$-98\cdot4$ dBc/Hz and the degradation of the 5 MHz MO is 20 log $(678/5) = 42\cdot6$ dB. The resultant values are plotted on 8 Figure 6, together with the VCO phase noise characteristic which is very similar to that of the VCO in Loop 1, modified by the slightly higher output frequency. The natural frequency of this loop was initially chosen at about 7 kHz but subsequently adjusted to 4 kHz which roughly approximates the intersection of the VCO characteristic with that of the noise floor and the MO. To illustrate the effect of f_n on the

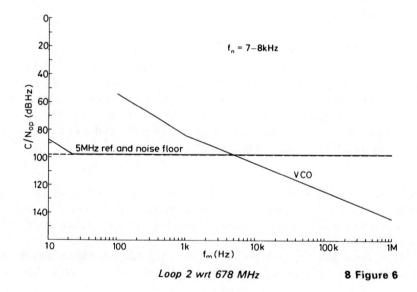

Loop 2 wrt 678 MHz   **8 Figure 6**

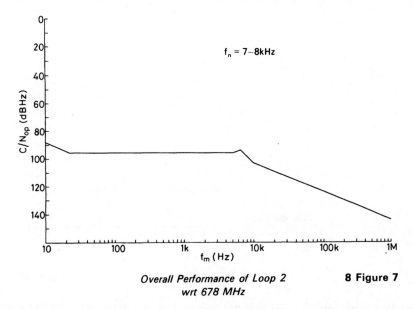

Overall Performance of Loop 2
wrt 678 MHz   **8 Figure 7**

performance, our calculations are carried out for $f_n = 7$ kHz and compared with measurements using an f_n of 4 kHz (see 8 Figure 9 and f below).

The overall performance of Loop 2 (derived from 8 Figure 6) is plotted on 8 Figure 7. Loop 2 output suffers neither multiplication nor division before being fed to a mixer in Loop 3.

(e) Loop 3

Loop 3 is the output loop and contains no frequency dividers and thus neither the PD noise floor, nor the input signals from Loops 1 and 2, suffer any phase noise degradation. Phase noise from Loop 2 adds to that of the VCO at the 10–12 MHz mixer output frequency and is then fed into one arm of the 10–12 MHz PD. The reference arm of the PD is fed from the output of Loop 1. The PD has a noise floor of about -150 dBc/Hz. The VCO performance is similar to that of Loop 2. All these phase noise parameters are plotted on 8 Figure 8. With an f_n of about 20 kHz the resultant overall performance of Loop 3 may readily be calculated (remembering the points made in Section 7.8) and the result is plotted on 8 Figure 9 which also shows the measured phase noise performance of one synthesiser of this design selected at *random* (Serial No. 106).

(f) Overall Performance

Examination of 8 Figure 9 shows that over the range of offset frequencies from about 600 Hz to 100 kHz the agreement between theory and measurement is good. Note the evidence that f_n for Loop 2 is nearer 5 kHz than the 7–8 kHz postulated. The difference below 600 Hz is due to the use of a poorer MO as

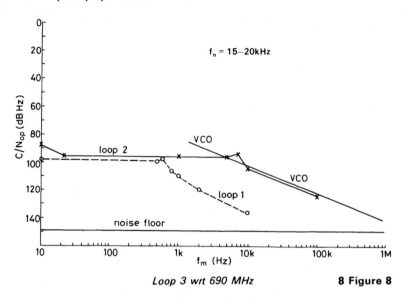

Loop 3 wrt 690 MHz   **8 Figure 8**

part of the measuring equipment. The poorer measured performance (still very much better than the specification) at $f_m > 150$ kHz appears to be due to some added noise in the Loop 3 VCO or in the following amplifier. As the input to the final amplifier is of the order of $+10$ dBm we may calculate the effective noise figure of the output amplifier necessary to produce $N_{op}/C = -125$ dB/Hz by using equation (6.23) which gives an added amplifier contribution as

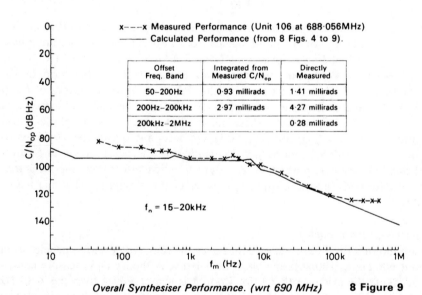

Overall Synthesiser Performance. (wrt 690 MHz)   **8 Figure 9**

$$\frac{N_{op}}{C} = \frac{F_A kT}{2C}$$

$$\therefore \quad F_A = \frac{N_{op}}{C}\frac{2C}{kT} \qquad\qquad (8.1)$$

with

$$\frac{N_{op}}{C} = -125 \text{ dB/Hz}$$

$$C = +10 \text{ dBm}$$

$$kT = -174 \text{ dBm}$$

This gives $F_A = 62$ dB.

Although such an amplifier is quite adequate to meet the required specification with a large margin, and although the amplifier intentionally limits, it seems unlikely that the noise figure is as high as this. The effect may be due to residual high frequency noise on the power supply to the oscillator. In an engineered design there is no point in inflating the cost to limit power supply contributions below -125 dBc/Hz unless this is a requirement.

The integrated phase jitter over the offset frequency bands 50–200 Hz and 200 Hz to 200 kHz has been calculated by integrating the *measured* phase noise performance over these bands using a Texas T159 programmable calculator programmed as described in Sections 8.4 and 9.3 below. The results are also shown on 8 Figure 9. The integrated phase jitter over the three specified offset frequency bands was also measured directly using the method outlined in Section 8.7 using an active filter switched in turn to pass the specified basebands:

> 50–200 Hz
> 200 Hz–200 kHz
> 200 kHz–2 MHz

It will be noted that the directly measured phase jitter is, in both cases, greater than that obtained by integrating the measured N_{op}/C curve. This is usually to be expected due to contributions to integrated phase jitter from discrete spurious outputs. These will show up on an N_{op}/C curve, if it is very carefully plotted using a high resolution analyser, but the Nop/C graph shown in 8 Figure 9 has been smoothed.

8.3.2 An X-band Synthesiser

(a) Specification
An X-band synthesiser (courtesy Microwave Associates) was designed to meet a very similar specification to the UHF synthesiser previously described, with a change in the integrated phase jitter requirements to those given below when

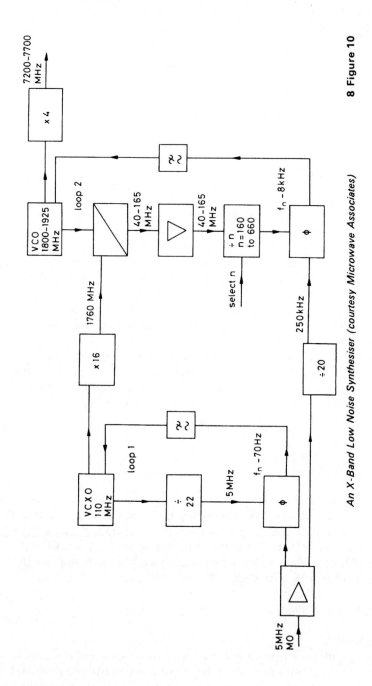

An X-Band Low Noise Synthesiser (courtesy Microwave Associates)

8 Figure 10

8 Table 1 *Integrated Phase Jitter Specification*

Offset frequency range	Max. integrated phase jitter
50–200 Hz	11 milliradians
200 Hz–200 kHz	12 milliradians
200 kHz–2 MHz	8·5 milliradians

operated in conjunction with the same 5 MHz MO. In this case the specified minimum frequency increment was 1 MHz.

(b) Overall Description of Synthesiser

Several versions were specified with different output frequency ranges. The version to be described has a required output frequency range of 7·2–7·7 GHz.

A simplified block diagram laid out in such a way as to highlight those factors which influence the phase noise performance, is given in 8 Figure 10. It will be seen that the synthesiser uses two phase lock loops. Loop 1 locks a 110 MHz VCXO to the 5 MHz MO. In Loop 2, a VCO operating over the range 1800–1925 MHz is phase locked to the 5 MHz MO. The output frequency of the Loop 2 VCO is multiplied by 4 to give the final output at 7200–7700 MHz.

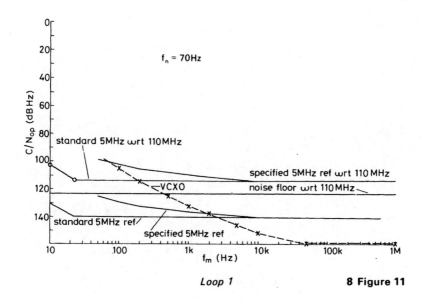

Loop 1 **8 Figure 11**

(c) *Loop 1*
The division ratio in this loop has a fixed value of 22. This will degrade both the 5 MHz MO performance and the noise floor of the PD by $20 \log_{10} 22$ ($= 26.8$ dB), when referred to the output frequency of 110 MHz.

The basic noise floor of the 5 MHz high performance PD is in the region of -150 dBc/Hz. The performance of the specified 5 MHz MO and that of the 'standard' MO (that also used in calculating the performance of the UHF synthesiser) are plotted in 8 Figure 11, referred both to 5 MHz and the 110 MHz output frequency of the PLL. The VCXO performance and the noise floor are both plotted wrt 110 MHz. It will be noted that the performance of the specified MO crosses that of the VCXO at $f_m = 70$ Hz. For this reason the natural frequency (f_n) of Loop 1 was chosen to be 70 Hz.

The overall performance of Loop 1 will follow that of the MO (wrt 110 MHz) up to 70 Hz and will then follow the VCXO phase noise performance, at higher offset frequencies. Using the 'standard' MO, the resultant curve is plotted in 8 Figure 12 both with respect to the 110 MHz output frequency and wrt 1760 MHz, the frequency at which it is fed into Loop 2.

(d) *Loop 2*
As a result of the mixer operation, phase noise at 1760 MHz from Loop 1 is added to the phase noise of the 1800–1925 MHz VCO at a maximum intermediate frequency of 165 MHz. The frequency of this IF signal is divided by 660 times and fed to a PD operating at 250 kHz. The reference signal into the PD is derived directly from the 5 MHz MO after division by 20. The PD

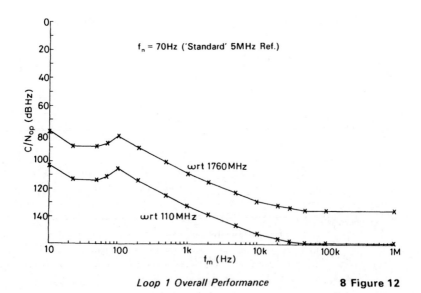

Loop 1 Overall Performance   **8 Figure 12**

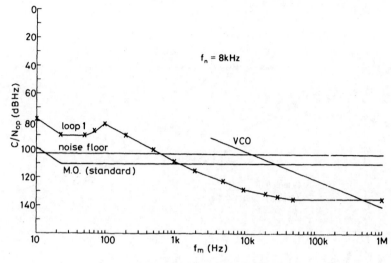

X-Band Synthesiser: Loop 2 wrt 1925 MHz   **8 Figure 13**

noise floor at 250 kHz is about -160 dBc/Hz. Both the PD noise floor and the 250 kHz reference signal suffer a degradation of $20 \log_{10} 660 = 56.4$ dB: the 5 MHz MO suffering a net degradation of $20 \log_{10} (660/20) = 30.4$ dB.

The PD noise floor and the MO phase noise performance wrt 1925 MHz are plotted on 8 Figure 13 together with the phase noise characteristics of the 1760 MHz signal from Loop 1 and those of the VCO. It is an interesting exercise for the reader to check that for the VCO,

$$\left(\frac{N_{op}}{C}\right)_{10\,kHz} = -102 \text{ dB/Hz when:}$$

$$F = 30 \text{ dB}$$

$$C = +19 \text{ dBm}$$

$$Q = 60$$

$$f_0 = 1925 \text{ MHz}$$

(e) Overall Performance

It should be noted that the VCO characteristic crosses the noise floor, which is not precisely known, at $f_m = 13$ to 20 kHz. To reduce MO enhancement (see Section 8.5.3) Loop 2 f_n has been chosen as 8 kHz. The overall performance of Loop 2 at 1925 MHz will then follow the Loop 1 performance (wrt 1760 MHz) out to about 600 Hz and will thereafter follow the noise floor out to about 8 kHz above which frequency it will be determined by the phase noise performance of the VCO.

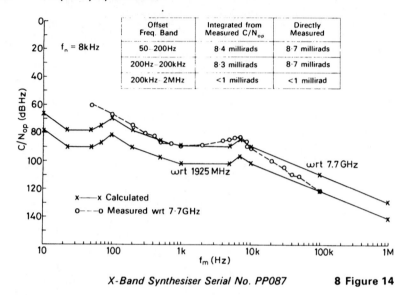

X-Band Synthesiser Serial No. PP087   **8 Figure 14**

The overall performance of Loop 2 wrt 1925 MHz, calculated in this way, is shown on 8 Figure 14. The frequency of the Loop 2 VCO is multiplied by 4 (12 dB degradation) before being fed to the output port of the synthesiser. This output (wrt 7·7 GHz) is also plotted on 8 Figure 14, together with the measured performance using a different 5 MHz MO in the test equipment. It is considered that the agreement between theory and measurement is quite good. This shows that the many possible causes of performance degradation, such as power supply noise, for example, have been successfully dealt with in this design.

Using the Texas TI 58/59 Program described in Sections 8.4 and 9.3, the integrated phase jitter over each of the specified offset frequency bands has been calculated from the *measured* curve of N_{op}/C. The integrated phase jitter has also been directly measured. A tabulation of the results is also given on 8 Figure 14.

8.4 A TI 58/59 Program to Integrate Phase Jitter over Specific Frequency Bands

8.4.1 General

Consider a graph showing $(N_{op}/C)_{dB}$ plotted against offset frequency f using a log scale for the f axis.

For any shape of graph this may be approximated by a series of straight line segments: the accuracy of the approximation improving as the number of segments in a given interval increases.

Consider a single segment between offset frequencies f_a, f_b ($f_a < f_b$).

The straight line approximation will be represented by a power law with a single valued exponent n.

For example, if the value of N_{op}/C falls by 20 dB per decade increase in frequency then $n = 2$.

8.4.2 Theory

$$n = \frac{\log\left(\frac{N_{op}}{C}\right)_a - \log\left(\frac{N_{op}}{C}\right)_b}{\log\left(\frac{f_b}{f_a}\right)} \tag{8.2}$$

$$\therefore \quad n = \frac{\frac{1}{10}\left(\frac{N_{op}}{C}\right)_a (\text{dB}) - \frac{1}{10}\left(\frac{N_{op}}{C}\right)_b (\text{dB})}{\log\left(\frac{f_b}{f_a}\right)} \tag{8.3}$$

$$\left(\frac{N_{op}}{C}\right)_f = \left(\frac{N_{op}}{C}\right)_a \left(\frac{f_a}{f}\right)^n \tag{8.4}$$

$$\overline{\phi_0^2}(f) = \left(\frac{2N_{op}}{C}\right)_f \tag{8.5}$$

$$\overline{\phi^2}(f_a \to f_b) = \int_a^b \left(\frac{2N_{op}}{C}\right)_f df \tag{8.6}$$

$$\therefore \quad \overline{\phi^2}(f_a \to f_b) = \int_a^b \left(\frac{2N_{op}}{C}\right)_a \left(\frac{f_a}{f}\right)^n df \tag{8.7}$$

$$\therefore \quad \overline{\phi^2}(f_a \to f_b) = \left(\frac{2N_{op}}{C}\right)_a f_a^n \frac{1}{1-n}[f^{(1-n)}]_a^b \quad (\text{except when } n = 1) \tag{8.8}$$

$$\overline{\phi^2}(f_a \to f_b) = \left(\frac{2N_{op}}{C}\right)_a f_a^n \frac{1}{1-n}[f_b^{(1-n)} - f_a^{(1-n)}] \tag{8.9}$$

$$\overline{\phi^2}(f_a \to f_s) = \phi^2(a \to b) + \phi^2(b \to c) + \phi^2(c \to d) \cdots \tag{8.10}$$

$$\phi_{rms}(f_a \to f_s) = [\phi^2(f_a \to f_s)]^{1/2} \tag{8.11}$$

8.4.3 The Program

Equations (3), (9), (10) and (11) have been programmed to run on a TI 58 or TI 59 calculator. With this program there are no restrictions on the total number of points which may be entered (the number of segments used). In addition special steps have been taken to circumvent the singularity that occurs normally when $n = 1$. For user instructions and program listing see Section 9.3.2, which describes a program which may be used either to evaluate integrated phase jitter as described above or to evaluate the frequency stability of the source given the phase noise density characteristics.

8.5 Discrete Spurious Outputs

8.5.1 Production of Discrete Spurious Outputs
Discrete spurious outputs from a synthesiser may be produced in a variety of ways. These include:

(a) Discrete spurious outputs from the Master Oscillator itself which pass through the synthesiser to the output, often being enhanced in the process.
(b) Direct breakthrough to the output, of signals which drive the various phase detectors, together with harmonics of these signals.
(c) Unwanted harmonics of the signal driving any frequency multipliers which may be used.
(d) Harmonics of the output frequency.
(e) Effects due to power supply ripple or vibration.

Before discussing these various effects in turn it is necessary to consider the relationship between the levels of discrete spurious signals and the equivalent phase jitter.

8.5.2 Phase Jitter due to Discrete Spurious Outputs

8.5.2.1 Contributions due to a Single Signal
Often the spectral width of a discrete spurious line is not known. In practice the width will depend upon the mechanism which produced the spurious signal. For the cases of mechanisms (b), (c) and (d) above it is possible in principle, though laborious, to calculate the line width. In the case of mechanism (e) it may also be possible, but usually not worthwhile. In all cases other than vibration effects, the energy bandwidth of the discrete spurious line is likely to be very narrow. A conservative assumption is that it occupies a bandwidth of 1 Hz. This assumption corresponds to the discussion and use of equation (5.53) in Section 5.9.4. Thus we shall assume a 1 Hz line width, which will over-estimate the phase jitter due to discrete signals of given level. It is usually the case that the ultimate criterion is the phase jitter, due to all causes, integrated over a specific offset frequency band. If this is so, an over-estimate of the contribution due to discrete spurious signals of a given level will lead to a conservative specification for the allowable level of individual spurious signals.

Thus, if the peak power in a spurious line is N_d, the peak power to carrier ratio is N_d/C. Now the nature of N_d may not be known except that it is deterministic and discrete: in particular it may be either an isolated line or one half of a conformable signal. In the latter case its 'phase' relation to the carrier may be such as to constitute PM or AM or a mixture of both. If our ultimate criterion is phase jitter then if its character is unknown, it is again a conservative assumption to assume that it is a PM sideband.

With our two assumptions:

(i) Its bandwidth is 1 Hz.
(ii) It is a PM sideband.

We may write:

$$\frac{N_d}{C} = \frac{N_{op}}{C} \tag{8.12}$$

and

$$\phi^2 = \left(\frac{2N_{op}}{C}\right) = \frac{2N_d}{C} \text{ rads}^2. \tag{8.13}$$

If it is known to be a single isolated line at $(f_0 + \delta f)$; no line being present at $(f_0 - \delta f)$ then, assuming 1 Hz bandwidth, it is analogous to SSB superposed noise.

$$\frac{N_d}{C} = \frac{N_0}{C} \tag{8.14}$$

In this case equation (3.46) gives:

$$\phi^2 = \left(\frac{N_0}{2C}\right) = \left(\frac{N_d}{2C}\right) \text{ rads}^2. \tag{8.15}$$

8.5.2.2 Integration of Contributions due to a Number of Spurious Signals

The phase jitter variance due to each *pair* of conformable lines may be zero if their 'phase' relative to the carrier is such as to produce AM: if it is such as to produce PM, the phase jitter variance due to each *pair* of sidebands is given by equation (8.13) and the total phase jitter variance due to a number of *pairs* of such lines is the sum of the individual variances.

$$\text{i.e. } \phi_c^2 = \phi_1^2 + \phi_2^2 + \cdots \phi_j^2 \tag{8.16}$$

where ϕ_c^2 is the total variance due to all conformable spurious lines and $(\phi_1^2, \phi_2^2, \rightarrow \phi_j^2)$ are the variances due to individual pairs of lines.

The phase jitter variance due to each single non-conformable line may be evaluated by equation (8.15). Let the results for lines $(k \rightarrow s)$ be written:

$$\phi_k^2, \phi_l^2, \phi_m^2 \cdots \phi_s^2$$

where the suffixes $(k \rightarrow s)$ refer to all lines in the offset frequency range of interest *both* below and above the carrier frequency. Let ϕ_n^2 be the total phase jitter variance due to all non-conformable lines on both sides of the carrier over the offset frequency range of interest. Then:

$$\phi_n^2 = (\phi_k^2 + \phi_l^2 + \phi_m^2 \ldots) \tag{8.17}$$

The total phase jitter due to both types of spurious signals ϕ^2 is:

$$\phi_t^2 = \phi_c^2 + \phi_n^2 \tag{8.18}$$

8.5.3 Enhancement of MO Spurious Outputs

Although a Master Oscillator will have a very good performance it may still produce discrete spurious signals at certain specific offset frequencies. For a well designed MO the performance will be very good at low offset frequencies (in the absence of vibration effects) both with regard to statistical noise and deterministic spurious line outputs at specific frequencies. At higher offset frequencies, particularly above 100 kHz the performance may cease to be outstanding both with regard to noise-like and deterministic outputs. In addition there will inevitably be some output at multiples of the MO frequency; that is the oscillator, however good, must give some output at harmonics of its basic frequency. A *good* 5 MHz MO may have a specification which states that at offset frequencies greater than 100 kHz all discrete spurious signals will be at least 120 dB below the carrier level.

In normal synthesiser applications the MO will be used as the reference input to a PLL, either directly or after an initial frequency division. In either case, the MO phase jitter density at the loop output frequency will be given by:

$$\phi_0(f_m)_{f_2} = n_1[H_1(j\omega)]\phi_0(f_m)_{f_1} \tag{8.19}$$

where f_2 is the loop output frequency, f_1 is the MO frequency and $n_1 = f_2/f_1$ and $H_1(j\omega)$ is the PLL transfer function.

This follows from 7 Figure 11(b) as modified by the discussion in Section 7.6.

In many synthesisers the output signal of this PLL (say PLL 1) may be fed into another PLL (say PLL 3) where it is used as the PD reference.

Let the output frequency of PLL 3 be f_3.

$$\therefore \quad \phi_0(f_m)_{f_3} = n_1 n_3 [H_1(j\omega)][H_3(j\omega)]\phi_0(f_m)_{f_1} \tag{8.20}$$

where

$$n_3 = \frac{f_3}{f_2}$$

Assume that the output frequency of the synthesiser is that of PLL 3 i.e. f_3.

$$n_1 n_3 = \frac{f_3}{f_1} = \frac{f(\text{output})}{f(\text{MO})} = n \tag{8.21}$$

8 Table 2   *MO Enhancement—an example*

Spurious Signal	100 kHz	200 kHz
Input level	− 120 dBc	− 120 dBc
n^2	+ 46 dB	+ 46 dB
$[H_1(j\omega)]^2$	− 40 dB	− 46 dB
$[H_3(j\omega)]^2$	− 20 dB	− 26 dB
Overall enhancement	− 14 dB	− 26 dB

Now the transfer functions of PLLs 1 and 3, $[H_1(j\omega)]$ and $[H_3(j\omega)]$ may each be evaluated using equation (7.33).

If the damping factor is not too different from 0·707 a good approximation is to assume that:

$$H(j\omega) = 1 \text{ for } \omega < \omega_n$$

$H(j\omega)$ falls from unity by 20 dB per decade increase in frequency from 1·5 f_n.

Alternatively we may evaluate $H(j\omega)$ more accurately by using Figure 2.3 of Ref. 24.

Thus we find that for MO phase jitter due to discrete spurious signals at baseband frequencies less than the natural frequency of both PLLs 1 and 3, the MO phase jitter will be *enhanced* by n times at the output of the synthesiser and its phase noise power to carrier ratio (N_{op}/C) will be enhanced by n^2 ($20 \log_{10}n$).

Consider, as an example a synthesiser with a maximum output frequency of 1 GHz and with this PLL configuration (any other PLLs being arbitrarily assumed to produce negligible enhancement, as is often true). The 5 MHz MO is assumed to have spurious lines at 100 kHz and 200 kHz at a level − 120 dB with respect to the carrier. The natural frequencies of PLLs 1 and 3 are:

$$f_{n_1} = 660 \text{ Hz} \qquad f_{n_2} = 6\cdot6 \text{ kHz}$$

$$20 \log_{10}n = 20 \log_{10} \frac{1000}{5} = 46 \text{ dB}$$

The resultant performance is shown in 8 Table 2.

It should be noted that, had f_{n1} and f_{n2} each been 10 times greater, the overall enhancement would have been worsened by 40 dB and would have been positive at both 100 kHz and 200 kHz.

8.5.4 Breakthrough of PD Signals

It is perfectly possible to derive a general formula for the output spurious signal level of a synthesiser of given design, due to direct breakthrough of PD signals. Nevertheless the factors involved are sufficiently complicated that it is felt that a step by step calculation for one specific example will be much more

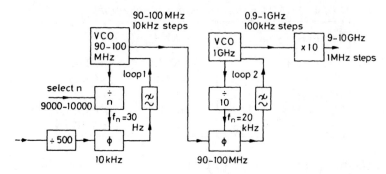

A Hypothetical Synthesiser   **8 Figure 15**

illuminating. The reader may then readily apply the same method to any other synthesiser being confident of the physical meaning of each step.

For an example we shall take a fairly simple and poorly designed synthesiser, where the effects are relatively large. A block diagram of the synthesiser we shall consider is given in 8 Figure 15.

The VCO of Loop 1 tunes over 10 MHz (say) with 20 volts variation on the varactor.

$$\therefore \quad \text{mean value of } K_0 = \frac{10^7}{20} \times 6\cdot28 \text{ rads/sec/V.}$$

Assume max. value of $K_0 = 10^7$ rads/sec/V.   (8.22)

Assume the use of an edge triggered digital PD (TTL)

From Section 7.2.2(c)

$$K_d = 0\cdot32 \text{ rads/v}$$

$$f_n = 30 \text{ Hz}$$

$$\therefore \quad \omega_n = 6\cdot28 \times 30 \text{ rads/sec.}$$

$$\therefore \quad \omega_n \simeq 2 \times 10^2.$$

From (7.51)

$$\omega_n^2 = \frac{K_0 K_d}{\tau_1 n}$$

$$\therefore \quad \tau_1 = \frac{K_0 K_d}{\omega_n^2 n}$$

$$\therefore \quad \tau_1 = \frac{0\cdot32 \times 10^7}{4 \times 10^8} \simeq \frac{1}{10^2} \text{ secs}$$ (8.23)

From (7.53)

$$\zeta = \omega_n \frac{\tau_2}{2} \qquad \text{Assume} \quad \zeta = 0.707$$

$$\therefore \quad \tau_2 = \frac{2 \times 0.707}{\omega_n} = \frac{2 \times 0.707}{2 \times 10^2}$$

$$\therefore \quad \tau_2 = \frac{0.707}{10^2} \text{ secs} \tag{8.24}$$

Now the output from the PD in Loop 1 will consist of pulses at a 10 kHz rate; the width of the pulses depending on the phase difference between the two input signals. For small duty ratios the fundamental component of a square wave of unity peak to peak amplitude is θ/π; so that, for a fairly large initial phase error of (say) 1×10^{-2} radians, the fundamental component of the 10 kHz output from the PD would be below the peak TTL output level by about:

$$\frac{1}{\pi \times 10^2} \simeq -50 \text{ dB} \tag{8.25}$$

The pulses, with a 10 kHz component of -50 dB wrt TTL level, will be smoothed by the active filter and pass to the control line of the Loop 1 VCO. Thus, to assess the magnitude at the VCO, we must consider the transfer function of the loop filter for a 10 kHz sinusoidal input.

The transfer function of the loop filter is given by (see equation (7.11)):

$$F_3(s) = -\frac{s\tau_2 + 1}{s\tau_1}$$

To calculate the residual 10 kHz component at the VCO we need to evaluate $F_3(s)$ for a 10 kHz sinusoidal input: From (7.11)

$$F_3(j\omega) = -\frac{j\omega\tau_2 + 1}{j\omega\tau_1} = -\frac{\omega^2\tau_1\tau_2 - j\omega\tau_1}{\omega^2\tau_1^2}$$

$$\therefore \quad F_3(j\omega) = -\frac{\omega\tau_2 - j}{\omega\tau_1}$$

The modulus of $F_3(j\omega)$ is given by:

$$|F_3(j\omega)| = \frac{(\omega^2\tau_2^2 + 1)^{1/2}}{\omega\tau_1} \tag{8.26}$$

Now for

$$\omega = 6 \cdot 28 \times 10^4, \ \tau_2 = \frac{0 \cdot 707}{10^2} \text{ and } \tau_1 = \frac{1}{10^2}$$

$$|F_3(j\omega)|_{10 \text{ kHz}} \simeq \frac{(\omega^2 \tau_2^2)^{1/2}}{\omega \tau_1} = \frac{\tau_2}{\tau_1}$$

$$\therefore \quad |F_3(j\omega)|_{10 \text{ kHz}} = 0 \cdot 707 = -3 \text{ dB} \tag{8.27}$$

Thus for 1×10^{-2} radians initial phase error, the equivalent sinusoidal signal at 10 kHz 'breaking through' from the PD to the VCO will have a peak value -53 dB, and an rms value -56 dB, wrt TTL level. Of the 53 dB attenuation of the peak breakthrough signal 3 dB is due to the active filter of Loop 1. The remainder is due to the assumed initial phase error.

Now the natural frequency of Loop 1 is 30 Hz so that it will not produce any correction of 10 kHz phase jitter of the Loop 1 VCO. Thus, for a TTL peak to peak level of 2 volts, and an initial error of 1×10^{-2} radians the peak voltage at 10 kHz reaching the VCO will be 4·5 mV. The rms signal will be 3·2 mV.

Now

$$K_0 = \frac{10^7}{6 \cdot 28} \text{ Hz/V} \qquad \text{(see 8.22)}$$

$$\therefore \quad \delta f \text{ (10 kHz)} = \frac{10^7}{6 \cdot 28} \times \frac{3 \cdot 2}{10^3} \text{ Hz}$$

$$\therefore \quad \delta f \text{(10 kHz)} = 5 \text{ kHz} \tag{8.28}$$

Now

$$\phi_{(10 \text{ kHz})} = \frac{\delta f}{f_m}$$

$$\therefore \quad \phi_{(10 \text{ kHz})} = 0 \cdot 5 \text{ radians rms} \tag{8.29}$$

Reference to 8 Figure 15 shows that this phase jitter at 10 kHz is present on the reference input to the PD of Loop 2. The natural frequency of this loop is 20 kHz. The phase jitter at 10 kHz at the output of Loop 2 will be:

$$\phi_{2 \text{ (10 kHz)}} = 0 \cdot 5 \left[H_2(j\omega) \right] \text{ radians rms} \tag{8.30}$$

where $H_2(j\omega)$ is the transfer function of Loop 2. For $\zeta = 0 \cdot 707$ and $f = f_n/2$, $|H(j\omega)| = +1$ dB (Ref. 24 Figure 2–3)

$$\therefore \quad \phi_{2 \text{ (10 kHz)}} = 0 \cdot 5 \times 1 \cdot 1 \text{ radians} \tag{8.31}$$

The output of Loop 2 is frequency multiplied by 10 to give the synthesiser output. Let $\phi_{3\,(10\,\text{kHz})}$ be the spurious phase jitter at the synthesiser output.

$$\therefore \quad \phi_{3\,(10\,\text{kHz})} = 5 \cdot 5 \text{ radians} \tag{8.32}$$

This is quite catastrophic.

The best way round this problem, whilst retaining the basic synthesiser configuration, would be to provide a 10 kHz bandstop filter in Loop 1 between the PD and the VCO. Alternatively such a filter might be inserted between the Loop 1 VCO and the Loop 2 PD. If such a filter provided 60 dB attenuation of the 10 kHz component then:

$$\phi_{3\,(10\,\text{kHz})} = 5 \cdot 5 \text{ millirads}$$

Arbitrarily assuming the line width to be 1 Hz

$$\frac{N_{op}}{C} = \frac{\phi_o^2}{2} = -48 \text{ dBc/Hz} \tag{8.33}$$

8.5.5 Unwanted Harmonics of Frequency Multipliers
As an example consider the synthesiser shown in 8 Figure 10. If it were poorly designed, the unwanted 15th and 17th harmonics of the 110 MHz VCXO might be coupled directly to the output to give output signals at 1650 MHz and 1870 MHz, or possibly, if they coupled to the input circuit of the final multiplier they might give spurious output signals at 6600 MHz and 7480 MHz.

Another possible mechanism is that 3rd or 5th harmonics of the final multiplier might be insufficiently suppressed at the synthesiser output.

8.5.6 Harmonics of the Output Frequency
The output VCO of any synthesiser, being imperfectly linear, will inevitably produce some power at harmonics of its operating frequency. In some instances it may be necessary to provide filtering at the output to attenuate these harmonics.

8.5.7 Effects due to Power Supply Ripple or to Vibration
Power supply imperfections or vibration may be responsible for discrete spurious output signals. These mechanisms were previously discussed in Section 5.9 and are further considered in Section 8.6.

8.6 Engineering Design Requirements

8.6.1 Power Supplies

The parts of a synthesiser which are most sensitive to ripple or noise on the power supply rails are the oscillators, particularly any wide tuning range VCO's. As an example a fairly typical tuning slope of 15 MHz/V represents 15 Hz rms per μV of rms ripple. At 50 Hz this represents an rms phase deviation ϕ of 0·3 radians rms. An rms ripple component of 1 μV at 150 Hz, gives

$$\phi_{150} = 0\cdot1 \text{ rads}$$

$$\phi^2 = 0\cdot01 = \left(\frac{2N_{op}}{C}\right)_{150} \text{ rads}^2$$

$$\therefore \quad \left(\frac{N_{op}}{C}\right)_{150} = -23 \text{ dBc/Hz}$$

It is true that in a synthesiser application the VCO may be phase locked, ultimately to the master oscillator, which, being a crystal oscillator, will be less sensitive to power supply ripple. However any noise or ripple on the power supplies at frequencies above the natural frequency of the PLL concerned will produce phase jitter which will appear at the output of the VCO.

Consider the permissible power supply noise voltage per Hz at (say) 1 kHz if the VCO noise performance due to this effect alone is not to be worse than -103 dBc/Hz at 1 kHz offset frequency.

$$\left(\frac{2N_{op}}{C}\right) = \bar{\phi}_0^2 = -100 \text{ dB rads}^2$$

$$\therefore \quad \phi_0 = \frac{1}{10^5}$$

$$\delta f_0 = \phi_0 f_m = \frac{10^3}{10^5} = 0\cdot01 \text{ Hz rms.}$$

At 15 MHz/V

$$Vn_0 = \frac{0\cdot01}{15 \times 10^6} = \frac{1}{1\cdot5 \times 10^9} \text{ volts.}$$

Thus it is apparent that, from the point of view of both noise and ripple, any power supply feed directly to the tuning control of a high slope VCO cannot be countenanced. The power supply must be connected to an operational amplifier which has a large power supply ripple rejection (PSRR) and, simultaneously, low noise characteristics Assume a PSRR of 86 dB ($=20{,}000$).

For an unchanged phase noise requirement the permissible noise density or ripple (at 1 kHz) of the power supply feeding this amplifier is

$$\frac{2 \times 10^4}{1\cdot5 \times 10^9} = 13\cdot3 \ \mu V.$$

The power supply purity requirements at other frequencies can easily be calculated in a similar manner after considering what the VCO phase noise specification should be at the corresponding offset frequency. In deciding the VCO phase noise specification the possible relaxation due to PLL action must be considered.

It is apparent from the foregoing that the design of power supply stabilisers used in synthesisers, is critical. This is particularly so for the VCO's and active filter amplifiers, each of which must normally be provided with its own low noise, low ripple stabiliser located as close as possible.

8.6.2 Earthing
Coupling of power supply noise or ripple, and various types of signal, from one part of a synthesiser to another by common earth coupling must be avoided at all costs.

The special steps taken to provide low ripple power supplies for each of the VCO's may readily be negated in practice by common earth path coupling. Consider the case of a synthesiser which consumes 1 amp at 20 volts. Assume the input power supply to the unit, to have a ripple voltage of 10 mV = 0·05% and that a path of 1 milliohm impedance is common to the main power supply return and a VCO control voltage return line. The ratio of ripple current to DC current will be of the same order as the ratio of ripple voltage to supply voltage: for purposes of illustration consider it to be 0·05% of 1 ampere; which equals 5×10^{-4} amps. Across 1 milliohm this will develop 5×10^{-7} volts = 0·5 μV. Such a ripple voltage applied directly to the tuning control of a wide range VCO would be catastrophic (see equation (5.48) and 5 Figure 9).

In a similar way common earth paths may couple signals from one part of a synthesiser to another, partially bypassing the filtering normally provided by PLL operation. At UHF or SHF frequencies a very small common inductance will provide considerable coupling. At 1 GHz, for example, 0·01 μH has an impedance of 62·8 ohms.

Summarising; common earth paths should be avoided completely. The same care is needed as in the design of very high gain VHF, IF amplifiers and the same techniques can be used to detect and measure the magnitude of unwanted couplings in an existing design (See Ref. 29).

8.6.3 Screening and Filtering
It is apparent from the previous discussions that power supply lines must be protected against pick up of unwanted, hum, noise and signals. To achieve this

it is necessary that all such lines should be adequately filtered where they enter or leave any critical compartment.

Protection against electrostatic pick-up can be achieved by screening the power supply leads: the chassis should not be used as a power supply earth return. Magnetic pick-up at low frequencies can be avoided by careful routing of leads and the avoidance of loops. If a synthesiser is to operate satisfactorily it is very important that signals from one part of the unit should not couple directly to another part, thus bypassing the filtering provided by PLLs and in other ways. For example the whole discussion of Section 8.5.4 is irrelevant if either of the 10 kHz signals at PD1 (8 Figure 15) couple directly to any part of Loop 2.

To prevent such effects each of the critical parts of the synthesiser must be separately screened using screening techniques adequate to the frequency range involved (Reference 29 Chapter 8).

8.6.4 Vibration

The parts of a synthesiser most subject to vibration effects are the oscillators. The effects of vibration on crystal oscillators were discussed in Section 5.9.4. Sensitivity of a unit to external vibration can be greatly reduced except at low vibration frequencies, by the provision of suitable anti-vibration mounts.

It is also important to note that with only one exception, low frequency vibration induced phase modulation of synthesiser oscillators, will be reduced by the operation of PLLs within the synthesiser. The exception, which in any case is often external to the synthesiser, is the Master Oscillator. This forms the phase reference for the whole synthesiser and any vibration induced phase modulation cannot be reduced by PLL action. In fact in most synthesisers MO phase modulation will be enhanced by $20 \log_{10} n$ dB: where n is the synthesiser output frequency divided by the MO frequency.

A knowledge of the likely vibration environment of the synthesiser, after allowing for the transmissibility characteristics of any vibration mounts, (Reference 30) and the natural frequency of each of the PLLs, will make it possible to decide whether or not any of the oscillators should be provided with its own anti-vibration mounts.

If the system specification demands operation of the complete equipment in a severe vibration environment then it may be necessary to use active vibration cancellation techniques, at least for the Master Oscillator (Reference 31).

8.7 Some Important Aspects of Phase Noise Measurements

8.7.1 A Simple Method of Measuring Phase Noise Density and/or Integrated Phase Jitter

In principle it would seem that the simplest means of measuring the phase noise characteristics of a signal source would be to feed it into a phase

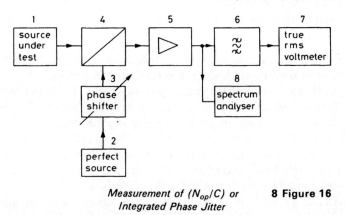

Measurement of (N_{op}/C) or   **8 Figure 16**
Integrated Phase Jitter

detector; the other input to the PD being a 'perfect' signal source *at the same frequency.* The baseband output, measured over any specific baseband frequency range, would then be proportional to the phase jitter of the signal source integrated over this particular range of baseband frequencies.

If the output of the PD were fed into a tuneable baseband filter of 1 Hz bandwidth, or a suitable low frequency spectrum analyser, then an output proportional to $(N_{op}/C)_f$ would be obtained.

This procedure, for which a block diagram is given as 8 Figure 16, is in fact a successful and popular method of measurement.

For measurement of low noise synthesisers it is apparent that the noise floor of the mixer/amplifier combination in 8 Figure 16 must be good. The mixer must be driven at an optimum level say $+10$ dBm for one signal and -10 dBm for the other. In addition a low noise amplifier must follow the mixer. For methods of calculating the noise floor see Section 7.7.1. It is also necessary to use low noise power supplies, to provide adequate screening and to avoid earth loops.

8.7.2 Methods of Calibration
In practice it will be necessary to measure the phase jitter of a synthesiser at a number of different output frequencies. The 'perfect' source (Item 2 in 8 Figure 16) must therefore be capable of providing outputs at any of these frequencies. It must therefore be a synthesiser, which will not, in fact, contribute negligible phase noise. If the synthesiser under test is a production model, then two synthesisers of the same type may be used in the test configuration. Although the phase noise performance of the two synthesisers may not be identical, they are likely to be very similar unless one is faulty. This can easily be checked if a third synthesiser is available. Assuming:

(*a*) The performance of Item 2 to be identical to that of the synthesiser under test and

(b) The performance of the two MO's (if these are separate from the synthesisers) also to be identical;

then the phase noise outputs of the two synthesisers will be incoherent and the total phase noise density will be 3 dB greater than either alone.

Initial calibration is carried out by setting the baseband filter (the characteristics of which must be accurately known) to pass the baseband frequency range over which it is desired to measure the integrated phase jitter. The synthesiser (Item 2) is set to give an output frequency different from that of the synthesiser under test by a frequency well within the filter passband. After allowing the two synthesisers to settle, the r.m.s output voltage is carefully measured. Denote this value by V_c.

Synthesiser 2 is then switched to give the same output frequency as synthesiser 1. Adjust the phase shifter 3 to give maximum output. The r.m.s output voltage is measured. Correct this value to allow for the known filter characteristics. Denote the resultant by V.

8.7.3 Derivation of Calibration Constants

Let the PD characteristic be:

$$v = K \sin \theta \tag{8.34}$$

where v is the instantaneous output voltage, K is a PD parameter and θ is the instantaneous phase error. When calibrating with two different input frequencies the phase error $\theta(t)$ varies sinusoidally.

$$\theta(t) = \theta \sin pt$$

The peak value of θ is $\pi/2$. Thus the *peak* voltage output from the PD during calibration (denoted V_{CP}) is:

$$V_{CP} = K \sin \frac{\pi}{2} = K \tag{8.35}$$

$$\therefore \quad v = V_{CP} \sin \theta \tag{8.36}$$

$$\therefore \quad \frac{dv}{d\theta} = V_{CP} \cos \theta \tag{8.37}$$

For $\theta \ngtr 0.1$ radians $\cos \theta = 1$ with an error $\ngtr 2\%$

$$\therefore \quad \frac{dv}{d\theta} = V_{CP} \text{ peak volts per peak radian} \tag{8.38}$$

when measuring sinusoidal phase jitter of the order of 100 millirads peak or less.

Integrating equation (8.38)

$$v = V_{CP}\theta \tag{8.39}$$

If θ is the peak phase deviation, v is the peak voltage. Dividing both sides of (8.39) by $\sqrt{2}$ gives

$$V = V_{CP}\,\phi$$

where ϕ is the rms phase error and V is the rms output voltage. If the calibration voltage is an rms voltage $V_c = V_{CP}/\sqrt{2}$

$$\phi = \frac{V}{V_{CP}} = \frac{1}{\sqrt{2}}\frac{V}{V_c} \tag{8.40}$$

$$\therefore\quad \phi^2 = \frac{1}{2}\left(\frac{V}{V_c}\right)^2$$

For noise induced phase jitter, the mean of a large number of measurements would give:

$$\overline{\phi^2} = \frac{1}{2}\left(\frac{V}{V_c}\right)^2 \tag{8.41}$$

Equation (8.41) shows that a single sample of the phase jitter variance is 3 dB less than the measured ratio $(V/V_c)^2$. If the result is normalised to a 1 Hz bandwidth

$$\overline{\phi_0^2} = \frac{1}{2}\left(\frac{V}{V_c}\right)^2\frac{1}{b} \tag{8.42}$$

where b is the baseband or offset frequency bandwidth used for the measurement.

Now

$$\overline{\phi_0^2} = \frac{2N_{op}}{C}$$

$$\therefore\quad \left(\frac{N_{op}}{C}\right) = \frac{1}{4}\left(\frac{V}{V_c}\right)^2\frac{1}{b} \tag{8.43}$$

Equation (8.43) shows that the phase noise density to carrier ratio is 6 dB less than the ratio of the measured power (in a 1 Hz bandwidth) to the calibration power. Strictly this gives a single sample of (N_{op}/C) and it is also an average value over the measurement bandwidth.

An intuitive understanding of the reason for the 6 dB correction factor in equation (8.43) may be obtained by remembering that phase noise sidebands equidistant from the carrier are conformable. As shown in Section 2.8 conformable sidebands add in phase when coherently demodulated, so that the baseband power is 4 times that of one of them.

If the signal source numbered '2' in 8 Figure 16 is not perfect, but has been *proved* to have identical phase noise characteristics to the source under test, then the corrections to $(V/V_c)^2$ given by (8.41) and (8.43) should be increased by 3 dB.

For a brief discussion of an alternative method of carrying out the above measurement whilst dispensing altogether with any reference signal source, that is using only the source under test, see the end of Section 10.1.2.

Another Misunderstanding

It has been the author's experience that some engineers use a calibration correction of 3·9 dB rather than 3 dB when calculating $\overline{\phi^2}$ from measured values of $(V/V_c)^2$.

It is argued that the ratio of the peak slope of the PD characteristic (sin θ) at the origin, to that of the mean slope from the origin to $\pi/2$ radians, is $2/\pi$. The peak slope is used for measurement and the mean slope for calibration. Hence the correction is stated to be 20 log $2/\pi = -3·9$ dB. This argument neglects the fact that we are concerned with *both* a sinusoidal PD characteristic and a sinusoidal phase error. Careful perusal of (8.36) and (8.39) should resolve any initial doubt. The necessary 3 dB correction is due solely to the fact that the calibration voltage is an rms voltage, not a peak voltage.

8.7.4 Statistical Aspects

Assume that we have set the filter, shown as item 6 in 8 Figure 16, at 1 Hz bandwidth. The true rms voltmeter (item 7) may be regarded as a detector. Let it have a time constant of t seconds. If $t = 1$ second then the output voltage will vary from second to second because the value of n_{op}/C varies from second to second due to the statistical characteristics of phase noise. (See Section 3.2).

Rewriting (8.43) for a specific second:

$$\left(\frac{V}{V_c}\right)^2 = \left(\frac{4n_{op}}{C}\right) \tag{8.44}$$

This is only a single sample of the phase noise density to carrier ratio. We need an estimate of the long term mean value.

Suppose that t is greater than one second: the meter will indicate the average value of V^2 over t seconds.

Consider now the more general case where the filter (item 6) is set to a bandwidth b Hz (the phase noise density being considered constant over this limited bandwidth (b)). At the output of the filter there will be b independent samples per second and bt independent samples in t seconds. That is the number of independent samples m is:

$$m = bt \tag{8.45}$$

The square of the reading of the true rms voltmeter will indicate the average power of m independent samples. In this case:

$$\left(\frac{V}{V_c}\right)^2 = 4b\left(\frac{\overline{n_{op}}}{C}\right) \tag{8.46}$$

If bt is sufficiently large $(\overline{n_{op}}/C)$ will be a reasonable estimate of (N_{op}/C).

If then we repeated the measurement n times and took the average power of these n results we should get an average over $m \times n$ samples (say) n' samples. Let each of the individual indicated powers (n in number) be $x_1, x_2 \rightarrow x_n$. The mean value of these n indicated powers will be

$$\bar{x} = \sum_{i=1}^{n} \frac{x_i}{n} \tag{8.47}$$

This will be an unbiassed value (Reference 32). To establish confidence limits that it lies within a given range we need to know the standard error of this mean.

Now the standard deviation of the mean of a sample of n items is

$$\frac{\sigma}{\sqrt{n}}. \tag{8.48}$$

If we knew σ (the standard deviation of the basic population) we should know the standard deviation of the sample. Let μ be the mean value of the total population, which is unknown but is what we are seeking.

Then:

$$Z = \frac{(\bar{x} - \mu)}{\sigma/\sqrt{n}} \text{ (Reference 32)} \tag{8.49}$$

where Z is the standardised normal deviate.

What this means in words is that the difference between our measured mean ($\bar{x}$) and the real mean is so many standard deviations for a selected value of Z, (the parameter of the cumulative normal distribution). For a confidence level of 68%, $|Z| \le 1$. If we select a confidence level of 80% then $|Z| \le 1\cdot28$. In this case we have an 80% confidence level that μ does not lie more than $\pm 1\cdot28$ standard deviations from our measured value $\bar{x}$, where the standard deviation is $\sigma/\sqrt{n}$.

Unfortunately we do not know σ. All we do know, or can readily calculate from our measured results, is $\hat{s}$ which is a best estimate of σ from our sample of n measurements.

$$\hat{s}^2 = \frac{1}{n-1} \sum_{i=1}^{n} (x_i - \bar{x})^2 \tag{8.50}$$

Many hand held calculators have statistical facilities which can be used to evaluate (8.50) merely by keying in the successive values of x_i (our measured powers).

A best estimate of $\sigma(\hat{s})$ differs from σ itself (the standard deviation of an infinitely large population) in that, being an estimate, it lies within a specified range with specified confidence limits.

If, in equation (8.49) we replace σ with its estimate $\hat{s}$, the standardised error of the mean is given by:

$$t = \frac{(\bar{x} - \mu)}{\hat{s}/\sqrt{n}} \qquad (8.51)$$

which is not normally distributed as $\hat{s}$ itself has confidence limits. Thus we have a special distribution for t (Student's t Distribution) which is tabulated for different degrees of freedom. In our case the number of degrees of freedom is $(n - 1)$. For a very large sample $\hat{s}$ is very close to σ and so the 't Distribution' approaches the Normal Distribution.

The 80% confidence limits for μ are shown in 8 Table 3. For example if we had taken 10 measurements $(n = 10)$ then $t = 1\cdot383$ and the maximum error in μ is $\pm 1\cdot383\,\hat{s}/\sqrt{10} = \pm 0\cdot437\,\hat{s}$.

Any prior effective integration due to the time constant of the power meter (combined with that of the observer) as described by equation (8.45) will of course reduce the value of $\hat{s}$ by $1/\sqrt{m}$ and thus increase the overall accuracy.

8 Table 3 *80% Confidence Limits for the Mean*

n	Degrees of freedom	t	$\lvert \bar{x} - \mu \rvert$
2	1	3·078	2·176 $\hat{s}$
3	2	1·886	1·089 $\hat{s}$
5	4	1·533	0·686 $\hat{s}$
7	6	1·440	0·544 $\hat{s}$
9	8	1·397	0·466 $\hat{s}$
10	9	1·383	0·437 $\hat{s}$
12	11	1·363	0·393 $\hat{s}$
15	14	1·345	0·347 $\hat{s}$
18	17	1·333	0·314 $\hat{s}$
20	19	1·328	0·297 $\hat{s}$
25	24	1·318	0·264 $\hat{s}$
30	29	1·311	0·239 $\hat{s}$
31	30	1·310	0·235 $\hat{s}$
∞	∞	1·282	0

A further improvement in accuracy occurs when the measured values of N_{op}/C are plotted over the offset frequency band. Abrupt changes in the phase noise performance of the source under test, with a small change in offset frequency, are most unlikely, except in the presence of discrete spurious signals. Drawing a smooth curve through the measured plot (which is usually made using a pen recorder) will in effect average a number of adjacent samples and give a further reduction in inaccuracy, often by as much as $1/\sqrt{10}$.

In conclusion we may say that if we take advantage of all the possibilities of statistical smoothing, phase noise density plots can be made to give a good approximation to the real long term average value N_{op}/C. With reasonable care an accuracy of ± 1 dB may readily be obtained.

8.8 A Generalisation of the Concept of Phase Noise Density

In this book, up until the present point, $N_{op}(f)$ has been regarded as the phase noise power in a 1 Hz bandwidth at an offset frequency f. It is necessary to extend the definition for very low offset frequencies. For example at 0.1 Hz offset frequency the phase noise power density will normally vary greatly over a 1 Hz bandwidth. Let δN_p be the phase noise power in a small increment (δf) of offset frequency.

The phase noise power density is then given by

$$N_{op} = \frac{\delta N_p}{\delta f} .$$

In the limit N_{op} is the derivative of the phase noise power with respect to offset frequency i.e.,

$$N_{op} = \left(\frac{dN_p}{df}\right) \tag{8.52}$$

CHAPTER 9

THE RECIPROCAL RELATIONSHIPS BETWEEN PHASE NOISE AND FREQUENCY STABILITY*

(Frequency Domain to Time Domain Transformations and their Inverses)

9.1 Introduction

. For sinusoidal modulation at a single frequency the rms frequency deviation of an angle modulated signal is related in a simple way (equation 4.3) to the rms phase deviation. For phase modulation indices due to random noise modulation of a carrier ($\theta' \ngtr 0.1$ radians) we may use the linear approximation, treating each of the sidebands separately. If we consider a number of 1 Hz (conformable) sidebands on either side of the carrier we may add their effects on a power basis, as they are at different frequencies.

Thus it would be expected that, if the phase noise characteristics of a signal were measured, it should be possible to calculate the square of the frequency deviation, and hence the frequency stability. Conversely, if the frequency deviation were measured, it would seem likely that the phase noise characteristics could be calculated.

We shall consider the subject from both these points of view. We shall find that if the density, of either the phase noise or the frequency deviation, is known as a function of offset frequency, over the offset frequency range of interest, then both its own integrated value and all the characteristics of the other parameter may be calculated. If only the integrated value of one parameter is known (over the offset frequency range of interest) then, except in special cases, it is not possible to calculate its own density characteristics or any of the characteristics of the other parameter. This situation is illustrated in 9 Figure 1.

---

* In accordance with widespread useage, 'fractional frequency instability' is referred to as frequency 'stability'.

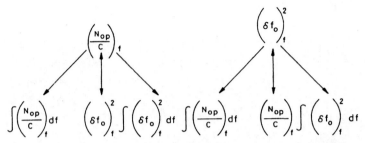

The Relationship between Phase Noise and   **9 Figure 1**
Frequency Stability

In some cases the availability of test equipment may make it necessary to measure one parameter (phase noise or frequency deviation) when a knowledge of the other is required. Thus the ability to transform from frequency domain (phase noise) results to time domain (frequency deviation) results, and vice versa, is important.

9.2 The Reciprocal Relationships between Phase Noise Density and Frequency Deviation Density

Equation (4.7) is repeated, for convenience

$$(\delta f_0)^2_{f_m} = \left(\frac{2N_{op}}{C}\right)_{f_m} f^2_m \tag{9.1}$$

Thus, if the value of $(N_{op}/C)_{f_m}$ is known over the complete range of relevant offset frequencies; $(\delta f_0)^2_{f_m}$ may be calculated for each of these offset frequencies.

Conversely if $(\delta f_0)^2_{f_m}$ is known, $(N_{op}/C)_{f_m}$ may be calculated (on a point by point basis).

9.3 The Calculation of Integrated Frequency Jitter from known Phase Noise Density Characteristics

9.3.1 The Theoretical Relationship
From equation (9.1)

$$(\delta f)^2 = \int_{f_1}^{f_2} \left(\frac{2N_{op}}{C}\right)_{f_m} f^2_m \, df_m \tag{9.2}$$

As in Section 8.4.1 the graph of $(N_{op}/C)_f$ plotted on log/log scales, is approximated by a series of straight line segments: the accuracy of the approximation improving as the number of segments in a given interval increases.

The straight line approximation is represented by a power law with a single valued exponent n. Consider a single segment between offset frequencies f_a and f_b ($f_a < f_b$).

From (8.3)

$$n = \frac{\frac{1}{10}\left(\frac{N_{op}}{C}\right)_a (dB) - \frac{1}{10}\left(\frac{N_{op}}{C}\right)_b (dB)}{\log\left(\frac{f_b}{f_a}\right)} \qquad (9.3)$$

$$\left(\frac{N_{op}}{C}\right)_f = \left(\frac{N_{op}}{C}\right)_{f_a}\left(\frac{f_a}{f}\right)^n \qquad (9.4)$$

$$\therefore \quad (\delta f)^2(f_a \to f_b) = \int_{f_a}^{f_b} \left(\frac{2N_{op}}{C}\right)_f f^2 \, df \qquad (9.5)$$

$$(\delta f)^2(f_a \to f_b) = \int_{f_a}^{f_b} \left(\frac{2N_{op}}{C}\right)_{f_a}\left(\frac{f_a}{f}\right)^n f^2 \, df \qquad (9.6)$$

$$\therefore \quad (\delta f)^2(f_a \to f_b) = \left(\frac{2N_{op}}{C}\right)_{f_a} f_a^n \int_{f_a}^{f_b} f^{(2-n)} \, df \qquad (9.7)$$

Except when $n = 3$:

$$(\delta f)^2(f_a \to f_b) = \left(\frac{2N_{op}}{C}\right)_a f_a^n \left[\frac{f^{(3-n)}}{3-n}\right]_{f_a}^{f_b} \qquad (9.8)$$

$$\therefore \quad (\delta f)^2(f_a \to f_b) = \left(\frac{2N_{op}}{C}\right)_a f_a^n \frac{1}{3-n}\left[f_b^{(3-n)} - f_a^{(3-n)}\right] \qquad (9.9)$$

$$(\delta f)^2(f_a \to f_s) = \delta f^2(a \to b) + \delta f^2(b \to c) + \delta f^2(c \to d) + \cdots \qquad (9.10)$$

$$\therefore \quad \delta f \text{ rms} = [\delta f^2(f_a \to f_s)]^{1/2} \qquad (9.11)$$

9.3.2 A TI 58/59 Program

Equations (9.3), (9.9), (9.10) and (9.11) have been programmed to run on a TI 58/59 calculator. This program may also be used to evaluate integrated phase jitter using equations (8.3), (8.9), (8.10) and (8.11), as derived in Section 8.4.2.

With this program there are no restrictions on the total number of points which may be entered. In addition special steps have been taken to eliminate the singularities which occur when $(1 - n)$ or $(3 - n) = 0$.

(a) User Instructions to Evaluate Integrated Phase Jitter

1. Press E—Display 21
   This step clears all previous results and initialises the program.
2. Enter f_a (the lowest frequency) Press A—Display f_a.
3. Enter $\left(\frac{N_{op}}{C}\right)_a$ (dB). This may be entered with or without the

   minus sign: the program takes care of this aspect.
   Press B Wait—Display (21).

4. Enter f_b—Press A—Display f_b.

5. Enter $\left(\dfrac{N_{op}}{C}\right)_b$ (dB) Press B

   Wait—Display $\phi^2(a \to b)$

6. Enter f_c Press A $\to$ Display f_c

7. Enter $\left(\dfrac{N_{op}}{C}\right)_c$ (dB) Press B

   Wait—Display $\Sigma\phi^2(a \to c)$

$8 - N$. Repeat for all required points up to (say) f_s and $\left(\dfrac{N_{op}}{C}\right)_s$

   Display shows $\Sigma\phi^2(a \to s)$

$N + 1$. Press C display $\phi_{rms}(a \to s)$

(b) User Instructions to Evaluate Frequency Stability
1. Press E—Initialise
2. Press D—Sets program to evaluate frequency stability.
3. Enter f_a—Press A
4. Enter $\left(\dfrac{N_{op}}{C}\right)_a$ (dB) Press B—wait
5. Enter f_b—Press A
6. Enter $\left(\dfrac{N_{op}}{C}\right)_b$ (dB) Press B—wait

   Display $(\delta f)^2(a \to b)$
7. Enter f_c Press A
8. Enter $\left(\dfrac{N_{op}}{C}\right)_c$ dB Press B

   Wait $\to$ Display $\Sigma(\delta f)^2(a \to c)$

$9 - N$. Repeat for all required points up to (say) f_s and $\left(\dfrac{N_{op}}{C}\right)_{f_s}$

   Display shows $\Sigma(\delta f)^2(a \to s)$
10. Press C Display $\delta f_{(rms)}(a \to s)$ Hz.
11. Enter carrier frequency f (Hz). Press 2nd A.
    Display fractional frequency stability.

(c) Program Listing

000	76	LBL	005	01	01	010	01	01	015	02	02
001	15	E	006	25	CLR	011	25	CLR	016	29	CP
002	47	CMS	007	01	1	012	02	2	017	91	R/S
003	22	INV	008	01	1	013	01	1	018	76	LBL
004	86	STF	009	42	STO	014	42	STO	019	11	A

020 72 ST*	064 95 =	108 95 =	152 91 R/S
021 01 01	065 55 ÷	109 42 STO	153 76 LBL
022 69 OP	066 43 RCL	110 08 08	154 10 E'
023 21 21	067 04 04	111 25 CLR	155 43 RCL
024 91 R/S	068 95 =	112 43 RCL	156 12 12
025 76 LBL	069 42 STO	113 06 06	157 42 STO
026 12 B	070 05 05	114 55 ÷	158 11 11
027 50 I × I	071 43 RCL	115 43 RCL	159 43 RCL
028 94 +/−	072 11 11	116 07 07	160 22 22
029 95 =	073 45 Y$^{\times}$	117 65 ×	161 42 STO
030 72 ST*	074 43 RCL	118 43 RCL	162 21 21
031 02 02	075 05 05	119 08 08	163 00 0
032 25 CLR	076 95 =	120 95 =	164 42 STO
033 02 2	077 42 STO	121 28 LOG	165 12 12
034 02 2	078 06 06	122 65 ×	166 00 0
035 32 X:T	079 87 IFF	123 01 1	167 42 STO
036 43 RCL	080 01 01	124 00 0	168 22 22
037 02 02	081 48 EXC	125 95 =	169 01 1
038 67 EQ	082 43 RCL	126 85 +	170 02 2
044 45 Y$^{\times}$	083 05 05	127 43 RCL	171 42 STO
040 69 OP	084 94 +/−	128 21 21	172 01 01
041 22 22	085 85 +	129 95 =	173 25 CLR
042 91 R/S	086 01 1	130 85 +	174 02 2
043 76 LBL	087 95 =	131 03 3	175 02 2
044 45 Y$^{\times}$	088 42 STO	132 95 =	176 42 STO
045 43 RCL	089 07 07	133 55 ÷	177 02 02
046 12 12	090 76 LBL	134 01 1	178 25 CLR
047 55 ÷	091 49 PRD	135 00 0	179 43 RCL
048 43 RCL	092 29 CP	136 95 =	180 09 09
049 11 11	093 67 EQ	137 22 INV	181 91 R/S
050 95 =	094 44 SUM	138 28 LOG	182 76 LBL
051 28 LOG	095 43 RCL	139 95 =	183 13 C
052 42 STO	096 12 12	140 44 SUM	184 43 RCL
053 04 04	097 45 Y$^{\times}$	141 09 09	185 09 09
054 25 CLR	098 43 RCL	142 25 CLR	186 34 $\sqrt{X}$
055 43 RCL	099 07 07	143 01 1	187 95 =
056 21 21	100 75 −	144 03 3	188 91 R/S
057 75 −	101 53 (	145 32 X:T	189 76 LBL
058 43 RCL	102 43 RCL	146 43 RCL	190 44 SUM
059 22 22	103 11 11	147 01 01	191 43 RCL
060 95 =	104 45 Y$^{\times}$	148 67 EQ	192 07 07
061 55 ÷	105 43 RCL	149 10 E'	193 85 +
062 01 1	106 07 07	150 43 RCL	194 93 .
063 00 0	107 54)	151 09 09	195 00 0

196	00 0	209	94 +/−	222	76 LBL	*Labels*	
197	00 0	210	85 +	223	16 A′		
198	01 1	211	03 3	224	42 STO	001	15 E
199	95 =	212	54)	225	03 03	019	11 A
200	42 STO	213	42 STO	226	35 1/X	026	12 B
201	07 07	214	07 07	227	65 ×	044	45 Y^x
202	61 GTO	215	61 GTO	228	53 (	091	49 PRD
203	49 PRD	216	49 PRD.	229	43 RCL	154	10 E′
204	76 LBL	217	76 LBL	230	09 09	183	13 C
205	48 EXC	218	14 D	231	34 $\sqrt{X}$	190	44 SUM
206	53 (	219	86 STF	232	54)	205	48 EXC
207	43 RCL	220	01 01	233	95 =	218	14 D
208	05 05	221	91 R/S	234	91 R/S	223	16 A′

Memories

01	Ind. Addressing for stores containing *f*.	09	$\Sigma\phi^2(a-k)$ or $\Sigma\delta f^2(a-k)$ current value.
02	Ind. Addressing of stores containing N_{op}/C		
03	Carrier frequency	11	Current value f_a
04	Log f_b/f_a	12	Current value f_b
05	*n*	21	Current value
06	f^n (current values)		$-\left(\dfrac{N_{op}}{C}\right)_a dB$
07	$(1-n)$ or $(3-n)$ current value.		
08	$[f_b^{(1-n)}-f_a^{(1-n)}]$ or $[f_b^{(3-n)}-f_a^{(3-n)}]$ current value.	22	Current value $-\left(\dfrac{N_{op}}{C}\right)_b dB$

9.3.3 The Relevant Offset Frequency Range

It will be apparent that to be able to use the TI 58/59 program described above to assess the frequency stability it is initially necessary to decide the offset frequency range over which to integrate.

(a) The Lower Frequency Limit

The lower frequency limit (say f_{c1}) is determined by the period (T) over which the frequency stability is specified. Any apparent difficulties in integrating down to zero frequency are purely mathematical in nature: the mathematical model does not represent a physically possible system if it is extrapolated down to zero offset frequency. This is because no real oscillator operates for an infinite period of time.

If the system specification lays down a value of T, this forms a basic starting point. If T is not specified, short term frequency stability, as distinct from close to carrier phase noise, is rarely a real operational requirement except in those systems which are used for making accurate time measurements or where frequency synchronisation of two or more systems is required. In other cases the real requirement is normally that the drift, together with all other contributions to frequency inaccuracies including any doppler shifts, shall not, in a period shorter than the planned interval between routine adjustments, exceed the pull-in range of the equipment. This often results in T having a value of the order of 1 month.

Even when T is known, the exact value to be chosen for f_{c1} is not immediately apparent. Intuitively it might seem that it should be chosen equal to $1/T$ or possibly $1/2T$. Actually it is more complicated than this. A full analysis is given in Section 9.4.5 where it is shown that for a $1/f^3$ phase noise density law which extends over an offset frequency range from zero to f_c, the mean square of the *measured* peak frequency jitter using a measurement time T is given by equation (9.68) which has been evaluated in (9.73). However putting $f_{c1} = 1/2T$ is adequate for purely theoretical purposes.

(b) The Upper Frequency Limit

The upper frequency limit (f_{c2}) should be stated in the system specification. Unfortunately this is not always done, and in any case the writer of the system specification cannot make an arbitrary decision and must therefore understand the factors involved in choosing it. To illustrate the necessity of specifying f_{c2}, consider different signal sources (a) and (b) with the following phase noise characteristics at large offset frequencies:

(a) white phase noise density
(b) phase noise density falling as $1/f^2$

Now

$$(\delta f_0)^2 = \left(\frac{N_{op}}{C}\right)_{f_m} f_m^2$$

Thus, in case (a), $(\delta f_0)^2$ is rising by 20 dB per decade increase in frequency. If we increase f_{c2} by a decade we shall greatly increase the calculated frequency jitter both because of the greater bandwidth and the greater density.

In case (b) the frequency jitter density will be white but an increase in bandwidth by a decade will still result in an appreciable increase in the value of the integral. In both these cases, which are fairly typical, a frequency stability specification without a top frequency limit is strictly meaningless. The phase noise density would have to be falling faster than $1/f^3$ with increasing frequency for such a specification to have even an approximate meaning.

We shall consider the factors which determine the choice of the upper offset frequency limit for signal sources serving two different functions as follows:

(i) A complete signal source which in practice is probably a synthesiser.
(ii) A high quality crystal oscillator which is to be used as a reference for a synthesiser.

In case (i) it is apparent that contributions to frequency jitter from phase noise at offset frequencies greater than the maximum RF or IF signal bandwidth of the system in which the source is to be used, are irrelevant. Phase noise sidebands at such offset frequencies will be rejected by IF filters which will have been chosen with the minimum bandwidth compatible with passing the signal spectrum with no more than acceptable levels of amplitude and group delay distortion. If the IF filters have not been specified at the time when it is necessary to specify the frequency stability of the signal source(s) then the spectral width of all signals to be handled by the system should be considered. In most cases the spectrum of the signal may be truncated so that 99% of the signal energy is inside the truncated band. Thus the upper offset frequency limit for a complete signal source is readily determined.

In case (ii) the upper offset frequency limit for an MO to be used as reference for a synthesiser is dependent on the design of the synthesiser. The overall phase noise performance of the synthesiser will only be determined by the MO phase noise characteristics up to an offset frequency determined by the natural frequency of one of the phase lock loops in the synthesiser. Depending on the choice of damping factor (ζ) the transfer function will fall by 20 dB per decade increase in frequency from $1.5f_n$ for $\zeta = 0.707$, and by 20 dB per decade from $2.5f_n$, for $\zeta = 1.5$. As a reasonable approximation we might then take $1.5f_n$ or $2.5f_n$ as the relevant upper offset frequency in calculating MO frequency stability. Above this frequency the synthesiser phase noise is likely to be independent of that of the MO.

Thus, in all cases, sufficient data is available to specify an upper offset frequency limit.

(c) Some General Comments

If the TI 58/59 program previously described were used to derive frequency stability from known phase noise density characteristics, integrated over a specific offset frequency range, then phase noise components with frequencies below and above this range would make (strictly) zero contribution to the calculated frequency stability. The theoretical procedure is thus equivalent to an ideal bandpass filter with an infinitely sharp cut-off at the lower and upper cut-off frequencies. No real filter can have such characteristics. In particular if measurements of frequency stability are made in the time domain, the equivalent filter in the frequency domain will have a sin x/x frequency representation rather than sharp cut-off characteristics. Thus it is true that the theoretical calculation of frequency stability, from measured phase noise density characteristics, must yield a different answer from that obtained by direct measurement of integrated frequency jitter. There is a very real sense in which density, is the fundamental concept.

If frequency jitter (stability) is *defined* as the result of measurements of frequency shift over a given time interval we have another concept which may have great practical value but differs from the theoretical concept which is based on the integration of phase noise density (or more directly frequency jitter density) over a precisely defined frequency band. This difference between the two concepts remains even if purely statistical errors have been adequately controlled.

Nevertheless the difference between the two concepts is essentially one of the different (frequency) transfer functions: if both could be corrected to a 1 Hz bandwidth then the two concepts would be seen to be fundamentally identical.

However, as is apparent from 9 Figure 1, it is not always possible to calculate the density, given only an integrated value: although in some cases this is possible if additionally the *shape* of the phase noise density curve is already known (see Section 9.6).

9.3.4 Statistical Aspects
Being based on a smoothed $(N_{op}/C)_f$ curve, the frequency deviation, evaluated by the TI 58/59 program over a specific offset frequency range, will represent a reasonable estimate of the statistical rms value. (See Section 8.7.4 for a more detailed discussion).

9.4 The Measurement of Integrated Frequency Jitter

Although it is not the primary aim of this book to discuss measurement techniques, some of the concepts and definitions associated with frequency stability are so closely related to measurement techniques that some discussion is unavoidable.

9.4.1 Initial Review of Concepts
Frequency measurements are normally carried out using a digital frequency counter, which, subject to some inaccuracies which will be discussed later, measures the average $(\bar{f})$ of the instantaneous frequency (f) over one specific time gate period τ. The result (say $\bar{f}_1$) represents one sample of the gate frequency. Due to the effects of phase noise on f itself, and measurement inaccuracies of $\bar{f}$, a second measurement would be likely to give a different result $(\bar{f}_2)$ and so on. A large number of measurements would be necessary to establish a statistically valid estimate. Consider one of a set of frequency measurements to be denoted $\bar{f}_k$.

We might attempt to assess the frequency stability in either of two ways. Firstly we might calculate the frequency difference (Δf_k) between two successive measurements $(\bar{f}_{k+1}$ and $\bar{f}_k)$ and carry out a statistical assessment of a large number of pairs of such measurements. However if the process is random and stationary (long term drift not being present or having been eliminated in the calculations) then the average value $\langle \Delta f \rangle$ of Δf will be zero; negative and positive frequency excursions Δf being equally likely. Our real interest is in the mean average frequency $\langle \bar{f} \rangle$ and in the variance of the frequency $\bar{f}$ from this mean rather than in Δf itself. However, as will become apparent, Δf may be used as one way of calculating the required parameters.

Secondly we might operate directly with the values of $\bar{f}_k$ rather than Δf_k. If we took a very large number M of frequency *measurements* $\bar{f}_k$, not frequency differences, we could find the mean:

$$\langle \bar{f} \rangle = \sum_{k=1}^{M} \frac{\bar{f}_k}{M} \tag{9.12}$$

This would be an unbiassed estimate of μ, the measured frequency $\bar{f}_k$ population mean, (Reference 32).

$$\text{Let us put} \quad \mu = f_0 \tag{9.13}$$

f_0 is the long term mean frequency: it may differ from the nominal frequency of the signal source although, for a very high quality source, this difference should be small. If we knew f_0, a single sample measurement of average frequency over time τ, $(\bar{f}_k)$, would permit calculation of a single sample of the frequency deviation from f_0 (δf_k).

$$\delta f_k = (\bar{f}_k - f_0) \tag{9.14}$$

(of zero long term mean, $\langle \delta f_k \rangle = 0$)

The technique used to measure δf_k will in fact determine the offset frequency band over which it represents an integral. This point will be discussed later (see Sections 9.4.3 and 9.4.4).

$$\text{Let} \quad y_k = \frac{\bar{f}_k}{f_0} \quad \text{and} \quad \delta y_k = \frac{\delta f_k}{f_0} \tag{9.15}$$

(Also of zero long term mean $\langle \delta y_k \rangle = 0$).

Then δy_k is the result of one measurement of the *fractional* frequency deviation, from the real mean frequency, assuming initially that the long term mean frequency (f_0) is known.

Our real interest is in the standard deviation (σ_y) of y or its square σ_y^2, which is the variance of the normalised frequency (y_k).

If we took M frequency measurements this would enable us to obtain an unbiassed estimate of f_0 (see equations 9.12 and 9.13), which would approach the true mean for large values of M. In addition we could obtain an estimate (s_y^2) of the real variance of the population (σ_y^2).

$$s_y^2 = \frac{1}{M} \sum_{k=1}^{M} (\delta y_k)^2 \tag{9.16}$$

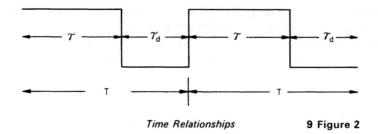

Time Relationships   **9 Figure 2**

This is a biassed estimate (Reference 32). An unbiassed estimate is given by:

$$\hat{s}_y^2 = \frac{1}{M-1} \sum_{k=1}^{M} (\delta y_k)^2 \tag{9.17}$$

The question of the statistical accuracy or the confidence limits to be attached to our estimate $\hat{s}_y^2$ of the population variance will be discussed in Section 9.4.6 below.

Now consider how σ_y^2 might be estimated from measurements of Δf, which is the difference between two frequency measurements both taken over a time τ but not overlapping in time. For example we might have a measurement sequence as illustrated in 9 Figure 2.

τ_d is the dead space (if any).
T is the time from the beginning of one measurement to the
   beginning of the next.

If the average frequency measured during the first period ($\bar{f}_1$), is deducted from the average frequency measured during the second period ($\bar{f}_2$), the resultant is one sample of the frequency deviation Δf over the time shift T.

i.e. $\Delta f_1 = (\bar{f}_2 - \bar{f}_1)$ \hfill (9.18)

If then $\bar{f}_2$ is deducted from $\bar{f}_3$ we have

$$\Delta f_2 = (\bar{f}_3 - \bar{f}_2) \tag{9.19}$$

In general:

$$\Delta f_k = (\bar{f}_{k+1} - \bar{f}_k) \tag{9.20}$$

It should be carefully noted that Δf_k is defined in a different way from δf_k. δf_k is simply the difference between the measurement $\bar{f}_k$ and the fixed mean, f_0. Thus Δf_k is not equal to δf_k although there is a simple relationship between $\langle \Delta f_k^2 \rangle$ and $\langle \delta f_k^2 \rangle$, as will shortly become apparent.

In order to compare the two approaches to estimating the variance of the frequency stability, consider that we take a number M of measurements of average frequency over time τ where the dead space τ_d as shown in 9 Figure 2 may be zero. We process these results in two different ways as follows:

(a) We calculate the sample mean frequency, calculate the deviation δf_k of each measurement from the sample mean, sum all M values of δf_k^2 and divide by $M - 1$. (See Reference 32).

In this case:

$$\hat{s}_f^2 = \frac{1}{M-1} \sum_{k=1}^{M} (\delta f_k)^2 \tag{9.21}$$

But (9.15 and 9.17)

$$\hat{s}_y^2 = \frac{\hat{s}_f^2}{f_0^2} \tag{9.22}$$

which is an estimate of $\langle (\delta y_k)^2 \rangle = \sigma_y^2$. When the dead space τ_d is zero, $\hat{s}_y^2$ will be written as $\hat{s}_y^2(\tau, \tau)$ to emphasise the fact that the measurement time is τ and the interval between measurements is τ.

(b) We calculate the mean of $(\Delta f_k)^2$. From (9.20)

$$\langle \Delta f_k^2 \rangle = \langle (\bar{f}_{k+1} - \bar{f}_k)^2 \rangle \tag{9.23}$$

$$\therefore \quad \left\langle \left(\frac{\Delta f_k}{f_0}\right)^2 \right\rangle = \langle (\bar{y}_{k+1} - \bar{y}_k)^2 \rangle \tag{9.24}$$

To find the relationship between our estimates

$$\hat{s}_y^2 \quad \text{of} \quad \sigma_y^2 \quad \text{and} \quad \left\langle \left(\frac{\Delta f}{f_0}\right) \right\rangle^2$$

we proceed as follows.

$$(\Delta f_k)^2 = (\bar{f}_{k+1} - \bar{f}_k)^2$$

$$(\bar{f}_{k+1} - \bar{f}_k)^2 = [(\bar{f}_{k+1} - f_0) - (\bar{f}_k - f_0)]^2$$

$$\therefore \quad (\Delta f_k)^2 = (\delta f_{k+1} - \delta f_k)^2 \tag{9.25}$$

$$\therefore \quad (\Delta f_k)^2 = (\delta f_{k+1}^2 - 2\delta f_{k+1}\delta f_k + \delta f_k^2)$$

$$\therefore \quad \langle (\Delta f_k)^2 \rangle = [\langle \delta f_{k+1}^2 \rangle - 2\langle \delta f_{k+1}\delta f_k \rangle + \langle \delta f_k^2 \rangle] \tag{9.26}$$

It is assumed that there is no correlation between two successive frequency deviations δf_{k+1} and δf_k. Thus the *mean* product of two successive deviations is zero. That is:

$$\langle \delta f_{k+1} \delta f_k \rangle = 0 \tag{9.27}$$

The long term statistics are also assumed to be stationary. Thus:

$$\langle \delta f_{k+1}^2 \rangle = \langle \delta f_k^2 \rangle \tag{9.28}$$

Substituting values given by (9.27) and (9.28) into (9.26) gives:

$$\langle(\Delta f_k)^2\rangle = 2\langle(\delta f_k)^2\rangle \tag{9.29}$$

$$\therefore \quad \langle(\Delta f_k)^2\rangle = 2\sigma_f^2 \tag{9.30}$$

Dividing through by f_0^2

$$\left\langle\left(\frac{\Delta f_k}{f_0}\right)^2\right\rangle = 2\sigma_y^2$$

$$\therefore \quad \sigma_y^2 = \tfrac{1}{2}\langle(y_{k+1} - y_k)^2\rangle \tag{9.31}$$

and, as an estimate

$$\hat{s}_y^2 = \tfrac{1}{2}\langle(y_{k+1} - y_k)^2\rangle_{\text{estimated}} \tag{9.32}$$

From (9.32) and (9.24)

$$\hat{s}_y^2 = \frac{1}{2}\left\langle\left(\frac{\Delta f_k}{f_0}\right)^2\right\rangle_{\text{estimated}} \tag{9.33}$$

Summarising in words: we may estimate the variance of the frequency deviation from a series of average frequency measurements $\bar{f}_k$ each in time τ with a constant dead space (which may be zero) between measurements, in either of two ways. Firstly we may calculate the variance from the individual measurements. Secondly we may take half of the mean of the squared difference between two successive measurements. After normalisation, that is dividing by the mean frequency squared, both approaches will provide an estimate of σ_y^2.

It should be noticed that this statement, and in particular the analysis on which it is based, is subject to the validity of the following assumptions:

(a) The frequency deviations, including measurement errors are random with zero means and finite variances.

(b) Successive measurements are not correlated.

(c) The statistical population is a stationary one. This would not be true if there were a steady drift in centre frequency. However the process *may* still prove to be stationary if the steady drift component is deducted from each measurement before statistical calculations (including calculation of the mean) are carried out.

9.4.2 *The Allan Variance (References 33 and 34)*

The Allan Variance is a widely accepted measure of frequency stability.

It is calculated by taking half the mean of the squared difference between two successive measurements of normalised frequency and estimated by equation (9.32). It is also defined with respect to a measurement sequence with zero dead time between measurements, each of which are taken in time τ. This ensures that measurements taken by different establishments on comparable signal sources give similar numerical results. This would not be the case if different laboratories selected arbitrary values for the dead space τ_d.

If M individual frequency measurements $(\bar{f}_k)$, or normalised frequency measurements $(y_k = \bar{f}_k/f_0)$, were taken, then there would be $(M - 1)$ differences Δf_k. Thus, for the case of zero dead space, (9.32) is a special case $m = 2$ of:

$$\hat{s}_y^2(m, \tau, \tau) = \frac{1}{M - 1} \frac{1}{2} \sum_{k=1}^{M-1} (y_{k+1} - y_k)^2$$

It will be noted that, for typographical convenience, we have *defined* y_k (9.15) without using a bar, although it represents the normalised frequency averaged over the gate time τ. It is conventional to write y_k as $\bar{y}_k$. Thus:

$$\hat{s}_y^2(2, \tau, \tau) = \frac{1}{2(M - 1)} \sum_{k=1}^{M-1} (\bar{y}_{k+1} - \bar{y}_k)^2 \qquad (9.34)$$

This is usually written (Reference 33):

$$\sigma_y^2(\tau) = \frac{1}{2(M - 1)} \sum_{k=1}^{M-1} (\bar{y}_{k+1} - \bar{y}_k)^2 \qquad (9.35)$$

Strictly the left hand side should be written $[\sigma_y^2(\tau)]$—estimated.

Equation (9.35) expresses the definition of the Allan Variance in succint form. Remember that zero dead space is implied. The Allan Variance has a major practical advantage over the direct use of equation (9.21) in that it is not necessary to have an accurate estimate of the mean frequency f_0 when normalising the frequency measurements to obtain $\bar{y}_k$ from $\bar{f}_k$. Use of the nominal frequency of the signal source to represent f_0 will involve negligible error.

9.4.3 An Important Relationship
Recalling equation (9.2)

$$(\delta f)^2 = \int_{f_1}^{f_2} \left(\frac{2N_{op}}{C}\right)_{f_m} f_m^2 \, df$$

In using this equation $(\delta f)^2$ is calculated from a known value of $(N_{op}/C)_{f_m}$.

Now (3.29) $N_{op} = N_0/2$ and (Section 3.2) N_0 is the *long term mean* noise power density [c.f. equations (3.3) and (3.4)].

Thus $(\delta f)^2$ as calculated from equation (9.2) is a long term mean value, which in the notation of Section 9.4.1 would be denoted by $\langle(\delta f)^2\rangle$.

Also, from (9.29)

$$\langle(\Delta f_k)^2\rangle = 2\langle(\delta f_k)^2\rangle \qquad (9.29)$$

Now $\Delta f = \sqrt{2}\,\delta f$ (see equations 4.2 and 4.3)

$$\therefore \quad (\Delta f)^2 = 2(\delta f)^2 \qquad (9.36)$$

Hence the simple $2:1$ numerical relationship between the squares of the peak and rms frequency deviations is identical to the relationship between $\langle(\Delta f_k)^2\rangle$ and $\langle(\delta f_k)^2\rangle$ established by statistical arguments in Section 9.4.1. This may seem a little surprising as Δf in Chapter 4 is defined as a peak modulation index whereas (Δf_k) in Section 9.4.1 is defined as the difference between two successive measurements of average frequency.

Nevertheless, the pair of parameters denoted by $(\Delta f)^2$ and $(\delta f)^2$ in Chapter 4 were defined as long term mean values and apart from typographical inconvenience could have been written $\langle(\Delta f)^2\rangle$ and $\langle(\delta f)^2\rangle$. In reality the only difference between the two pairs of parameters is due to the fact that in one case (the sinusoidal representation) integration is carried out over a (rectangular) bandwidth (from f_{c1} to f_{c2}) and in the other case, that of the time domain measurement of frequency jitter, over a bandwidth determined by the Fourier Transform of the Weighting Function of the time domain measurement. If this difference is nullified by referring both parameters to a 1 Hz bandwidth, (or more strictly to density) then the two pairs of concepts are identical. To emphasise this important basic identity, the same symbols (Δf and δf) have been used in both cases; although it must be remembered that unless referred back to density they may differ in numerical value.

9.4.4 'Transfer Function' to a Frequency Jitter Input

Consider the use of a digital frequency counter to measure $\bar{f}_k$. A conventional frequency counter counts the number of one way zero crossings of the waveform from the signal source being measured, in a time τ defined by an exact number of cycles of the counter clock. The counter thus only detects complete cycles of the source and it will therefore be limited in accuracy unless τ is made so long that an inaccuracy of up to practically 1 cycle of the source is a small fraction of the n cycles which occur in a time τ. For a fixed length of time τ, the fractional accuracy is at its worst when the source being measured has a low output frequency.

The accuracy may be improved by using a very high frequency clock oscillator, detecting counter clock zero crossings which correspond to an exact number of cycles of the source under test. The inaccuracy in this case is due to the fact that τ is only known to within plus or minus one cycle of the clock oscillator. For a 500 MHz clock, τ would be known to within 2 nanoseconds. This type of counter is known as a reciprocal frequency counter. For this type of counter, the worst fractional inaccuracy is $\Delta t / \tau$ in magnitude, where

$$\Delta t = \frac{1}{f_{\text{clock}}} \tag{9.37}$$

Apart from such inaccuracies and those due to frequency jitter of the clock oscillator itself, we may measure the number of cycles of the source under test which occur in a time τ. The *average* frequency of the source during *this period*

τ will then be given by:

$$\bar{f} = \frac{n}{\tau} \qquad (9.38)$$

where n is the measured number of cycles in time τ.

Frequency jitter of the source which occurs at a rate f_m very much *greater* than $1/\tau$ will have little effect on the average number of cycles in time τ and will therefore be heavily attenuated by the measurement.

Consider a carrier frequency modulated with a single sinusoidally represented noise component at an angular frequency $p(=2\pi f_m)$:

$$V(t) = \sin\left[\omega t + \frac{\Delta f_0}{f_m} \sin pt\right] \qquad (9.39)$$

At time t_1 the phase Φ_1 is:

$$\Phi_1 = \omega t_1 + \frac{\Delta f_0}{f_m} \sin pt_1 \qquad (9.40)$$

At time t_2 the phase Φ_2 is:

$$\Phi_2 = \omega t_2 + \frac{\Delta f_0}{f_m} \sin pt_2 \qquad (9.41)$$

$$\Phi_2 - \Phi_1 = \omega(t_2 - t_1) + \frac{\Delta f_0}{f_m}(\sin pt_2 - \sin pt_1) \qquad (9.42)$$

$\therefore$ total number of elapsed cycles 'n' in time $(t_2 - t_1) = \tau$ seconds is:

$$n = \frac{\omega}{2\pi}(t_2 - t_1) + \left(\frac{\Delta f_0}{2\pi f_m}\right)(\sin pt_2 - \sin pt_1) \qquad (9.43)$$

Divide by the elapsed time τ to find the average frequency $\bar{f}$.

$$\bar{f} = f + \frac{\Delta f_0}{p\tau}(\sin pt_2 - \sin pt_1) \qquad (9.44)$$

$$\therefore \quad \bar{f} = f + \Delta f_0 \cos\left[\frac{p(t_1 + t_2)}{2}\right]\left[\frac{\sin(p\tau/2)}{p\tau/2}\right] \qquad (9.45)$$

The peak value of the cosinusoidal frequency deviation (Δf_0) starts to fall when $p\tau$ exceeds the lowest value for which $\sin p\tau/2 = 1$. That is when:

$$p = \frac{\pi}{\tau}, \text{ giving } f_m = \frac{1}{2\tau} \qquad (9.46)$$

Thus for a given value of τ the *maximum* value of the average measured frequency falls by 20 dB per decade for constant frequency jitter density as f_m increases beyond

$$f_m = \frac{1}{2\tau}.$$

In a sense the averaging of frequency over a time τ represents a low pass filter with a cut off frequency f_{c2} given by:

$$f_{c2} = \frac{1}{2\tau} \tag{9.47}$$

However it is apparent from (9.45) that the real situation is somewhat more complicated than this.

Suppose now that the measurement of average frequency is repeated in another period τ which starts T seconds after the start of the first period as shown in 9 Figure 2.

We may now calculate the frequency change Δf as given by equation (9.18) repeated here for convenience.

$$\Delta f_1 = (\bar{f}_2 - \bar{f}_1) \tag{9.48}$$

This represents the frequency change in time T with heavy attenuation of very high frequency jitter components due to the averaging time τ. Also frequency jitter at very low offset frequencies will be heavily attenuated due to the differencing time T as is apparent from 9 Figure 3 which shows a very low frequency sinusoidal jitter component for which $1/f_m \gg T$. The measured value of Δf varies depending upon which part of the sinusoid is sampled. It is also apparent that if $1/f_m$ is reduced, that is if f_m is increased, relative to T, then the maximum value Δf can reach is the *peak to peak* value of the frequency deviation sinusoid reached when $f_m = 1/2T$.

It is obvious that the measurement process, when considered in the frequency domain, represents some sort of bandpass filter with a lower cut off

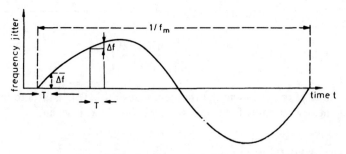

Low Offset Frequency Contribution to Δf   **9 Figure 3**

frequency (f_{c1}) which is some function of T and an upper cut off frequency (f_{c2}) which is some function of τ. For the case of a phase jitter input (the basic parameter) a full analysis is carried out in Section 9.4.5 below, under the special restriction that $T = \tau$. The reader is especially referred to equations (9.61) and (9.62).

The attenuation of the measurement process for frequency jitter components well below $1/2T$ ensures that the effective integrated value will converge even for (frequency jitter density)^2 which varies as $1/f_m$. This of course corresponds to phase noise density which rises as $1/f_m^3$ with falling offset frequency. Thus taking the difference of two frequency measurements separated by time T has ensured the convergence of the result even for this important practical case.

The effective upper cut off frequency f_{c2} also ensures convergence at the high frequency end for flat frequency jitter. However in many signal sources the phase noise density being flat at high offset frequencies due to thermal contributions, ensures that $(\delta f_0)^2$ rises as f_m^2. In this important case the measurement process will not converge at the high frequency end, unless an additional low pass filter is provided.

If such a low pass filter is not provided the result of the measurement will depend on the value of any accidental and uncontrolled upper cut-off frequency due, for example, to stray capacitances.

Thus, unless it is known that the phase noise density is falling continuously by more than 20 dB per decade, it is meaningless to refer to frequency stability without specifying an upper offset frequency limit. To obtain a meaningful measurement, a low pass filter with a suitably chosen cut off frequency, must be inserted between the source under test and the frequency counter. The need for a filter is the measurement analogue of the theoretical problem of choosing an upper frequency limit when calculating frequency stability from phase noise density measurements. This problem was adequately discussed in Section 9.3.3.

9.4.5 *Transfer Function to a Phase Noise Input* (Reference 35)

Consider a sequence of measurements of the average frequency of a signal source over time τ, using a frequency counter. Let there be zero dead time between measurements, so that τ is equal to T, the interval between measurements.

Initially consider the output of the signal source to be noise free apart from the fact that it is phase modulated by a single baseband noise component at an angular frequency p. Let the output voltage be:

$$V(t) = \sqrt{2C} \cos\left[\omega t + \theta_0 \cos(pt + \psi)\right] \tag{9.49}$$

The phase of this signal is

$$\phi(t) = \omega t + \theta_0 \cos(pt + \psi) \tag{9.50}$$

Let the first frequency measurement be carried out from $t = 0$ to $t = T$ and the second from $t = T$ to $t = 2T$. Now

$$\phi_0 = \theta_0 \cos \psi \tag{9.51.1}$$

$$\phi_1 = \omega T + \theta_0 \cos (pT + \psi) \tag{9.51.2}$$

$$\phi_2 = 2\omega T + \theta_0 \cos (2pT + \psi) \tag{9.51.3}$$

$$(\phi_1 - \phi_0) = \omega T + \theta_0 \cos (pT + \psi) - \theta_0 \cos \psi \tag{9.52}$$

$$(\phi_2 - \phi_1) = 2\omega T + \theta_0 \cos (2pT + \psi) - \omega T - \theta_0 \cos (pt + \psi) \tag{9.53}$$

$$\overline{f_1} = \frac{1}{2\pi T}(\phi_1 - \phi_0) \tag{9.54}$$

$$\overline{f_2} = \frac{1}{2\pi T}(\phi_2 - \phi_1) \tag{9.55}$$

$$(\Delta f)_0 = \overline{f_2} - \overline{f_1} = \frac{1}{2\pi T}[2\omega T + \theta_0 \cos (2pT + \psi) - \omega T$$
$$- \theta_0 \cos (pT + \psi)]$$
$$- \frac{1}{2\pi T}[\omega T + \theta_0 \cos (pT + \psi) - \theta_0 \cos \psi]$$

$$\therefore \quad (\Delta f)_0 = \frac{1}{2\pi T}[\theta_0 \cos (2pT + \psi) - \theta_0 \cos (pT + \psi)]$$
$$- \frac{1}{2\pi T}[\theta_0 \cos (pT + \psi) - \theta_0 \cos \psi] \tag{9.56}$$

Now

$$\cos B - \cos A = 2 \sin \frac{A + B}{2} \sin \frac{A - B}{2}$$

$$\therefore \quad (\Delta f)_0 = \frac{\theta_0}{2\pi T}\left[2 \sin \frac{(3pT + 2\psi)}{2} \sin \left(-\frac{pT}{2} \right) \right.$$
$$\left. - 2 \sin \frac{pT + 2\psi}{2} \sin \left(-\frac{pT}{2} \right) \right]$$

$$\therefore \quad (\Delta f)_0 = \frac{2\theta_0}{2\pi T}\left(-\sin \frac{pT}{2} \right)\left[\sin \frac{3pT + 2\psi}{2} - \sin \frac{pT + 2\psi}{2} \right]$$

$$\therefore \quad (\Delta f)_0 = -\frac{\theta_0}{\pi T} \sin \frac{pT}{2}\left[2 \cos (pT + \psi) \sin \frac{pT}{2} \right] \tag{9.57}$$

$$\therefore \quad (\Delta f)_0^2 = \left(\frac{2\theta_0}{\pi T} \right)^2 \sin^4 \frac{pT}{2} \cos^2 (pT + \psi) \tag{9.58}$$

Averaging with respect to ψ (which is uniformly distributed wrt time):

$$\overline{(\Delta f)_0^2} = \frac{1}{2}\left(\frac{2\theta_0}{\pi T}\right)^2 \sin^4\left(\frac{pT}{2}\right) \tag{9.59}$$

Now θ_0 is the peak phase jitter density

$$\therefore \quad \text{rms phase jitter density} = \frac{\theta_0}{\sqrt{2}}$$

i.e. $\overline{\phi_0^2} = \dfrac{\theta_0^2}{2}$

$$\therefore \quad \overline{(\Delta f)_0^2} = \frac{4\overline{\phi_0^2}}{(\pi T)^2} \sin^4\left(\frac{pT}{2}\right) \tag{9.60}$$

But $\overline{\phi_0^2}(f_m) = 2\left(\dfrac{N_{op}}{C}\right)_{f_m}$

$$\therefore \quad \overline{(\Delta f)_0^2}(f_m) = 8\left(\frac{N_{op}}{C}\right)_{f_m} \frac{1}{(\pi T)^2} \sin^4\left(\frac{pT}{2}\right) \tag{9.61}$$

The phase noise will in fact be due not to a single sinusoid in a 1 Hz baseband bandwidth but to the integrated effect of all noise components for offset frequencies from 0 to infinity.

The mean value of $(\Delta f)^2$ will therefore be:

$$\overline{(\Delta f)^2} = \int_0^\infty 8\left(\frac{N_{op}}{C}\right)_{f_m} \frac{1}{(\pi T)^2} \sin^4\left(\frac{2\pi f_m T}{2}\right) df_m \tag{9.62}$$

$\overline{(\Delta f)^2}$ is the variance of Δf, since its mean $\overline{\Delta f} = 0$, and is related to $\overline{(\delta f)^2}$ (see 9.29) as follows:

$$\overline{(\Delta f)^2} = 2\overline{(\delta f)^2} \tag{9.63}$$

Whether or not the integral (9.62) is convergent is very dependent on the characteristics of $(N_{op}/C)_{f_m}$.

Consider the case when the phase noise of the source is white: $(N_{op}/C)_{f_m}$ is a constant.

$$\therefore \quad \overline{(\Delta f)^2} = 8\left(\frac{N_{op}}{C}\right) \frac{1}{(\pi T)^2} \int_0^\infty \sin^4\left(\frac{2\pi f_m T}{2}\right) df_m \tag{9.64}$$

Now

$$\sin^4 x = \left[\frac{\cos 4x}{8} - \frac{\cos 2x}{2} + \frac{3}{8}\right] \tag{9.65}$$

Hence, apart from fluctuations, the integrand is also a non-zero constant. It is obvious that the integral will not converge to a finite value over the infinite frequency range. A low pass filter must therefore be provided at the output of the source to impose a finite upper frequency limit. This confirms the statement derived by means of a heuristic argument in Section 9.4.4.

Now consider the case where the phase noise density is proportional to $1/f^3$. No real signal source can maintain this characteristic out to very high offset frequencies. Hence this is a hypothetical case unless we provide a low pass filter at the output of the signal source to reject phase noise at all offset frequencies greater than f_c; where f_c is the highest offset frequency at which the $1/f^3$ law applies. That is, assume that

$$\left(\frac{N_{op}}{C}\right)_{f_m} = \left(\frac{N_{op}}{C}\right)_{f_c} \left(\frac{f_c}{f_m}\right)^3 \text{ for } f_m \leqslant f_c \tag{9.66.1}$$

and

$$\left(\frac{N_{op}}{C}\right)_{f_m} = 0 \text{ for } f_m > f_c \tag{9.66.2}$$

Equation (9.62) becomes

$$\overline{(\Delta f)^2} = \int_0^{f_c} \left\{ 8\left(\frac{N_{op}}{C}\right)_{f_c} \left(\frac{f_c}{f_m}\right)^3 \frac{1}{(\pi T)^2} \sin^4\left(\frac{2\pi f_m T}{2}\right) \right\} df_m \tag{9.67}$$

$$\overline{(\Delta f)^2} = 8\left(\frac{N_{op}}{C}\right)_{f_c} f_c^3 \int_0^{f_c} \left[\frac{1}{f_m^3 (\pi T)^2} \sin^4\left(\pi f_m T\right) \right] df_m \tag{9.68}$$

which converges at the origin, since the integrand is zero at $f_m = 0$. Using (9.65), and writing $x = (\pi f_m T)$, the integral may be written:

$$I = \int_0^{(\pi f_c T)} \left(\frac{\sin^4 x}{x^3}\right) dx = \int_0^{(\pi f_c T)} \left[\frac{1}{2}\left(\frac{1 - \cos 2x}{x^3}\right) - \frac{1}{8}\left(\frac{1 - \cos 4x}{x^3}\right) \right] dx \tag{9.69}$$

To avoid divergence of the integrals of the separate terms at the origin it is helpful to integrate from a lower limit ϵ and then find the limiting value of the integral as $\epsilon \to 0$. Now

$$\int_\epsilon^\infty \frac{(1 - \cos ax)}{x^3} dx = a^2 \int_{a\epsilon}^\infty \frac{(1 - \cos x)}{x^3} dx \tag{9.70}$$

putting $f_c = \infty$

$$I = \text{Lim} \left[\int_{2\epsilon}^\infty 2\left(\frac{1 - \cos x}{x^3}\right) dx - \int_{4\epsilon}^\infty 2\left(\frac{1 - \cos x}{x^3}\right) dx \right] \tag{9.71}$$

$$= \text{Lim} \left[2 \int_{2\epsilon}^{4\epsilon} \left(\frac{1 - \cos x}{x^3}\right) dx \right] = \text{Lim} \left[2 \int_{2\epsilon}^{4\epsilon} \left(\frac{\frac{1}{2} x^2}{x^3}\right) dx \right]$$

$$\therefore \quad I = \int_0^\infty \frac{\sin^4 x}{x^3} dx = \text{Lim} \left[\ln (x)\right]_{2\epsilon}^{4\epsilon} = \ln 2 \tag{9.72}$$

Thus, subject to f_c being a sufficiently large offset frequency that, even without the low pass filter, phase noise contribution above f_c would be completely negligible, so that substituting ∞ for f_c in I produces little error in equation (9.68):

$$\therefore \quad \overline{\Delta f^2} = 8 \left(\frac{N_{op}}{C} \right)_{f_c} f_c^3 \ln 2 \qquad\qquad (9.73)$$

But (see (9.29)

$$\overline{\delta f^2} = \frac{\overline{\Delta f^2}}{2}$$

$$\therefore \quad \overline{\delta f^2} = 4 \left(\frac{N_{op}}{C} \right)_{f_c} f_c^3 \ln 2 \qquad\qquad (9.74)$$

and

$$\sigma_y^2(\tau) = \frac{\overline{\delta f^2}}{f_0^2}$$

$$\therefore \quad \sigma_y^2(\tau) = \frac{4}{f_0^2} \left(\frac{N_{op}}{C} \right)_{f_c} f_c^3 \ln 2 \qquad\qquad (9.75)$$

Remember that equation (9.75) is only true if the phase noise follows a $1/f^3$ law over the whole of the effective offset frequency range.

Much of the preceding analysis in this sub-section using a different notation, was due to Dr. H. Smith formerly of Marconi Space and Defence Systems Ltd., and was reported in a technical note with a limited circulation L04/A/MS/05 (Reference 35).

9.4.6 Statistical Aspects

A pair of frequency measurements separated by a time T, provides one sample of frequency shift over time T. This sample is only a crude estimate of the frequency shift, both because of test equipment inaccuracies and more fundamentally because a single sample of a random process does not give much information as to the 'expected value' in a statistical sense. It will be assumed that measured values of the frequency (f) will follow a normal probability curve, and that long term drift has been eliminated.

If we took a number $(M$-1) measurements of frequency shift (Δf) (involving M frequency measurements) we might take the average but this would be negligible because (Δf) is normally distributed, and the long term frequency drift is small.

What we need to know (and what in fact is usually specified) is the standard deviation (σ) of the frequency f itself.

For data obtained from small samples the range of values from σ^2_{min} to σ^2_{max} within which the variance of the population (σ^2) may confidently be taken to lie is given by (References 32 and 36):

$$\sigma^2_{min} = \frac{Ms^2}{[\chi^2(v)]_{max}}; \qquad \sigma^2_{max} = \frac{Ms^2}{[\chi^2(v)]_{min}} \qquad (9.76)$$

where s^2 is the calculated variance of the sample of M measurements; i.e. where (Ms^2) is the sum of the square deviations from the sample mean. The normalised value of this sum, viz Ms^2/σ^2 is denoted by:

$\chi^2_{(v)} = $ Chi squared with

$v = (M\text{-}1)$ degrees of freedom,

and has standard statistics. Although σ^2 is unknown, max and min limits of the statistical variable χ^2 are tabulated for various confidence levels and degrees of freedom (Reference 32), yielding the corresponding limits σ^2_{min} and σ^2_{max} of σ^2.

We may carry out such a series of M measurements of frequency with either of two purposes in mind as follows:

(*a*) To determine whether or not we meet a specification such as that the one σ value of the frequency deviation in a time T shall not exceed (say) R Hz with a specific confidence level (say) 90%.
(*b*) In order to determine the phase noise characteristics of the source.

In case (*a*) we are concerned to determine the *maximum* value of σ at a specified confidence level.

In case (*b*), underestimates and over estimates of σ are equally undesirable. Thus, confidence limits specified are single sided in case (*a*) and two sided in case (*b*). To simplify the tabulation for specific confidence levels we shall assume that the requirements are 90% single sided and 80% two sided.

If we took a very large number of samples, for purposes of illustration say 1000, then we could calculate $\hat{s}$ (the best estimate of σ) as follows:

$$\hat{s}^2 = \sum_{k=1}^{M} \frac{(x_k - \bar{x})^2}{M\text{-}1} \qquad (9.77)$$

and $\hat{s}$ would be very close to the population standard deviation σ.

Here x_k is the result of the k_{th} measurement and $\bar{x}$ is the mean value calculated from our measurements.

In this case $\hat{s}$ is also very closely equal to s.

$$\hat{s} = \sqrt{\frac{M}{M\text{-}1}} \, s \qquad (9.78)$$

$$\hat{s} = \sqrt{\frac{1000}{999}} s \qquad (9.79)$$

Column 6 of 9 Table 1 shows for samples of different sizes the number of dB by which the measured value of the standard deviation must be increased to give with 90% confidence an upper limit that the real (population) value of σ will not exceed. Given M measurements of frequency, s may of course readily be calculated on any one of a variety of hand held calculators with statistical facilities.

As an example, assume that we have carried out 20 measurements of frequency and that we have calculated the standard deviation of these measurements to be s Hz. Then we have a 90% confidence level that σ (the standard deviation of the population) does not exceed $1\cdot31$ s Hz or $2\cdot35$ dB up on s Hz.

Column 10 is used in an analogous way for the lower limit.

9.5 Measurement of Density using Time Domain Methods

9.5.1 Measurement of Frequency Jitter Density
In principle frequency jitter *density* can be measured using a frequency counter.

The process (slightly modified) of measuring average frequency in each of two intervals, and subtracting one from the other, can be represented as a continuous linear operation performed on a voltage proportional to the instantaneous frequency of the source to yield, after each pair of measurements, a voltage proportional to this difference. In the time-domain, the transfer-function of this process becomes the weighting function waveform shown in 9 Figure 4. The amplitude A is a constant representing the process of averaging the frequency during one measurement; and is shown as negative for the process of averaging the frequency during the other measurement, as it is subtracted from the former measurement.

Each frequency measurement of the pair occupies a time τ and there is a 50% dead space $\tau/2$. As the measurement total cycle time is $2T$ seconds, the fundamental frequency of the waveform is given by:

$$f_m = \frac{1}{2T} = \frac{p}{2\pi} \qquad (9.80)$$

Let x be the angle in radians, representing normalised time.

$$x = pt \qquad (9.81)$$

The origin of x is chosen so that

$$x = 0, t = 0 \text{ at the centre of the waveform.}$$

9 Table 1   Statistical Relationships

$$\sigma^2_{max} = \frac{Ms^2}{\chi^2_{min}}: \quad \sigma^2_{min} = \frac{Ms^2}{\chi^2_{max}}$$ Double Sided 80%
Single Sided 90%

		Upper Limit 90%				Lower Limit 90%			
1	2	3	4	5	6	7	8	9	10
M	v	χ^2	σ^2_{max}	σ_{max}	dB/s	χ^2	σ^2_{min}	σ_{min}	dB/s
3	2	·211	$14\cdot22s^2$	$3\cdot77s$	$11\cdot53$	$4\cdot61$	$0\cdot65s^2$	$0\cdot81s$	$-1\cdot86$
5	4	1·064	$4\cdot7s^2$	$2\cdot17s$	$6\cdot72$	$7\cdot78$	$0\cdot64s^2$	$0\cdot8s$	$-1\cdot92$
8	7	2·833	$2\cdot82s^2$	$1\cdot68s$	$4\cdot5$	$12\cdot02$	$0\cdot66s^2$	$0\cdot82s$	$-1\cdot77$
10	9	4·168	$2\cdot4s^2$	$1\cdot55s$	$3\cdot8$	$14\cdot68$	$0\cdot68s^2$	$0\cdot82s$	$-1\cdot67$
11	10	4·865	$2\cdot26s^2$	$1\cdot5s$	$3\cdot54$	$15\cdot99$	$0\cdot69s^2$	$0\cdot83s$	$-1\cdot62$
16	15	8·547	$1\cdot87s^2$	$1\cdot37s$	$2\cdot72$	$22\cdot31$	$0\cdot72s^2$	$0\cdot85s$	$-1\cdot44$
20	19	11·65	$1\cdot72s^2$	$1\cdot31s$	$2\cdot35$	$27\cdot2$	$0\cdot74s^2$	$0\cdot86s$	$-1\cdot34$
25	24	15·66	$1\cdot59s^2$	$1\cdot26s$	$2\cdot03$	$33\cdot2$	$0\cdot75s^2$	$0\cdot87s$	$-1\cdot23$
30	29	19·77	$1\cdot52s^2$	$1\cdot23s$	$1\cdot81$	$39\cdot1$	$0\cdot77s^2$	$0\cdot88s$	$-1\cdot15$
31	30	20·6	$1\cdot5s^2$	$1\cdot23s$	$1\cdot77$	$40\cdot26$	$0\cdot77s^2$	$0\cdot88s$	$-1\cdot14$
41	40	29·05	$1\cdot41s^2$	$1\cdot19s$	$1\cdot5$	$51\cdot8$	$0\cdot79s^2$	$0\cdot89s$	$-1\cdot02$
51	50	37·69	$1\cdot35s^2$	$1\cdot16s$	$1\cdot31$	$63\cdot2$	$0\cdot81s^2$	$0\cdot9s$	$-0\cdot93$
61	60	46·46	$1\cdot31s^2$	$1\cdot15s$	$1\cdot18$	$74\cdot4$	$0\cdot82s^2$	$0\cdot91s$	$-0\cdot86$
71	70	55·33	$1\cdot28s^2$	$1\cdot13s$	$1\cdot08$	$85\cdot5$	$0\cdot83s^2$	$0\cdot91s$	$-0\cdot81$
81	80	64·28	$1\cdot26s^2$	$1\cdot12s$	$1\cdot0$	$96\cdot6$	$0\cdot84s^2$	$0\cdot92s$	$-0\cdot76$
91	90	73·29	$1\cdot24s^2$	$1\cdot11s$	$0\cdot94$	$107\cdot6$	$0\cdot85s^2$	$0\cdot92s$	$-0\cdot73$
101	100	82·36	$1\cdot23s^2$	$1\cdot11s$	$0\cdot89$	$118\cdot5$	$0\cdot85s^2$	$0\cdot92s$	$-0\cdot69$

Calculated from Normal Distribution (z)

$$\sigma_{max} = \hat{s} + \sigma_{(\sigma\,max)}z \quad \sigma_\sigma = \frac{\hat{s}}{\sqrt{2M}} \quad \hat{s} = s\sqrt{\frac{M}{M-1}}$$

M	$\hat{s}$	σ_σ	Z_{max}	σ_{max}	dB/s
101	$1\cdot005s$	$\cdot07s$	$1\cdot28$	$1\cdot095s$	$0\cdot79$
200	$1\cdot0025s$	$\cdot05$	$1\cdot28$	$1\cdot067s$	$0\cdot56$
500	$1\cdot001s$	$\cdot032s$	$1\cdot28$	$1\cdot04s$	$0\cdot35$
1000	$1\cdot0005s$	$\cdot022s$	$1\cdot28$	$1\cdot029s$	$0\cdot25$
10,000	$1\cdot00005$	$\cdot007s$	$1\cdot28$	$1\cdot009s$	$0\cdot08$
10^6	$1\cdot5 \times 10^{-7}$	$\cdot0007s$	$1\cdot28$	$1\cdot0009s$	$0\cdot008$

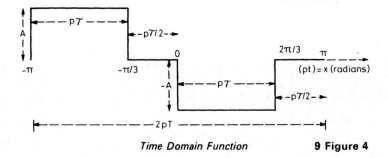

Time Domain Function   **9 Figure 4**

If the weighting function waveform repeats a large number of times $(N = M/2)$ then a Fourier Series representation may be used, valid for its whole duration.

$$\therefore \quad f(x) = \frac{a_0}{2} + \sum_{n=1}^{\infty} (a_n \cos nx + b_n \sin nx) \tag{9.82}$$

Now $f(x)$ is an odd function as

$$f(x) = -f(-x) \tag{9.83}$$

and hence the a_0 term and the cosine terms disappear (Reference 37).

$$\therefore \quad f(x) = \sum_{n=1}^{\infty} b_n \sin nx \tag{9.84}$$

$$b_n = \frac{1}{\pi} \int_{-\pi}^{+\pi} f(x) \sin nx \, dx \tag{9.85}$$

Reference to 9 Figure 4 shows that

$$\left. \begin{array}{ll} f(x) = A: & \text{from } x = -\pi \text{ to } x = -\dfrac{\pi}{3} \\[2ex] f(x) = 0: & \text{from } x = -\dfrac{\pi}{3} \text{ to } x = 0 \\[2ex] f(x) = -A: & \text{from } x = 0 \text{ to } x = \dfrac{2\pi}{3} \\[2ex] f(x) = 0: & \text{from } x = \dfrac{2\pi}{3} \text{ to } x = \pi \end{array} \right\} \tag{9.86}$$

$$\therefore \quad b_n = \frac{1}{\pi} \int_{-\pi}^{-\pi/3} A \sin nx \, dx - \frac{1}{\pi} \int_{0}^{2\pi/3} A \sin nx \, dx \tag{9.87}$$

$$\therefore \quad b_n = -\frac{A}{n\pi} [\cos nx]_{-\pi}^{-\pi/3} + \frac{A}{n\pi} [\cos nx]_{0}^{2\pi/3} \tag{9.88}$$

$$b_n = \frac{A}{n\pi}(\cos n\pi - \cos \frac{n\pi}{3} + \cos \frac{n2\pi}{3} - 1) \qquad (9.89)$$

$$b_n = \frac{A}{n\pi}\left[\left(2 \sin \frac{n\pi}{2} \sin\left(-\frac{n\pi}{6}\right) + \cos n\pi - 1\right)\right] \qquad (9.90)$$

When n is even $\sin n\pi/2 = 0$ and $\cos n\pi = 1$

$$\therefore \quad \text{when } n \text{ is even;} \quad b_n = 0$$

Thus all even harmonics disappear. From (9.89) when n is an odd multiple of 3.

$$b_n = \frac{A}{n\pi}[-1 + 1 + 1 - 1] = 0$$

Thus, all harmonics which are either even, or are multiples of 3, disappear. Hence, up to the 20th harmonic, the only outputs are for:

$$n = 1, n = 5, n = 7, n = 11, n = 13, n = 17 \quad \text{and} \quad n = 19.$$

From (9.81), (9.84) and (9.89) the overall measurement weighting function is:

$$f(pt) = \sum_{n=1}^{\infty} \frac{A}{n\pi}\left[\cos n\pi - \cos \frac{n\pi}{3} + \cos \frac{n2\pi}{3} - 1\right] \sin npt \qquad (9.91)$$

This corresponds to a measurement transfer function in the frequency domain comprising discrete lines at multiples (nf_m) of the frequency $f_m = 1/2T = p/2\pi$ (i.e. to a comb filter) where values of n which are even, or are multiples of 3, are omitted.

The first harmonic, or fundamental line of the comb filter in which we are primarily interested, is given in the time domain by the weighting function:

$$f(pt) = \frac{A}{\pi}\left[\cos \pi - \cos \frac{\pi}{3} + \cos \frac{2\pi}{3} - 1\right] \sin pt \qquad (9.92)$$

(the other non-zero lines falling off like $1/n$). Thus approximately

$$f(pt) = -\frac{3A}{\pi} \sin 2\pi f_m t \qquad (9.93)$$

If we take N pairs of frequency measurement the waveform exists for a total time of $2TN$ seconds.

The bandwidth of each line (β) will be given by

$$\beta = \frac{1}{2TN} \text{ Hz} \qquad (9.94)$$

but $\quad f_m = \frac{1}{2T} \text{ Hz} \qquad (9.95)$

$$\therefore \quad \beta = \frac{f_m}{N} \text{ Hz} \tag{9.96}$$

In other words the Q of the equivalent filter is given by:

$$Q = N \tag{9.97}$$

It is apparent from 9 Figure 4 that a single pair of frequency measurements gives the frequency difference between two average frequencies, each measured over a time τ, and separated by time T.

A sequence of N of these measurements gives the frequency shift at a rate f_m, that is at an offset frequency $f_m = 1/2T$ with a baseband bandwidth $= 1/2TN$. Such a measurement may easily be carried out at small offset frequencies and with a very narrow bandwidth.

The measurement yields Δf, the peak value of the pseudo sinusoid, at an offset frequency f_m. Knowing the effective bandwidth, the density $(\Delta f_0)^2$ of the square peak frequency deviation at f_m may readily be calculated, as may $(\delta f_0)^2$, the mean square of the frequency deviation density. This yields a value for (δf_0), the rms frequency deviation per $\sqrt{\text{Hz}}$.

Apart from the amount of computer calculation involved, and the consequent loss in operator visibility and understanding, this approach has the major disadvantage of unwanted responses at 5th, 7th, 11th, 13th etc. harmonics. If, as is unlikely, the *frequency deviation* density is falling very rapidly with increasing offset frequency the resultant error may not be too important.

In all other cases special steps must be taken to eliminate the errors. This may be done either by the provision of special filters, by calculating the correction involved or by weighting the samples with binomial coefficients (Reference 38).

9.5.2 Measurement of Phase Noise Density
Rewriting equation (9.1) for convenience

$$(\delta f_0)^2_{f_m} = \left(\frac{2N_{op}}{C}\right)_{f_m} f_m^2 \tag{9.1}$$

it is apparent that a measurement of δf_0 or $(\delta f_0)^2$ will also yield the phase noise density characteristics.

9.5.3 Statistical Aspects
The statistical aspects are closely analogous to those discussed in Section 9.4.6 above, see especially equation (9.76). However for a given accuracy the total number of measurements $(M = 2N)$ must be considerably greater than the required value of n (Equation 8.47) when carrying out phase noise density measurements by the method described in Section 8.7. This is because the inherent integration given by equation (8.45) is not available for the present method of measurement.

9.5.4 The Hadamard Variance

The Hadamard Variance (Reference 33) is:

$$\langle \sigma_H^2(M, T, \tau) \rangle = \langle (\bar{y}_1 - \bar{y}_2 + \bar{y}_3 \cdots - \bar{y}_M)^2 \rangle \tag{9.98}$$

where $\bar{y}_k$ is as defined in (9.15) (there called y_k).
In our notation this is:

$$\left\langle \left(\frac{\delta f}{f_0}\right)^2 (M, T, \tau) \right\rangle = \left\langle \left(\frac{\bar{f}_1}{f_0} - \frac{\bar{f}_2}{f_0} + \frac{\bar{f}_3}{f_0} \cdots - \frac{\bar{f}_M}{f_0}\right)^2 \right\rangle \tag{9.99}$$

If we put $T = 1 \cdot 5\tau$ (50% dead space) then equations (9.98) and (9.99) are just alternative descriptions of the process illustrated in 9 Figure 4, when the waveform is repeated many times, (with $M = 2N$). Only for large N is the covered offset frequency band narrow.

9.6 Calculation of Phase Noise Density Knowing Integrated Frequency Jitter

Assume that we have measured the variance of the integrated frequency jitter, effectively integrated over some broad offset frequency band determined by the measurement parameters T, τ and the cut off frequency of any associated low pass filter (see Section 9.4.4). It would be very useful if it were possible to use this information to calculate the *density* characteristics of both phase noise and frequency jitter variance. In general this is not possible. However if the *shape*, not necessarily the magnitude, of the phase noise density versus offset frequency curve is known, it becomes possible. We shall carry out the conversion for two special cases.

9.6.1 When $\dfrac{N_{op}}{C}$ follows a $\dfrac{1}{f^3}$ Law

Assume that the frequency jitter variance of a source has been measured with large values of T and τ and a low pass filter with a cutoff frequency of f_{c2} (say 20 Hz). In this case we know that there are negligible contributions to the frequency jitter variance from frequencies above 20 Hz. Assume the signal source to be a high quality 5 MHz crystal oscillator. In view of the analysis carried out in Sections 5.2 and 5.3 above, we can be pretty certain that the phase noise density of the signal source will follow *a* $1/f^3$ law in the offset frequency range of interest.

$$\left(\frac{N_{op}}{C}\right)_{f_m} = \left(\frac{N_{op}}{C}\right)_{f_{c2}} \left(\frac{f_{c2}}{f}\right)^3 \tag{9.100}$$

The lower offset frequency limit will be imposed by the interval between measurements T. At a frequency below $1/2T$ the response to *frequency jitter* will fall by 20 dB per decade. The density of the frequency jitter variance

$(\delta f_0)^2_{f_m}$ will fall by a further 20 dB per decade relative to the phase noise density [see (9.1)]. So that measured frequency jitter variance will fall by 10 dB per decade below $f_{c1} = 1/2T$ even for phase noise which is rising as $1/f^3$. Thus contributions from phase noise at offset frequencies below f_{c1} may normally be neglected.

$$\therefore \quad (\delta f)^2 \simeq \int_{f_{c1}}^{f_{c2}} 2\left(\frac{N_{op}}{C}\right)_{f_{c2}} \left(\frac{f_{c2}}{f_m}\right)^3 f_m^2 \, df_m \tag{9.101}$$

$$(\delta f)^2 \simeq 2\left(\frac{N_{op}}{C}\right)_{f_{c2}} f_{c2}^3[\ln f_{c2} - \ln f_{c1}] \tag{9.102}$$

Dropping the approximation sign and estimating δf^2 by $\overline{\delta f^2}$

$$\left(\frac{N_{op}}{C}\right)_{f_{c2}} = \frac{\overline{(\delta f)^2}}{2f_{c2}^3[\ln f_{c2} - \ln f_{c1}]} \tag{9.103}$$

$\overline{(\delta f)^2}$ has been measured, f_{c2} and f_{c1} are known and hence it is possible to calculate $(N_{op}/C)_{f_{c2}}$ from (9.103).

Knowing $(N_{op}/C)_{f_{c2}}$ it is then possible to calculate $(N_{op}/C)_{f_m}$ from (9.100) for any frequency less than or equal to f_{c2}.

As an example assume that the frequency stability of a 5 MHz oscillator, followed by a low pass filter with a 20 Hz cut off frequency has been measured with parameters $\tau = 25$ milliseconds and $T = \frac{1}{4}$ second. Then $f_{c2} = 20$ Hz.

$$f_{c1} = \frac{1}{2T} = 2 \text{ Hz}$$

Assume $\quad \sigma_y^2 = \left(\frac{3.84}{10^{12}}\right)^2$

$$\therefore \quad (\delta f)^2 = (\sigma_y^2)(f_0^2) = \left(\frac{1.92}{10^5}\right)^2$$

From (9.103)

$$\left(\frac{N_{op}}{C}\right)_{20} = \frac{(1.92)^2}{2 \times 20^3 \times 10^{10} \, (\ln 20 - \ln 2)}$$

$$\simeq 1 \times 10^{-14}$$

$$\therefore \quad \left(\frac{N_{op}}{C}\right)_{20} = -140 \text{ dB/Hz}$$

and

$$\left(\frac{N_{op}}{C}\right)_{2} = -110 \text{ dB/Hz}$$

9.6.2 For Other Known Laws

The values of $(N_{op}/C)_f$ may also be calculated in other cases where, for theoretical or other reasons, the shape of the phase noise density curve, when plotted in dB against log offset frequency, is known to have a constant slope with a known exponent over a known offset frequency range. Frequency jitter measurements, limited to this offset frequency range or to part of this range, may then be converted into phase noise density values using a method analogous to that used in Section 9.6.1 above.

The Special Case of White Phase Noise

Given an integrated value of $(\delta f)^2$ over an offset frequency band which extends from f_{m1} to f_{m2}, where $f_{m2} > f_{m1}$ the value of (N_{op}/C) may readily be calculated for this case. This approach is often relevant for calculating $\overline{\phi^2}$ for the case of thermal noise which is almost invariably white over the signal frequency band. From (4.10)

$$(\delta f)^2 = \frac{2N_{op}}{C} \left(\frac{f_{m2}^3 - f_{m1}^3}{3} \right)$$

$$\therefore \quad \frac{2N_{op}}{C} = \overline{\phi_0^2} = \frac{3\delta f^2}{f_{m2}^3 - f_{m1}^3} \tag{9.104}$$

Note that the value of $(\delta f)^2$ is not derived directly from the Allan Variance which has a non-uniform transfer function from f_{m1} to f_{m2}.

CHAPTER 10

SYSTEM PHASE NOISE REQUIREMENTS

It is important for a systems engineer to be able to calculate the phase noise requirements for any signal sources to be used in his system. To specify unnecessarily stringent requirements will result in complex and expensive signal sources. In some cases too lax a specification may degrade the system performance to the point where the operational performance cannot be met regardless of what adjustments are made to other parameters. In other cases, the degradation may be capable of being offset by adjusting other parameters but only with an excessive economic penalty. A satellite communication system is a good example of this, where a small performance degradation due to local oscillator phase noise may be offset by increasing the satellite transmitter power or the antenna diameter of all the ground stations, usually at great expense (see Section 1.1).

Both radar and communication systems will be discussed sufficiently to determine the requirements which must be imposed upon the spectral purity of transmitters and/or local oscillators.

The different types of radar are so numerous that an analytical discussion of each is impossible in the space available. Fortunately it is also not necessary, as non-coherent pulse radars impose no critical constraints in these respects. To a lesser degree this is also true of 'Chirp' radars. Any radar which relies upon radial velocity differences between target and ground clutter to achieve discrimination between target and clutter must necessarily rely on coherence between the transmitter and the local oscillator. Spurious angle modulation of the transmitter will be transferred to the delayed returns from the ground clutter and, at an offset frequency equal to the doppler frequency difference between target and clutter, the sidebands from the large clutter return will obscure the target signal. Thus, the greater the potential clutter rejection of a radar using doppler discrimination, the more severe will be the requirements for low phase noise transmitters and local oscillators.

Maximum doppler discrimination can be achieved in a pure CW radar system, so that this system has the most demanding requirements for a low level of spurious angle modulation. Because the transmitter and receiver operate simultaneously, a CW radar using a common antenna for transmission and reception also requires the lowest level of achievable spurious AM. A single dish CW radar system will therefore be fully analysed as it represents the most complete illustration of the constraints.

Analogous methods using the same equations may be used by the reader in analysing the spurious PM requirements for coherent MTI (Moving Target Indication) radars although the numerical values will be different due to velocity ambiguities and the resultant limitations to the achievable clutter rejection (see References 39 and 40).

Another type of radar which would seem to rely heavily on coherence between the transmitter and the receiver local oscillator is a range measuring FM radar. One extreme illustration of such a system will be considered. Cases intermediate between this example and that of a pure CW radar may then readily be analysed by the reader using an adaptation or combination of the methods described.

In the case of communication systems little comment is necessary on the reasons for selecting particular systems for analysis, as most of the interesting systems other than Single Sideband and Delta modulation are fully covered.

10.1 A Single Dish Pure CW Radar System

10.1.1 General Description

Consider a doppler radar system transmitting a continuous wave carrier without any intentional modulation of this carrier. The received signal from a target with relative radial velocity will suffer a doppler frequency shift and is distinguished by this means. Assume that a common antenna is used both for transmission and reception.

To clarify ideas we will assume the radar to be operating at 8·5 GHz. The doppler shift is given by (Reference 39)

$$f_D = 89 \cdot 4 \, \frac{v_r}{\lambda} \tag{10.1}$$

For convenience f_D will be denoted by d

where d is in Hz

v_r is in mph

and λ is in cms

This gives $d = 944$ Hz for 60 kilometres per hour.

A block diagram showing a radar of this type is given in 10 Figure 1. Only those items relevant to an assessment of the spectral purity are shown. This radar was one for the design of which the author was responsible.

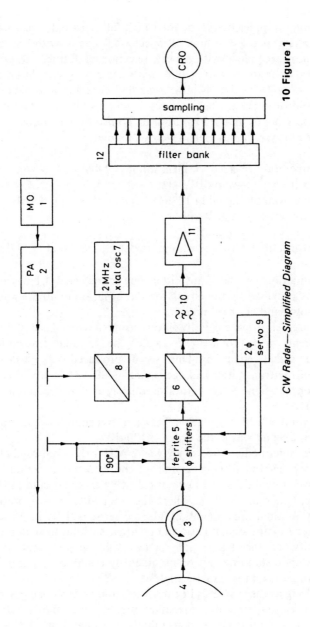

CW Radar—Simplified Diagram

10 Figure 1

The master oscillator was a Ferranti low noise two cavity klystron, the PM and AM noise performance of which are shown at (e) in 6 Figure 3 and 6 Figure 4 respectively.

The minimum doppler frequency for which full sensitivity was required was 1 kHz. The crystal band stop filter 10 rejected all signals with less than 1 kHz shift from centre frequency with a peak rejection of 120 dB. Doppler signals were identified in frequency by a filter bank consisting of a large number of filters, each of 100 Hz bandwidth, together covering the whole doppler frequency band of interest. The same result might of course now be obtained by digital processing using the Fast Fourier Transform technique.

The important system parameters were as follows:

(a) Noise temperature referred to mixer input = 2900°K
(b) Noise density N_0 = -194 dBW/Hz
(c) Minimum detectable signal in 100 Hz bandwidth = -174 dBW.
(d) Transmitter power = 200 Watts
$ = +23$ dBW
(e) Ratio of transmitter power to minimum detectable signal = 197 dB.

Some attenuation of the transmitter power reflected from the antenna mismatch was provided by the circulator. An isolation of 40 dB was obtained over the narrow bandwidth of interest.

A two phase servo system (autoplexer) using ferrite phase shifters as RF control elements was provided to cancel the residual transmitter signal reaching the signal mixer 6. This servo produced an effective further attenuation of the transmitter leakage signal by 80 dB. To avoid cancellation of wanted doppler signals the natural frequency of the two phase servo was less than 1 kHz.

Thus the total effective isolation between transmitted signal power and receiver leakage signal was of the order of 120 dB.

The receiver intermediate frequency was 2 MHz: the local oscillator for the mixer (6) being derived from a sample of the transmitter output frequency shifted by 2 MHz by means of a low noise 2 MHz crystal oscillator (7).

Unlike a pulsed radar system, where the transmitter is inoperative during reception, a CW radar must maintain full receiver sensitivity in the presence of the transmitter leakage signal. It should be apparent that in such a system the spectral purity of the transmitter signal will be of great importance. Considerable understanding can be obtained by a simple physical discussion prior to carrying out a more precise analysis.

Firstly, as long as the system is linear, overloading effects being avoided, it is apparent that transmitter noise sidebands offset by less than 1 kHz from the carrier will not be important, as detectability of signals with less than 1 kHz Doppler shift is not required. Thus the discussion will be limited to transmitter noise sidebands at offset frequencies of 1 kHz or greater.

The permissible spurious modulation of the transmitter will be found to be determined by five effects as follows:

(*a*) Local breakthrough of the transmitter signal to the receiver mixer will be one factor determining the permissible transmitter amplitude modulation at offset frequencies covering the required doppler frequency range. In this case the received 'signal' is delayed relative to the transmitted signal by only a very short time. In all cases τ will denote the delay in seconds.

(*b*) Reflections from ground clutter will impose a constraint on the permissible transmitter AM. This constraint is usually less severe than that due to (*a*) above.

(*c*) The permissible spurious AM at offset frequencies differing from the receiver first intermediate frequency by the range of wanted doppler frequencies, is dependent upon the permitted level of the transmitter breakthrough signal to the receiver mixer and also directly upon the local oscillator noise suppression achieved in the balanced mixer.

(*d*) The permissible spurious phase modulation of the transmitter will be partly determined by the level and delay of the transmitter breakthrough signal. This requirement is normally less severe than (*e*) below because of the short delay.

(*e*) The permissible PM will be determined by its effects in producing spectrum spreading of delayed returns from ground clutter.

Initially a general analysis will be carried out to determine the effects of delay on a transmitter signal which has unwanted AM and PM sidebands. This will permit us to evaluate the effects of both short and long delays. It will also assist in providing a basis for considering the effects of transmitter AM at an offset frequency equal to the receiver IF.

10.1.2 Delayed Reflection of a Signal with Spurious AM and PM

To simplify the analysis we shall assume that spurious modulation of the transmitter is sinusoidal in nature, that we are in fact dealing with the noise components in a 1 Hz bandwidth at an angular offset frequency p. To simplify the equations we shall not initially use a suffix zero to denote a density. Thus $\theta_0(t)$ and $M'_0(t)$ will be written $\theta(t)$ and $M(t)$. Only when we reach the stage where it is necessary to consider effects integrated over some other bandwidth shall we introduce the more precise notation. $\theta'(t)$ and $M'(t)$ will be used to denote the IF modulation indices and should not be confused with the transmitter modulation indices $\theta(t)$ and $M(t)$.

Let the transmitter waveform be:

$$e(t) = A(t) \exp j[\omega t + \theta(t)] \qquad (10.2)$$

where $A(t)$, the modulus of the waveform is

$$A(t) = A(1 + M \sin pt) \qquad (10.3)$$

M is the amplitude modulation index at an angular rate p. The carrier angular frequency is ω, $\theta(t)$ represents spurious phase modulation

$$\theta(t) = \theta \sin (pt + \psi) \tag{10.4}$$

Assume the receiver local oscillator to be derived from the transmitter by beating a 'clean' IF oscillator with a sample of the transmitter in a mixer and taking the filtered sum frequency to the signal mixer as a local oscillator. The receiver local oscillator will then have the same spurious phase modulation as the transmitter and, assuming no limiting, it will also have the same spurious AM. To simplify the argument, with only a trivial loss in generality, we shall assume negligible time delay or phase shift in the path from the transmitter, via the local oscillator generation process to the local oscillator feed to the signal mixer. The inputs to the signal mixer will then form the reference point from which other time delays are calculated. Any real fixed time delay in the local oscillator path (τ') may easily be allowed for by an appropriate modification to the signal delay: our real interest is relative delay between the returned signal and the local oscillator, which will be denoted τ.

If the transmitter signal is reflected from a distance of r metres, the delay τ in the received signal will be:

$$\tau = \frac{2r}{c} \tag{10.5}$$

where c is the velocity of propagation, which will be assumed equal to the free space EM wave velocity. Corrections may readily be made for parts of the circuit which may have lesser propagation velocities.

Let q be the angular frequency of the IF. Then the local oscillator signal will be:

$$A(t) \exp j[(\omega + q)t + \theta(t)] \tag{10.6}$$

If the relative delay of a reflected transmitter signal, on arrival at the mixer, is τ seconds; the local oscillator signal at this time $(t + \tau)$ seconds, will be:

$$A(t + \tau) \exp j[(\omega + q)(t + \tau) + \theta(t + \tau)] \tag{10.7}$$

The 'signal' input to the mixer, the reflected signal, will be that generated at time t, with possibly a fixed phase shift β, due to the phase characteristic of the reflecting object, and it will have suffered a total attenuation a, to give:

$$\frac{A}{a}(t) \exp j[\omega t + \theta(t) + \beta] \tag{10.8}$$

The amplitude of the mixer output will be:

$$\frac{KA}{a}(t) \tag{10.9}$$

where K is the mixer gain constant.

The phase of the IF output will be:

$$\gamma(t) = q(t + \tau) + \theta(t + \tau) - \theta(t) + \beta' \tag{10.10}$$

where $\beta' = \omega\tau - \beta$.

Thus the IF output of the mixer will be:

$$\frac{KA}{a}(t) \exp j[q(t + \tau) + \theta(t + \tau) - \theta(t) + \beta'] \tag{10.11}$$

We shall now consider the amplitude and the phase separately and in more detail in order to assess the magnitudes of the AM and PM sidebands.

(a) The AM Output

Using (10.9) and (10.3) the IF amplitude is

$$\frac{KA(t)}{a} = \frac{KA}{a}(1 + M \sin pt) \tag{10.12}$$

where K, the mixer gain constant, will apply equally to wanted signals and unwanted reflections.

Comparing this with the expression for an amplitude modulated signal (see, for example, Section 2.1), it is obvious that the amplitude modulation index of the IF signal (M') is equal to the AM index of the transmitter (M) and is independent of the signal delay

$$M' = M \tag{10.13}$$

(b) Phase Modulation of the IF Output

Repeating (10.10) for convenience:

$$\gamma(t) = q(t + \tau) + \theta(t + \tau) - \theta(t) + \beta'$$

Substituting for $\theta(t)$ from (10.4)

$$\gamma(t) = q(t + \tau) + \theta \sin [p(t + \tau) + \psi] - \theta \sin (pt + \psi) + \beta' \tag{10.14}$$

$$\therefore \quad \gamma(t) = q(t + \tau) + 2\theta \cos \left[\frac{2pt + p\tau + 2\psi}{2}\right] \sin \frac{p\tau}{2} + \beta'$$

$$\therefore \quad \gamma(t) = q(t + \tau) + 2\theta \sin \frac{p\tau}{2} \cos \left[pt + \frac{p\tau}{2} + \psi\right] + \beta' \tag{10.15}$$

Comparison with (2.3.1) shows that the peak phase modulation index of the IF output is:

$$\theta'(\tau) = 2\theta \sin \left(\frac{p\tau}{2}\right) \tag{10.16}$$

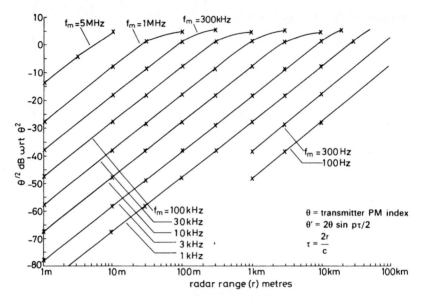

Effect of Radar Range Delay on IF Phase   **10 Figure 2**
Modulation Index

Using (10.5), and $p/2\pi = f_m = c/\lambda_m$

$$\therefore \quad \theta'(\tau) = 2\theta \sin\left(\frac{2\pi r}{\lambda_m}\right) \tag{10.17}$$

Equation (10.17) has been evaluated for various radar range distances (r), for different modulating frequencies (f_m). The results are plotted on 10 Figure 2. The ordinate shows the attenuation of the transmitter spurious phase modulation power density, expressed in dB, for a radar system with a local oscillator derived from the transmitter. From (10.16) when

$$\frac{p\tau}{2} < 0.5 \text{ radians: } \sin\left(\frac{p\tau}{2}\right) \simeq \frac{p\tau}{2}.$$

For such cases, which are widely applicable for the normal ranges covered by f_m and τ:

$$\theta'(\tau) = \theta p\tau = \theta 2\pi f_m \tau \tag{10.18}$$

Multiplying both sides of (10.18) by f_m, the equivalent IF peak frequency deviation is:

$$\Delta'f = \Delta f 2\pi f_m \tau \tag{10.19}$$

A Brief Digression on a Method of Measuring Phase Jitter Without a Reference Source

Two outputs are taken from the single signal source which it is desired to measure; one output being fed directly to the phase detector and the other to the phase detector via a long delay, τ. The phase jitter may be measured as described in Section (8.7). The results give a measure of ϕ' and hence of θ'. θ and ϕ may then readily be calculated using (10.16).

(c) Comparison of the Effects of Delayed Reflections for Transmitter AM and for Transmitter PM

For AM a signal delay does not affect the modulation index.

For PM, the effect of a delay of the reflected signal is to multiply the transmitter modulation index by 2 sin $p\tau/2$. For short delays the effective modulation index is very small. It will reach a maximum of twice that of the transmitter when $p\tau/2$ is an odd multiple of $\pi/2$.

10.1.3 Permissible AM

10.1.3.1 Transmitter Breakthrough

The IF output of the mixer due to transmitter breakthrough has a peak amplitude

$$\frac{KA}{a}(1 + M \sin pt) \qquad \text{see (10.12)}$$

A, is the peak amplitude of the transmitter $= \sqrt{2C}$ where C is the transmitter power. From equation (2.2) the peak amplitude of each sideband is half $(KA/a)M$.

$$\therefore \quad \text{amplitude of each sideband} = \frac{K\sqrt{2C}}{2a} M \qquad (10.20)$$

$$\text{Power (SSB)} = \frac{K^2 C}{4a^2} M^2 \qquad (10.21)$$

If we now consider M_0 to be the amplitude modulation index of the transmitter due to noise in a 1 Hz bandwidth at a baseband angular frequency p, then the AM noise density at the IF output, due to transmitter breakthrough is:

$$N_{oa}(\text{IF}) = \frac{K^2 C}{4a^2} M_0^2 \qquad (10.22)$$

Let N_0 be the receiver noise power density at the mixer input due to receiver noise temperature.

$\quad \therefore \quad$ *receiver* noise density at mixer output

$$N_0(\text{IF})(\text{Rec}) = K^2 N_0 \qquad (10.23)$$

If the AM contribution due to the transmitter breakthrough (10.22) is equal to that due to receiver noise, then the overall noise level will rise by 3 dB when the transmitter is switched on. This is true because a single sideband receiver cannot distinguish between pure noise, AM noise and PM noise. Ideally the contribution due to the transmitter breakthrough should be 20 dB below receiver thermal noise.

$$\therefore \quad N_{oa}(\text{IF}) = \frac{K^2 N_0}{100} \tag{10.24}$$

$$\therefore \quad K^2 M_0^2 \frac{C}{4a^2} = \frac{K^2 N_0}{100}$$

$$\therefore \quad M_0^2 = \frac{4N_0}{100} \frac{a^2}{C} \tag{10.25}$$

But, at the transmitter (see 2.2)

$$N_{oa} = \frac{C M_0^2}{4} \tag{10.26}$$

$$\therefore \quad \frac{N_{oa}}{C} = \frac{N_0}{100} \frac{a^2}{C}$$

$$\therefore \quad \left(\frac{N_{oa}}{C}\right)_{f_m} = \frac{a^2}{100} \frac{N_0}{C} \tag{10.27}$$

Equation (10.27) relates the permissible transmitter AM noise density to the ratio of receiver noise density to the transmitter breakthrough power C/a^2.

Now, for the radar we are considering:

$$N_0 = -194 \text{ dBW/Hz} \tag{10.28}$$

$$C = 200 \text{ watts} = +23 \text{ dBW} \tag{10.29}$$

$$\frac{N_0}{C} = -217 \text{ dB/Hz} \tag{10.30}$$

If the transmitter/receiver isolation were only 40 dB.

$$\left(\frac{N_{oa}}{C}\right)_{f_m} = (40 - 10 \log_{10} 100 - 217) \text{ dB/Hz}$$

$$= -197 \text{ dB/Hz} \tag{10.31}$$

This is an impossibly severe requirement and the only solution is to reduce the level of the transmitter breakthrough. Given a knowledge of the best value of $(N_{oa}/C)_{f_m}$ which may be economically achieved (or be dictated by other factors) the minimum value of a^2 is determined from (10.27).

$$a^2 = \left(\frac{N_{oa}}{C}\right)_{f_m} \frac{100C}{N_0} \tag{10.32}$$

for

$$\left(\frac{N_{oa}}{C}\right) = -150 \text{ dB/Hz at 1 kHz} \tag{10.33}$$

which is achieveable (see 6 Figure 4e).

$$a^2 = (-150 + 20 + 217) \text{ dB}$$

$$\therefore \quad a^2 = 87 \text{ dB} \tag{10.34}$$

Thus in this radar, transmitter AM demands a minimum transmitter/receiver isolation of 87 dB.

10.1.3.2 Ground Clutter

For an airborne radar even stationary targets such as ground clutter will experience a doppler shift except in the very limited number of cases where the radial velocity with respect to the radar is zero. Due to the finite antenna beamwidth ground clutter returns will occur over a range of doppler frequencies. A target flying over the ground at a low velocity may produce the same doppler frequency as ground clutter from another part of the antenna beamwidth. Even a radar with perfect spectral purity cannot improve the target/clutter ratio for that small fraction of the total clutter power with the same frequency shift as the signal. Improvement in performance in this respect can only come by narrowing the antenna beamwidth so that groundclutter with the same radial velocity as the target is heavily attenuated by the antenna polar diagram. We shall not further consider this special case as it has no impact on the specification of radar spectral purity.

Due to the extended area of the ground intercepted by the antenna beamwidth of a high flying airborne interception (AI) radar, the total power of the ground clutter returns, irrespective of frequency, will be very much greater than that of a low flying aircraft target. Typically the total power of the ground clutter might be as much as 90 dB above the target level. Spectral impurity of the radar transmitter will ensure that every frequency component of the ground clutter return has associated with it some sideband energy at the frequency of the target signal. To simplify the argument imagine the antenna polar diagram to give uniform response over a certain beamwidth and zero response outside. This is analogous to approximating the frequency response of a real filter by that of a rectangular filter. Let the ground clutter return in 100 Hz bandwidth (the receiver discrimination) be x times higher than the signal return. Let there be m such clutter signals; that is the frequency spread of the clutter is 100 m Hz. The total clutter power (I) will be mx times greater than the signal power S, i.e.

$$\frac{I}{S} = mx = \text{(say) } 90 \text{ dB} \tag{10.35}$$

If the AM noise of the transmitter were white, that is if $(N_{oa}/C)_{f_m}$ were constant over a large offset frequency range, then each ground clutter component would make equal sideband noise power contribution within the single filter containing the signal. To be (say) 20 dB below the signal, the sideband power in 100 Hz bandwidth would have to be 110 dB below the carrier. This implies that $(N_{oa}/C)_{f_m}$ is constant at -130 dB/Hz.

It will be realised that the above discussion is very approximate, partly because of the complexity of the operational situation: variables include:

(a) radar altitude
(b) radar beam width
(c) target altitude and velocity
(d) antenna beam pointing direction
(e) the assumption that the transmitter AM sidebands are white
(f) an unspecified false alarm rate.

Nevertheless it gives an understanding of the factors involved in a full analysis and the numerical values are fairly typical. It shows that really bad transmitter AM noise sidebands could not be tolerated even if the effects of direct transmitter breakthrough were in some way eliminated. In most cases, however, the breakthrough problem determines the transmitter AM requirements and this is amenable to a reasonably accurate analysis (see Section 10.1.3.1 above).

10.1.3.3 At an Offset Frequency Close to the Receiver IF

Denote the angular signal frequency by ω, the angular local oscillator frequency by h and that of the IF by q.

Let the derived local oscillator be above the signal frequency so that:

$$h = (\omega + q) \tag{10.36}$$

The wanted output signal will then occur at an angular frequency:

$$(h - \omega) = q \tag{10.37}$$

Unwanted AM noise sidebands of the local oscillator at offset frequencies $(h - q)$ and $(h + q)$ will represent local oscillator noise components at ω and $(\omega + 2q)$. The first of these is the signal frequency and the second is the image frequency (See 10 Figure 3).

Assuming no image rejection within the mixer itself, both components will produce noise at the IF output. In practice a balanced mixer will partially

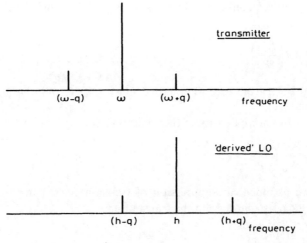

10 Figure 3
The Effects of AM Noise, Offset by the IF

cancel local oscillator noise; the attenuation being at least 20 dB and probably 30 dB. (See Appendix II).

The process of generating the local oscillator from the transmitter in a frequency shifting mixer may itself reduce the LO AM sideband to carrier ratio below that of the transmitter, if the transmitter sample is large compared with the IF oscillator input. This is not always convenient. Thus for a local oscillator power of P watts it is conservative to assume that $(N'_{oa}/P)_q$ for the radar local oscillator is identical to that of the transmitter. The local oscillator level at the radar mixer will be of the order of $+3$ dBm. The two contributions to unwanted noise at ω and $(\omega + 2q)$ were originally due to two conformable sidebands $(N'_{oa}/P)_q$ of the local oscillator signal with angular frequency h. They were therefore equal. On further beating with the local oscillator $(\omega + q)$ in the receiver mixer they will both be converted to the same frequency q. The phase relations are such that the two components will add on a voltage basis, rather than a power basis at the mixer output. This is closely analogous to the phase relationships existing in a coherent detector, where sidebands add on a voltage basis at baseband (see Section 2.8).

Thus the total power will be 4 times that of one sideband and as it is centred on a single frequency it must now be described as 'noise' and not as 'AM noise'.

Thus the power output density of the mixer due to the two LO noise sidebands together will be:

$$(N_0)_{IF} = 4P \left(\frac{N'_{oa}}{P} \right)_q L \tag{10.38}$$

where L is the mixer conversion loss. As an equivalent noise level referred to the mixer input:

$$N_0 = 4P \left(\frac{N'_{oa}}{P} \right)_q \tag{10.39}$$

To distinguish it from receiver thermal noise density, N_0 due to LO noise will be denoted LO.

The ratio of LO noise density to thermal noise density will be:

$$\frac{LO}{N_0} = \frac{4P}{N_0} \left(\frac{N'_{oa}}{P} \right)_q \tag{10.40}$$

As we have assumed no suppression of transmitter AM in the process of deriving the local oscillator from the transmitter:

$$\left(\frac{N'_{oa}}{P} \right)_q = \left(\frac{N_{oa}}{C} \right)_q \tag{10.41}$$

$$\therefore \quad \frac{LO}{N_0} = \frac{4P}{N_0} \left(\frac{N_{oa}}{C} \right)_q \tag{10.42}$$

Assuming the local oscillator AM noise is multiplied by $1/M$ due to the use of a balanced mixer.

$$\frac{LO}{N_0} = \frac{4P}{N_0} \left(\frac{N_{oa}}{C} \right)_q \frac{1}{M} \tag{10.43}$$

Let it be required that the LO noise contribution be $1/x$ of that due to receiver thermal noise, then

$$\frac{1}{x} = \frac{4P}{N_0} \left(\frac{N_{oa}}{C} \right)_q \frac{1}{M} \tag{10.44}$$

$$\therefore \quad \left(\frac{N_{oa}}{C} \right)_q = \frac{M}{x} \frac{N_0}{4P} \tag{10.45}$$

As an example assume:

$$N_0 = -194 \text{ dBW/Hz (see 10.28)} \tag{10.46}$$

$$M = 30 \text{ dB} \tag{10.47}$$

$$x = 20 \text{ dB minimum} \tag{10.48}$$

$$P = +3 \text{ dBm} = -27 \text{ dBW} \tag{10.49}$$

$$\frac{q}{2\pi} = 2 \text{ MHz} \tag{10.50}$$

∴   It is necessary that

$$\left(\frac{N_{oa}}{C}\right)_{2\,\text{MHz}} = [+30 - 20 - 194 - 6 + 27]\ \text{dB/Hz} \qquad (10.51)$$

$$= -163\ \text{dB/Hz} \qquad (10.52)$$

From 6 Figure 4 (*e*) it is apparent that the Ferranti transmitting klystron oscillator is likely to meet this requirement although extrapolation of the graph is necessary.

10.1.4 Permissible P.M.

10.1.4.1 Transmitter Breakthrough

Assume as before that the IF oscillator adds completely negligible PM noise to the transmitter at the derived local oscillator frequency, $1/2\pi(\omega + q)$. That is for the local oscillator

$$\left(\frac{N_{op}}{P}\right)_f = \left(\frac{N_{op}}{C}\right)_f$$

for the transmitter.

In the case where there is a relative time delay of zero between the transmitter breakthrough signal and the local oscillator signal, over the two paths from the transmitter to the two inputs to the receiver mixer, both signals will have identical phase jitter. The output of this mixer will then be an IF signal with zero phase jitter. Physically, rather than mathematically, this is most easily understood by considering two signals which are both varying in frequency but at every instant of time have a frequency difference of exactly the intermediate frequency. The output of the mixer will be the difference frequency exactly: the IF signal will be spectrally pure CW. Thus for zero delay the level of transmitter spurious PM is unimportant.

When the relative delay of the two paths from the transmitter to the two inputs to the mixer is τ seconds, the peak phase jitter at IF is given by (10.17), which is repeated below for convenience.

$$\theta'(r) = 2\theta\,\sin\left(\frac{2\pi r}{\lambda_m}\right) \qquad (10.53)$$

where θ is the peak phase jitter of the transmitter.

Now $\sin x = x$ with an error hardly greater than 1% up to $x = 0\cdot25$ radians.

$$\therefore\quad \text{for}\ \frac{2\pi r}{\lambda_m} \ngtr 0\cdot25 \qquad (10.54)$$

$$\theta'(r) \simeq 2\theta\,\frac{2\pi r}{\lambda_m} \qquad (10.55)$$

when

$$r = \frac{0 \cdot 25 \lambda_m}{2\pi} \qquad (10.56)$$

$$\theta'(r) = 0 \cdot 5\theta \qquad (10.57)$$

and the phase noise sidebands of the IF signal are only 6 dB below those of the transmitter.

For local reflections internal to the radar r is very unlikely to reach, let alone exceed, 30 metres. Phase noise sidebands at offset frequencies greater than the highest doppler frequency of interest, are unimportant as they will be rejected by the receiver filtering or digital processing. Assume the highest frequency of interest to be 100 kHz, for this case $\lambda_m = 3000$ metres.

Substituting these values in (10.55) or using 10 Figure 2:

$$\theta'(r) \simeq 0 \cdot 1256\theta \qquad (10.58)$$

Thus even for this extreme case the phase jitter at IF due to transmitter breakthrough will be 18 dB below that of the transmitter. At lower offset frequencies the factor by which the transmitter phase jitter is reduced at the intermediate frequency will increase by 20 dB per decade fall in offset frequency below 100 kHz.

In most cases transmitter breakthrough is not the factor which determines the permissible transmitter phase jitter. Large signals reflected with long delays, such as those from ground clutter normally impose a much more severe constraint. For this reason, and because the interested reader may obtain full information from 10 Figure 2, this discussion will be terminated.

10.1.4.2 Ground Clutter

For a CW radar using a transmitter of perfect spectral purity, targets with a radial velocity with respect to the radar different from that of any ground clutter components, would not be subject to ground clutter interference. However special processing, based on a knowledge of the velocity of the radar, the beam pointing direction and the antenna beamwidth, would be necessary to prevent ground clutter from being falsely identified as additional signals.

Spurious PM of the transmitter would appear as spurious PM sidebands on each of the clutter returns with a modulation index modified according to (10.17) by the delay applicable to each component of clutter return.

The unwanted PM sidebands of the clutter returns would include components at the target frequency and thus constitute interference.

Consider a single small clutter element with a single doppler frequency shift. To a very close approximation the PM sidebands of the returned clutter signal will suffer the same doppler shift as the clutter signal itself. This is because the sideband offset frequency is a very small fraction of the carrier frequency. The clutter PM sidebands will also suffer the same time delay τ as the clutter return due to the transmitter 'carrier'. Let the clutter 'carrier' be of an angular frequency $(\omega + 2\pi \, d_c)$, where d_c is the doppler frequency shift of this clutter element. Regardless of the delay τ this would beat with a spectrally pure local oscillator to give an IF output of angular frequency $(q - 2\pi \, d_c)$. Let the real wanted target have a doppler frequency shift d, which is (say) greater than d_c. Phase noise sidebands of the transmitter at an offset frequency of $(d - d_c)$, say δd, resulting from a modulation index $\theta(\delta d)$, will potentially interfere with the wanted signal.

The effective modulation index of the IF signal (in a 1 Hz bandwidth) due to the return from the clutter element with a delay τ, or at a distance r, will be denoted $\theta'_{oc}(\tau)$ and [(10.16) and (10.17)] will be given by:

$$\theta'_{oc}(\tau) = 2\theta_0(p) \sin \frac{p\tau}{2} \tag{10.59}$$

where

$$p = 2\pi \, \delta d \tag{10.60}$$

and $\theta_0(p)$ is the peak modulation index of the transmitter at an offset angular frequency p

$$\text{or} \quad \theta'_{oc}(\tau) = 2\theta_0(p) \sin \frac{2\pi r}{\lambda_{(\delta d)}} \tag{10.61}$$

The clutter element sideband power density interfering with the wanted target signal will be (3.1):

$$N_{opc} = \frac{C_1[\theta'_{oc}(\tau)]^2}{4} \tag{10.62}$$

where N_{opc} is the interfering phase noise power density and C_1 is the 'carrier' power of the elementary clutter return.

$$\therefore \quad N_{opc} = C_1\theta_0^2(p) \sin^2 \frac{p\tau}{2} \tag{10.63}$$

For most operational situations any group of clutter elements with a very small spread in doppler frequency will also produce signals with almost identical echo delays τ. The doppler frequency discrimination of the receiver will normally be good, such as 1 part in 500 or 1 part in 1000 of the total doppler frequency range. The receiver frequency discrimination bandwidth will be denoted b. In the radar described in Section 10.1.1, b was 100 Hz. Thus, to a good approximation, we may regard all doppler returns from ground clutter which differ in doppler frequency by no more than b Hz as having the same delay τ or being at the same range r. Let the total 'carrier' power of this small group of clutter sub-elements now to be called a clutter group element be denoted C_g. If the phase noise density of the transmitter may be regarded as constant over the bandwidth b about δd then the square of the peak modulation index due to the clutter group element will be:

$$\theta_c'^2(p) = \theta_{oc}'^2(p)b \tag{10.64}$$

Thus the phase noise power due to a single clutter group element which interferes with the signal will be:

$$N_{pc} = C_g \theta_0^2(p)b \sin^2 \frac{p\tau}{2} \tag{10.65}$$

In the total clutter return there will be a large number of such group elements (say l) in total, of significant amplitude within the beamwidth.

Equation (10.65) is written for a typical group element k:

$$N_{pck} = C_{gk} \theta_0^2(p_k)b \sin^2 \frac{p_k \tau_k}{2} \tag{10.66}$$

The total clutter power within the signal bandwidth due to the transmitter phase modulation producing spectral spreading of the ground clutter return denoted I_c is given by:

$$I_c = \sum_{k=1}^{l} C_{gk} \theta_0^2(p_k)b \sin^2 \frac{p_k \tau_k}{2} \tag{10.67}$$

The signal return power is denoted S. Thus the clutter to signal ratio is:

$$\frac{I_c}{S} = \frac{1}{S} \sum_{k=1}^{l} C_{gk} \theta_0^2(p_k)b \sin^2 \frac{p_k \tau_k}{2} \tag{10.68}$$

This is a difficult equation to evaluate accurately. To do so would require a knowledge of the power C_{gk}, the relative doppler frequency δd (and hence p) and the differential delays τ_k for each of the clutter group elements of significant magnitude. The variation of the transmitter phase modulation density θ_0 with offset frequency would also be required to be known.

However an upper limit to the value of I_c/S may readily be obtained.

Firstly the maximum value of $\sin^2 p_k \tau_k/2$ is unity. Secondly assume initially that the transmitter peak phase modulation density index $\theta_0(p)$ is in fact constant over the range of offset frequencies involved: denote this by θ_0. Then the clutter to signal ratio is:

$$\frac{I_c}{S} = \frac{1}{S} \theta_0^2 b \sum_{k=1}^{l} C_{gk} \tag{10.69}$$

Now $\sum_{k=1}^{l} C_{gk}$ is the sum of the power of all the clutter returns, denoted C_c.

$$\therefore \quad C_c = \sum_{k=1}^{l} C_{gk} \tag{10.70}$$

$$\therefore \quad \frac{I_c}{S} = \frac{C_c}{S} \theta_0^2 b \tag{10.71}$$

If the total clutter return power C_c is x times greater than the power of the signal S:

$$\frac{C_c}{S} = x \tag{10.72}$$

Then

$$\frac{I_c}{S} = x\theta_0^2 b \tag{10.73}$$

For unity clutter to signal ratio in the receiver bandwidth, neglecting direct clutter return at the same Doppler frequency as that of the target:

$$\theta_0^2 = \frac{1}{xb} \tag{10.74}$$

A typical ratio x of total clutter power to signal power might be 90 dB. For a receiver bandwidth b of 100 Hz this gives:

$$\overline{\theta_0^2} = -110 \text{ dB peak rads}^2$$

$$\therefore \quad \overline{\phi_0^2} = -113 \text{ dB rads}^2 \tag{10.75}$$

$$\therefore \quad \left(\frac{N_{op}}{C}\right) = -116 \text{ dB/Hz}$$

If a signal/clutter ratio of 20 dB were required

$$\left(\frac{N_{op}}{C}\right) = -136 \text{ dB/Hz} \tag{10.76}$$

Effectively this might be regarded as an average value of $(N_{op}/C)_f$ over the offset frequency range which includes all values of δd relevant to any of the returned ground clutter components. Nevertheless it is apparent that if the radar antenna beam centre is pointing directly at the target, which will be the case either during auto tracking or during search when the target return is a maximum, which by implication is the case we have been considering all along, then the major clutter returns will be relatively close in an angular sense to the target.

The worst case is probably that when the radar antenna is pointing vertically downwards to a target vertically below with a radial velocity of only 60 kph giving a target doppler shift of 1 kHz. The major part of the ground clutter return, near beam centre, will suffer an even smaller doppler shift. It will then be the transmitter phase noise sidebands near 1 kHz offset frequency which will determine the target to clutter ratio. For the case where the ground clutter return was 90 dB greater than the target return, and for a receiver bandwidth of 100 Hz, equation (10.76) would be a very good guide as to the maximum permissible transmitter phase noise sideband density at $f_m = 1$ kHz, i.e.

$$\left(\frac{N_{op}}{C}\right)_{1\text{ kHz}} \not> -136 \text{ dB/Hz} \tag{10.77}$$

Perusal of 6 Figure 3(e) will show that the Ferranti kystron was about 20 dB worse than this, resulting in a signal/clutter ratio close to unity. This would not give a reasonable target acquisition probability under these highly adverse conditions. The false alarm probability would also be large. Thus there was a case for improving the phase noise performance of even this superb transmitter.

10.1.5 Single Sideband Signals and Double Sideband Interference (Reference 22)
The doppler shift for a single target with a single relative radial velocity produces what is essentially a single sideband signal. From the analysis in Chapter 3 it is apparent that, for example, an upper sideband doppler signal is equivalent to four real or potential sidebands all of equal amplitude, consisting of a pair of conformable AM sidebands and another pair of conformable PM sidebands. In this case the lower AM and PM sidebands mutually cancel.

On the other hand a pure PM interfering signal has conformable sidebands, that is it is essentially double sideband. Similarly an interfering signal which is purely AM, is essentially double sideband.

Many sources of interference in a CW radar are essentially double sideband in nature, although unfortunately this is not true of ground clutter.

Consider firstly transmitter spectral impurities. The phase noise spectral density $(N_{op}/C)_f$ is usually greater than the AM noise spectral density $(N_{oa}/C)_f$. In many cases the AM noise may be further reduced by limiting in the transmitter channel. In other cases the transmitter output frequency may be achieved by high order frequency multiplication, when the phase noise sidebands may be completely dominant. Depending on details of the radar design, the effects of vibration on a single dish airborne CW radar may also produce primarily PM sidebands.

Relatively good cancellation of DSB interference may be achieved at the price of converting the wanted single sideband doppler signal into a double sideband signal. In some operational situations this is not a major embarrassment. The author has achieved 30 dB suppression of DSB signals, experimentally without loss in target sensitivity in a clutter free situation. In cases where all significant clutter returns and the target return have positive doppler frequency shifts, there is no loss in signal/clutter ratio. When clutter extends into the negative doppler frequency range such signals after conversion to DSB, add to the total clutter interference.

A simplified block diagram of the double sideband suppression equipment devised and made by the author many years ago is given in 10 Figure 4. The

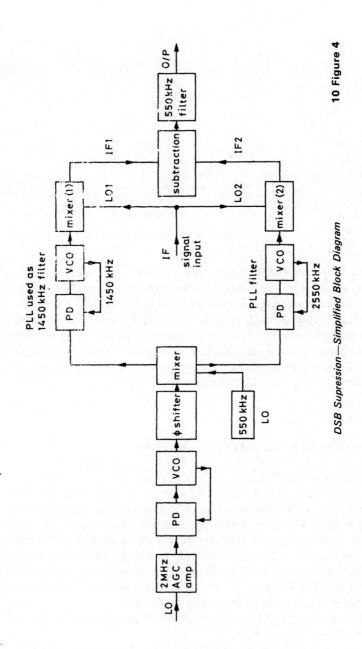

DSB Supression—Simplified Block Diagram

10 Figure 4

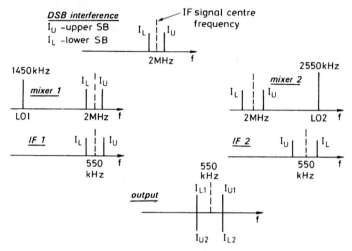

DSB Suppression—Mode of Operation   **10 Figure 5**

method of operation is illustrated qualitatively in 10 Figure 5. Although it is a good example of the application of the methods of analysis described in this book this is not a radar textbook and it is felt to be out of place to carry out the detailed mathematical analysis here. The procedure is straightforward, if somewhat lengthy, and the interested reader may easily carry it out for himself. A by-product of such an analysis is a quantitative specification for the required accuracy of phase and amplitude matching necessary for the achievement of a desired DSB suppression ratio.

10.2 A Radar using High Deviation Linear FM

10.2.1 Types of FM Radar

There is a variety of types of FM radar, using sinusoidal FM, symmetrical sawtooth and asymmetrical sawtooth modulation. In addition the frequency deviation may be large or small. The addition of low deviation FM to a pure CW radar has a relatively small impact on the *methods* of calculating spectral purity requirements already discussed. It will be interesting to explore the differences introduced by the addition of high deviation FM. The discussion will be limited to linear sawtooth modulation. Descriptions of systems using sinusoidal FM are given in References 41 and 42.

Two types of linear FM waveform, symmetric and asymmetric are shown in 10 Figure 6 and 10 Figure 7.

In both cases it is assumed that any receiver local oscillator used is 'derived' from the transmitter so that it is identically frequency modulated. Any signal

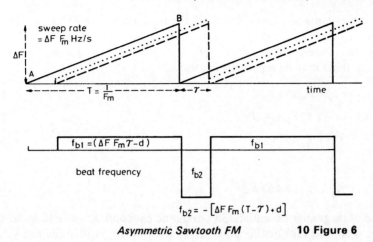

$f_{b1} = (\Delta F \, F_m \, T - d)$

beat frequency

$f_{b2} = -[\Delta F \, F_m \, (T - \tau) + d]$

Asymmetric Sawtooth FM   **10 Figure 6**

with zero delay will then possess a single line spectrum at the IF centre frequency.

Consider a single linear frequency sweep such as that labelled AB in 10 Figures 6 and 7. The sweep rate in the two cases is $\Delta F F_m$ Hz/sec and $2\Delta F F_m$ Hz/sec. In the absence of doppler shift, a signal delay τ will result in a constant beat frequency of $\Delta F F_m \tau$ and $2\Delta F F_m \tau$ Hz in the two cases. The values of the beat frequencies, including the effects of target doppler shift are shown on the figures. For a range of r metres the signal delay τ is:

$$\tau = \frac{2r}{c} \tag{10.78}$$

where c is the velocity of light $= 3 \times 10^8$ m/sec.

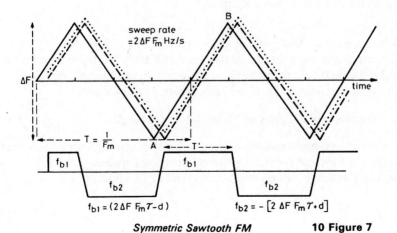

$f_{b1} = (2 \Delta F \, F_m \, T - d)$ $f_{b2} = -[2 \Delta F \, F_m \, T + d]$

Symmetric Sawtooth FM   **10 Figure 7**

For the asymmetric sawtooth of 10 Figure 6 there is no way in which the doppler frequency shift may be separated from the range frequency shift. In the case of the symmetric sawtooth of 10 Figure 7, the doppler and range frequency shifts may be separated as follows:

$$f_{b1} = 2\Delta FF_m\tau - d \tag{10.79}$$

$$f_{b2} = -(2\Delta FF_m\tau + d) \tag{10.80}$$

$$\therefore \quad f_{b1} + f_{b2} = -2d \tag{10.81}$$

and

$$f_{b1} - f_{b2} = 4\Delta FF_m\tau \tag{10.82}$$

Because of its greater potential, the symmetric sawtooth linear FM radar is the type we shall select for further consideration although parts of the analysis will be applicable with some modification, to the other system.

10.2.2 Required Frequency Deviation

If the smallest range increment, which we wish to distinguish is δr and the smallest beat frequency increment which we are able to distinguish is δf_b then:

$$\delta f_b = 2\Delta FF_m\frac{2\delta r}{c} \tag{10.83}$$

and

$$\Delta F = \frac{\delta f_b}{F_m}\frac{c}{4\delta r} \tag{10.84}$$

It remains to determine any constraints on δf_b; particularly if we wish to minimise ΔF in the interests of simplicity.

(a) Measurement of Beat Frequency Using a Frequency Counter

If only one target were present in the beam at a time, the radio altimeter is an example, it would be possible to use a frequency counter to measure f_b.

Let f_b remain constant for T' seconds, where T' is fractional.

The number of cycles in T' seconds $= f_b T' = N$ $\quad$ (10.85)

A conventional zero crossing counter will determine only an integer part of N i.e. $(N - \delta N)$ where δN is a fractional part lying between ± 1.

The real frequency, or real number of cycles in 1 second, is:

$$f_b = \frac{N}{T'} \tag{10.86}$$

The number of cycles in one second, calculated from the counter reading over T' seconds is:

$$f'_b = \frac{N - \delta N}{T'} \qquad (10.87)$$

The error

$$= \frac{\delta N}{T'}$$

But maximum value of $|\delta N|$ is 1 cycle.

$$\therefore \quad \text{maximum value of frequency error} = \frac{1}{T'} \text{ Hz.} \qquad (10.88)$$

From 10 Figure 7 it is apparent that:

$$T' = \left(\frac{T}{2} - \tau \right) \qquad (10.89)$$

It is convenient to write

$$T' = \frac{1}{n} \frac{T}{2} \qquad (10.90)$$

where

$$n = \frac{\dfrac{T}{2}}{\dfrac{T}{2} - \dfrac{2r}{c}} \qquad (10.91)$$

Now

$$\frac{T}{2} = \frac{2r_{max}}{c} \qquad (10.92)$$

where r_{max} is the maximum non-ambiguous range.

$$\therefore \quad n = \frac{\dfrac{2r_{max}}{c}}{\dfrac{2r_{max}}{c} - \dfrac{2r}{c}} = \frac{r_{max}}{r_{max} - r} \qquad (10.93)$$

$1/n$ is then the fraction of a single frequency sweep for which a constant beat frequency persists.

$$\therefore \quad \delta f_b = \frac{1}{T'} = \frac{2n}{T} = 2nF_m \qquad (10.94)$$

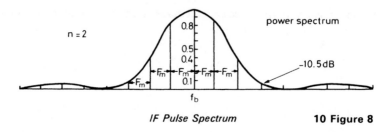

IF Pulse Spectrum   **10 Figure 8**

This limitation is a fundamental one for a single measurement using a conventional frequency counter. It may, of course, be improved upon by taking an average over a number of 'pulses' of f_b. It might also be thought possible to achieve greater accuracy by using a reciprocal counter or alternatively by using a bank of narrow band IF filters which would have the additional advantage of providing a multi-target capability.

(b) The Use of Narrow Band IF Filters
From 10 Figure 7 it is apparent that, for a constant range point target, pulses of frequency f_{b1} occur with a pulse recurrence frequency (PRF) given by

$$PRF = F_m(Hz) \tag{10.95}$$

and the duration of a single pulse is:

$$T' = \frac{1}{n}\frac{T}{2} = \frac{1}{2nF_m} \tag{10.96}$$

The line spectrum of these pulses is shown in 10 Figure 8 for the simple case where $n = 2$.

If we attempt to use filters narrower than (say) $4F_m$, then more than one range filter will be excited by a single point target. The difficulty is now one of range ambiguity. To avoid this, for the case $n = 2$, it would be necessary to adopt minimum filter bandwidths at least as great as $4F_m$.

In the general case the minimum IF filter bandwidths are such that:

$$\delta f_b = n2F_m \tag{10.97}$$

If no signal or clutter returns are possible beyond half the maximum non ambiguous range r_{max}, then:

$$\delta f_b = 4F_m \tag{10.98}$$

The restriction as to the accuracy with which we may measure f_b using a counter given by equation (10.94) is identical to that applicable when we use an IF filter bank (10.97).

(c) Minimum Frequency Deviation

For a required incremental range discrimination δr, the minimum value of ΔF is (10.84):

$$\Delta F = \frac{\delta f_b}{F_m} \frac{c}{4\delta r}$$

But (10.94) and (10.97)

$$\delta f_b = n2F_m$$

$$\therefore \quad \Delta F = \frac{n2F_m}{F_m} \frac{c}{4\delta r} \tag{10.99}$$

$$\therefore \quad \Delta F = \frac{nc}{2\delta r} \tag{10.100}$$

It is interesting to note that, apart from the factor n the spectral width of the transmitted signal must be equal to or greater than $1/\delta\tau$ where $\delta\tau$ is the incremented time delay over a distance equal to the range resolution. This is true even for pulse radar.

10.2.3 Spurious AM and PM

(a) Spurious AM

Spurious transmitter AM will have to be such that at an offset frequency equal to the range frequency shift for a target at the minimum target range of interest the noise sidebands from transmitter breakthrough are (say) 20 dB below receiver noise. This is analogous in principle to the requirements for a CW radar (see Section 10.1.3) but for a high deviation system the requirements are much less severe because the minimum interesting offset frequency is very much larger.

(b) Spurious PM

In order to discriminate against heavy clutter it is important that returns from clutter except from clutter with exactly the same frequency (range shift and doppler shift) as the target, shall not have sidebands which extend into the IF filter excited by the target itself. More precisely the sidebands of the clutter which do excite the target filter must be of a level which is below receiver noise by an appropriate amount (say 20 dB). This must be true after allowing for the much greater amplitude of the clutter 'carrier' and changes in the peak PM modulation index due to range delays, see equations (10.16) and (10.17).

This is perhaps best illustrated by taking an example. The same method of analysis, using the same formulae will be applicable to any other particular cases.

Consider a system which must meet the following requirements:

1. Maximum non-ambiguous range = 10 km
2. Maximum target or clutter range = 5 km
3. Minimum clutter range 100 metres.
4. Required range discrimination = 10 metres

$$\Delta F = \frac{nc}{2\delta r} \text{ see (10.100)}$$

For this case with values of 5 km for the maximum range of any returned signals and a maximum non-ambiguous range of 10 km:

$$n = 2$$

$$\therefore \quad \Delta F = \frac{2c}{2\delta r} = \frac{2 \times 3 \times 10^8}{20} = 30 \text{ MHz}$$

$$\frac{T}{2} = \frac{2r_{max}}{c} = \frac{1}{2F_m} \text{ see (10.92)}$$

$$\therefore \quad F_m = \frac{c}{4r_{max}} \tag{10.101}$$

$$\therefore \quad F_m = \frac{3 \times 10^8}{4 \times 10^4} = 7 \cdot 5 \text{ kHz}$$

$$\therefore \quad \delta f_b = 4 \times 7 \cdot 5 \text{ kHz} = 30 \text{ kHz}$$

This means that the individual filter bandwidths will be 30 kHz and that we cannot distinguish smaller frequency differences, regardless of whether these are due to range or radial velocity. Thus the price we have paid for a good range discrimination is an inability to distinguish doppler frequencies which are separated by less than 30 kHz.

It is simple to show that, having lost fine doppler discrimination, the close to carrier phase noise is relatively unimportant compared with the CW radar discussed in Section 10.1.4. For example, even when the radar is operating at an altitude of only 100 metres, the minimum range beat frequency of the clutter returns is 300 kHz (i.e. $r/r_{max} \times 30$ MHz).

The range shift of a target 50 metres above the ground would be 150 kHz. Even with an adverse relative doppler shift of 50 kHz between target and ground clutter, the nearest ground clutter returns would be 100 kHz from those of the target. This represents about three filter bandwidths. Thus in this case it would be necessary to control PM noise sidebands in the region of 100 kHz from the carrier, but little closer. At higher radar altitudes and with greater range differences between the target and the ground clutter, the requirements would be less severe.

Consider then the spurious PM requirement at 100 kHz for ground clutter at 100 metres range. From (10.16) the effective modulation index with this delay is:

$$\theta'(\tau) = 2\theta \sin\left(\frac{p\tau}{2}\right)$$

For

$$p = 10^5 \times 2\pi \qquad \text{and} \qquad \tau = \frac{200}{3 \times 10^8}$$

$$\frac{p\tau}{2} = 0.209 \text{ radians} \tag{10.102}$$

For such a small value of $\left(\dfrac{p\tau}{2}\right)$,

$$\sin\frac{p\tau}{2} \simeq \frac{p\tau}{2}$$

$$\therefore \quad \theta'(\tau) = 2\theta\,\frac{p\tau}{2} = \theta 2\pi f_m \tau \tag{10.103}$$

$$\theta'(\tau) = 0.418\theta$$

Thus the effective modulation index or the PM noise sideband is still attenuated by 7.58 dB relative to the transmitter noise sideband to carrier ratio at 100 kHz offset frequency and 100 metres range.

If the clutter 'carrier' with a beat frequency within 30 kHz of the target return (i.e. in the same filter) is x dB above the target return, then the transmitter phase noise density to carrier ratio must be:

$$\left(\frac{N_{op}}{C}\right)_{100 \text{ kHz}} = [-x + 7.58 - 10\log_{10}(3 \times 10^4)] \text{ dB/Hz}$$

in order to be equal to the target signal. For a 20 dB signal/noise ratio

$$\left(\frac{N_{op}}{C}\right)_{100 \text{ kHz}} = (-x + 7.58 - 44.77 - 20) \text{ dB/Hz}$$

$$= (-x - 57.2) \text{ dB/Hz} = -(x + 57.2) \text{ dB/Hz} \tag{10.104}$$

As the competing clutter is only that within a 10 metre radial range differential, x would probably not be a very large number. Even if it were as high as 40 db, $(N_{op}/C)_{100 \text{ kHz}}$ would not have to be better than -97.2 dB/Hz.

Thus it may be seen that for high deviation linear FM systems the transmitter spectral purity requirements are not very demanding. Nevertheless the requirements should always be calculated and specified as the use of a transmitter of very poor spectral purity might easily degrade the signal/clutter ratio.

The linearity requirements for the transmitter FM waveform will not be discussed here as they are only very indirectly related to the subject of spectral purity. As a warning it will merely be stated that they are often critical.

10.3 Communication Systems using SCPC-FM

10.3.1 Introduction

The reason for the introduction of frequency modulation for sound broadcasting purposes was the great improvement in the audio signal/noise ratio achieved by using a fairly large frequency deviation. The maximum peak value used was 75 kHz. This resulted in a system which was highly resistive to noise contamination, whether due to thermal noise or local oscillator noise. With regard to thermal noise this was only true above the improvement threshold of about 10 dB IF signal/noise ratio. The price paid for these advantages was the use of a broad bandwidth of about 200 kHz and a resultant poor threshold input C/N ratio. In such a system local oscillator phase noise is not critical as the frequency deviation due to the signal modulation is high, the thermal noise density is constant and the local oscillator phase noise density at IF (N_{op}) falls proportionately with the carrier level as this falls towards the threshold level.

In the case of satellite communication systems with signals destined in at least one direction to small earth terminals, mobile or otherwise, the single channel per carrier (SCPC) mode of operation is often adopted. This has the advantage of only demanding simple, and hence cheap, demodulation equipment and it also results in economy in transmitters and transmitter modulators. However in such a case the threshold level is a critical economic factor. A reduction in the required threshold value of C/N_0 will reduce the requirements for satellite eirp (effective isotropic radiated power) per carrier or in earth terminal G/T ratio (figure of merit). For this economic reason and also to save spectral occupancy, which in the case of satellite systems is a scarce resource, it is natural to adopt medium deviation FM. The choice of channel bandwidth is a trade-off between maintaining an adequate threshold level and having a sufficiently large frequency deviation to obtain an adequate baseband S/N ratio. The widest application at present for systems of this type is to carry a single voice channel, for which the required audio bandwidth is 300 Hz $-$ 3·4 kHz and the required TTNR (test tone to noise ratio) is not worse than 50 dB. The RF channel bandwidth is given approximately by Carson's Rule $[Bc = 2(\Delta f + f_m)]$ where Δf is the peak frequency deviation.

There are various special measures which may be taken to improve the real or subjective S/N ratio. The two most important are 'pre-emphasis and de-emphasis' and 'companding'. The former is used for broadcasting as well as for communication links, although the wider audio bandwidth used for broadcasting involves different numerical factors.

A more detailed discussion of pre-emphasis and de-emphasis will be given in Section 10.3.3. 'Companding' is a rather specialised subject and will not be further discussed except to say that its only effect on the calculations which follow is to reduce the value of (C/N_0) required to achieve a specified baseband signal/noise ratio. The engineer who has studied companding will have no difficulty in doing this calculation. A full treatment is rather outside our present scope.

The procedure we shall adopt is to deal first with a simple system without pre-emphasis (Section 10.3.2) and then expand the treatment to include it (Section 10.3.3).

10.3.2 Phase Noise Requirements for a Simple FM System

Consider first the effects of signal and thermal noise alone. Assume the input thermal noise to be white, as will be approximately true over the relatively narrow signal bandwidth.

Let the input carrier/noise density ratio be C/N_0. Assume a good limiter to be used which strips off the AM noise.

$$N_{op} = \frac{N_0}{2} \text{ (see 3.29)} \tag{10.105}$$

Let the carrier be frequency modulated with a single sinusoidal tone, such as is used in measuring the TTNR and let the *peak* frequency deviation be Δf.

The signal output voltage of the FM demodulator will have a peak output voltage proportional to Δf and an rms output voltage proportional to $\Delta f/\sqrt{2}$. Without loss of generality, as our purpose is to determine *relative* signal and noise outputs, we may make the constant of proportionality equal to unity and assume an impedance level of 1 ohm. The baseband signal output power (S) is then:

$$S = \frac{(\Delta f)^2}{2} \tag{10.106}$$

The baseband filter will be designed to reject all output components with a frequency greater than the maximum value of the baseband frequency.

$$b \simeq f_m(\text{max})$$

Well above threshold, cross products between noise components will be negligible. The only noise terms in the baseband output will be those giving maximum frequencies of $\pm b$ after beating with the large carrier.

In the case of a white phase noise input:

$$(\delta f)^2 = \left(\frac{N_{op}}{C}\right) \int_{-b}^{+b} f_m^2 \, df_m$$

where δf is an rms value as N_{op}/C is a power ratio.

$$\therefore \quad (\delta f)^2 = \left(\frac{2N_{op}}{C}\right) \int_0^b f_m^2 \, df_m \qquad (10.107)$$

The noise power output is (10.105).

$$N_T = \left(\frac{N_0}{C}\right)_T \frac{b^3}{3} = \left(\frac{N_0}{C}\right)_T \frac{f_{m_{max}}^2 b}{3} \qquad (10.108)$$

where the suffix T has been added to denote the thermal noise contribution. Then the output signal/noise ratio is:

$$\frac{S}{N} = 3\left(\frac{\Delta f}{f_{m_{max}}}\right)^2 \frac{C}{2bN_{0T}} \qquad (10.109)$$

Now $(\Delta f/f_{m_{max}})$ is the peak FM modulation index M or θ at the highest baseband frequency f_m max.

$$\therefore \quad \frac{S}{N} = \frac{3M^2}{2b}\left(\frac{C}{N_0}\right)_T \qquad (10.110)$$

(a) The Case of Local Oscillator White Phase Noise

If the local oscillator phase noise were white over the full offset frequency band $f_0 \pm b$, then the specification of local oscillator phase noise requirements would be very simple. Firstly it would be necessary to decide the acceptable performance degradation due to local oscillator phase noise, considering the effects on baseband signal/noise ratio both at maximum and minimum working levels of input carrier power.

By analogy with (10.108)

$$N_L = \left(\frac{2N_{op}}{C}\right)_L \frac{b^3}{3} \qquad (10.111)$$

where the suffix L refers to local oscillator noise. Assume that it is necessary that the local oscillator noise contribution shall be a fraction $1/x$ of that due to thermal noise.

$$N_L = \frac{1}{x}N_T \qquad (10.112)$$

Using (10.111) and (10.108)

$$2\left(\frac{N_{op}}{C}\right)_L \frac{b^3}{3} = \frac{1}{x}\left(\frac{N_0}{C}\right)_T \frac{b^3}{3} \qquad (10.113)$$

$$\therefore \quad \left(\frac{N_{op}}{C}\right)_L = \frac{1}{2x}\left(\frac{N_0}{C}\right)_T \qquad\qquad (10.114)$$

As C changes from C_{min} to C_{max} $(N_{op}/C)_L$ does not change in value. This is because the power in a single phase noise sideband (N_{op}) due to a fixed value of phase jitter is proportional to C (consult equation 2.11 if in doubt).

However N_0 is constant; being due to the input design of the earth terminal, especially the low noise amplifier. Thus $(N_0/C)_T$ falls with increasing input carrier power.

Except apparently in the case where operational requirements specify varying baseband S/N ratios at different carrier levels, and this is sometimes so, the required value $(N_{op}/C)_L$ is calculated from:

$$\left(\frac{N_{op}}{C}\right)_L = \frac{1}{2x}\left(\frac{N_0}{C_{max}}\right)_T \qquad\qquad (10.115)$$

Even in the case where the baseband S/N is given specified values for different values of C, equation (10.115) may often be used to advantage. If the degradation due to local oscillator noise at high carrier levels is chosen to be 0·4 dB (that is $x = 10$), then at lower carrier levels down to threshold, where the performance is usually most critical, the degradation due to L.O. phase noise will be less.

So far we have only been considering the simple case of white phase noise, which may not be typical over an offset frequency range such as 300 Hz to 3·4 kHz. To complete this simplified case it only remains to determine a reasonable value for x and to deal with a typical numerical example.

Consider a one way complete satellite link. This will consist of a transmitting earth terminal, a satellite repeater and a receiving earth terminal. The first question is; how much *link* degradation should be allowed due to the overall phase noise contributions of all signal sources in this link? There is no precise answer to this question, which ultimately is an economic one.

In a single SHF satcom link, five equipments are likely to make major, and roughly equal, contributions to the total phase noise budget and to the resultant signal degradation. These are:

 (i) Local oscillators in the transmitting ground station
 (ii) High power amplifiers in the transmitting ground station
 (iii) Local oscillators in the satellite
 (iv) The satellite transmitter
 (v) Local oscillators in the receiving ground station.

Link degradation due to carrier phase jitter is independent of the carrier power level; whereas up-link thermal noise contributions are usually made small relative to the down-link contributions by increasing the up-link carrier power.

If each of the five contributions listed above is 1/50th (− 17 dB), of the phase noise component of down-link thermal noise, then the overall link degradation due to this cause is 0·4 dB. In a complete satellite communication system costing several hundreds of millions of pounds over a decade such a link degradation represents a financial loss of several tens of millions of pounds and the use of relatively complex low phase noise signal sources is thus economically justified. It is suggested that the contribution of any one of the five items listed above should be 20 dB below the phase noise component of thermal noise.

Now

$$\frac{N_{op}}{C} = \frac{N_0}{2C}$$

Hence it is recommended that:

$$\frac{N_{op}}{C} \text{L0 (dB)} \not> \left[\left(\frac{N_0}{C} \right) \text{dB} - 3 - 20 \right] \text{(dB)} \qquad (10.116)$$

This performance must be maintained so that the integrated value of the phase noise density over critical offset frequency bands is 23 dB below the integrated value of the thermal noise density over the same offset frequency bands.

Thus we shall asume that:

$$x = 20 \text{ dB} = 100 \text{ (see 10.112)} \qquad (10.117)$$

If we knew $(N_0/C)_T$ it would now be possible to use equation (10.114) to calculate the required value of $(N_{op}/C)_L$. If we took the minimum value of $(C/N_0)_T$ when doing this calculation we should normally be safe at all carrier levels. For example even if the carrier power rose by 6 dB the baseband S/N ratio, as limited by thermal noise, would rise by 6 dB (see 10.110) and *receiver* local oscillator phase noise contributions would rise from 20 dB below the thermal contribution to 14 dB below it, thus degrading the improved S/N ratio by only 0·17 dB.

An Example
As we have as yet taken no advantage of pre-emphasis or companding, it will be assumed that the required baseband S/N ratio is 30 dB, rather than the 50 dB normally specified. This will result in the other parameters having fairly typical values.

Let the RF channel bandwidth be 25 kHz, from Carson's Rule (Reference 43):

$$B_c = 2(\Delta f + f_m) \qquad (10.118)$$

For

$$f_m = 3\cdot4 \text{ kHz} \tag{10.119}$$

$$\Delta f = 9\cdot1 \text{ kHz} \tag{10.120}$$

From (10.110)

$$\left(\frac{C}{N_0}\right)_T = \frac{S}{N}\frac{2b}{3M^2} \tag{10.121}$$

Assume a 1 kHz test tone is used, with the same value of M.

$$\therefore \quad M = \frac{9100}{3400} = 2\cdot68 \tag{10.122}$$

$$\therefore \quad \left(\frac{C}{N_0}\right)_T (\text{dB Hz}) = (30 + 3 + 10 \log_{10} 3400 - 4\cdot77 - 20 \log_{10} 2\cdot68)$$

$$\therefore \quad \left(\frac{C}{N_0}\right)_T = 54\cdot98 \text{ dB Hz and } \left(\frac{N_0}{C}\right)_T = -54\cdot98 \text{ dB/Hz} \tag{10.123}$$

for $x = 20$ dB equation (10.114) gives

$$\left(\frac{N_{op}}{C}\right)_L = (-3 - 20 - 54\cdot98) \text{ dB/Hz} = -77\cdot98 \text{ dB/Hz}$$

The preceeding calculations have been based on the assumption of white phase noise. Thus, for the example chosen, it would be necessary that local oscillator phase noise density should be held equal to or better than:

$$\left(\frac{N_{op}}{C}\right) \leq -77\cdot98 \text{ dB/Hz} \tag{10.124}$$

over the offset frequency band 300 Hz to 3·4 kHz. In practice if a double superhet were used with IF's of (say 700 MHz and 70 MHz then the total phase noise of the two local oscillators together would have to satisfy (10.124). The phase noise budget would normally be such as to allocate most of this to the microwave L.O.

(b) The Case of Non White Phase Noise

In practice the local oscillator phase noise density is unlikely to be constant over an offset frequency range of 300 Hz – 3·4 kHz. The discussion in (a) above was also simplified by assuming that the noise bandwidth of the baseband filter was equal to the highest signal frequency which it was required to pass (3·4 kHz). In practice the noise bandwidth is likely to be some 10% wider than this. A bandwidth increase of 10% in a region where $(\delta f_0)^2$ is rising by 20 dB per decade would increase both thermal and white phase noise contributions by 30% which is very approximately 1 dB. Thus our real interest will be in phase noise density from perhaps 250 Hz – 3·8 kHz offset frequency.

In the example we chose before, equation (10.123) gave

$$\left(\frac{N_0}{C}\right)_T = -54 \cdot 98 \text{ dB/Hz}$$

If we assume conservatively that this should be improved by 1 dB to allow for a realistic noise bandwidth, then:

$$\left(\frac{N_0}{C}\right)_T = -55 \cdot 98 \text{ dB/Hz} \tag{10.125}$$

From (10.108) increasing b by 10%

$$(\delta f)_T^2 = \left(\frac{N_0}{C}\right)_T \frac{(1 \cdot 1b)^3}{3} \tag{10.126}$$

For local oscillator noise

$$(\delta f)_L^2 = \int_{250 \text{ Hz}}^{1 \cdot 1b} \left(\frac{2N_{op}}{C}\right)_{f_m} f_m^2 \, df_m \tag{10.127}$$

It is necessary that (10.127) should be x dB below (10.126).

It will always be the case either that $(N_0/C)_T$ is directly known as a basic system parameter or that it may be calculated, as was done in the example above, from a knowledge of the required baseband S/N ratio and the modulation parameters. Thus $(\delta f)_T^2$ may readily be calculated from (10.126).

The local oscillator phase noise requirement may then be expressed as:

$$(\delta f)_L^2 \not> \frac{(\delta f)_T^2}{x} \tag{10.128}$$

where $(\delta f)_L^2$ is the integrated value from (say 250 Hz) to $1 \cdot 1b$.

Thus the specification for the permissible random angular modulation of the local oscillator(s) in an earth terminal receiver or transmitter may be expressed in either of two forms as follows:

$$(\delta f)_L^2 \not> \frac{1}{x} \left(\frac{N_0}{C}\right)_T \frac{(1 \cdot 1b)^3}{3} \tag{10.129}$$

$$\int_{250 \text{ Hz}}^{1 \cdot 1b} \left(\frac{2N_{op}}{C}\right)_{f_m} f_m^2 \, df_m \not> \frac{1}{x} \left(\frac{N_0}{C}\right)_T \frac{(1 \cdot 1b)^3}{3} \tag{10.130}$$

Strictly, the expression for $(\delta f)_T^2$ should be slightly reduced to allow for the fact that the demodulator output filter rejects the contribution to $(\delta f)_T^2$ below 250 Hz. However as $(\delta f_0)_T^2$ falls by 20 dB per decade fall in f_m, this contribution is negligible. It is not always safe to make the same approximation in evaluating

$(\delta f)^2_L$, as $(N_{op}/C)_{f_m}$ might itself be rising rapidly with falling f_m. In fact for very small values of f_m, even a good local oscillator will have a phase noise density rising as $1/f^3_m$ in which case $(\delta f_0)^2_L(f_m)$ will be rising as $1/f$.

For an existing L.O, $(\delta f)^2_L$ may be measured with a calibrated discriminator and suitable baseband filter. However it is more usual to measure the phase noise *density* over this offset frequency band. Once this is known, the TI58/59 program described in Section 9.3.2 may be used to evaluate the integral in equation (10.127) to see whether or not the specification requirement is satisfied.

It should be noted [see (10.127), also 4 Table 1 and 4 Figure 1] that the main contribution to $(\delta f)^2_T$ tends to come from phase noise components at offset frequencies at the upper end of the range, that is in the region of $\pm b$. $(\delta f_0)^2_T$ (the density of the frequency variance) rises by 20 dB per decade relative to the phase noise density.

10.3.3 The Addition of Pre-emphasis and De-emphasis

If the frequency deviation of the transmitter signal were increased by 6 dB per octave (20 dB per decade) above a certain modulating frequency f_c, then in principle the distortion might be corrected at the receiver by providing a de-emphasis network after the FM demodulator, the response of which falls by 6 dB per octave above f_c. Due to the lower energy density of the spectrum of most audio signals above f_c, both speech for communication, and speech and music for broadcasting, pre-emphasis at the transmitter may be achieved without greatly increasing the peak deviation. A *small* reduction in the low frequency sensitivity of the FM modulator will then prevent over-deviation at any time.

The de-emphasis network at the receiver will reduce the noise power output of the FM detector by 6 dB per octave above f_c. This falling characteristic combined with the naturally rising response of an FM detector (6 dB per octave) with increase in baseband frequency will result in a reponse to a white phase noise input which is flat above f_c.

Thus, with de-emphasis, we may modify equation (10.107) for white thermal noise input to show the baseband output power (P_{bT}) as:

$$P_{bT} = \left(\frac{N_0}{C}\right)_T \int_0^{f_c} f^2_m \, df_m + \left(\frac{N_0}{C}\right)_T f^2_c(b - f_c) \tag{10.131}$$

$$P_{bT} = \left(\frac{N_0}{C}\right)_T \left[\frac{f^3_c}{3} + f^2_c(b - f_c)\right] \tag{10.132}$$

If greater accuracy is required, equations (10.131) and (10.132) may be modified in respect of upper and lower baseband filter cut-off frequencies to allow for the actual noise bandwidth.

The L.O. frequency density variance is given by:

$$(\delta f_0)_L^2 = \left(\frac{2N_{op}}{C}\right)_{f_m} f_m^2 \tag{10.133}$$

The baseband output power density $P_{0(bL)}$ is equal to $(\delta f_0)_L^2$ from (say) 250 Hz up to f_c. Above f_c the de-emphasis modifies this so that

$$P_{0(bL)} = \left(\frac{2N_{op}}{C}\right)_{f_m} f_c^2 . \tag{10.134}$$

Thus

$$P_{bL} = \int_{250}^{f_c} \left(\frac{2N_{op}}{C}\right)_{f_m} f_m^2 \, df_m + \int_{f_c}^{b} \left(\frac{2N_{op}}{C}\right)_{f_m} f_c^2 \, df_m \tag{10.135}$$

If (N_{op}/C) has been measured on a supposedly suitable L.O, equation (10.135) may readily be evaluated using the TI58/59 program for frequency jitter from $f_m = 250$ Hz to $f_m = f_c$ and the program for integrated phase noise from $f_m = f_c$ to $f_m = f_b$ after allowing for the term f_c^2.

The specification requirement is that

$$P_{bL} \ngtr \frac{P_{bT}}{x} \tag{10.136}$$

i.e. (10.132)

$$P_{bL} \ngtr \frac{1}{x} \left(\frac{N_0}{C}\right)_T \left[\frac{f_c^3}{3} + f_c^2(b - f_c)\right] \tag{10.137}$$

All the terms in (10.137) will be known so that a numerical value may be found for the maximum permissible value of P_{bL}.

Different L.O. designs may have different distributions of (N_{op}/C) over the offset frequency range of $f_m = (250$ to $b)$ Hz but in all cases P_{bL} may quickly be calculated using the TI58/59 program. There is probably no unique solution to the design of a suitable local oscillator but many solutions can be devised using the techniques described in the earlier chapters of this book.

One particular point is worth stressing. If a local oscillator is only required for this application it is a waste of effort to aim for an outstanding phase noise performance below 250 Hz offset frequency. The only exception is if no frequency acquisition circuit is provided and frequency drift is excessive. In this case long term frequency stability requirements may determine the close to carrier phase noise. This aspect was adequately treated in Chapter 9.

In writing a L.O. specification it may be convenient to express it in terms of $(\delta f)^2$ below f_c and in terms of integrated single sideband phase noise or phase jitter (DSB phase noise) above f_c.

Thus

$$(\delta f)^2 \not> \frac{1}{x}\left(\frac{N_0}{C}\right)_T \frac{f_c^3}{3} \text{ for } f_m = 250 \text{ Hz} \rightarrow f_c$$

and

$$\overline{\phi^2} \not> \frac{1}{x}\left(\frac{N_0}{C}\right)_T (b - f_c) \text{ for } f_m = f_c \rightarrow b$$

(10.138)

An Example

Consider a maritime satellite system achieving an audio S/N ratio of 50 dB by means of companding and pre-emphasis/de-emphasis. Such a system might operate at $(C/N_0)_T = 52$ dB. Assume x to be set at 20 dB and $f_c = 1$ kHz. Then

$$(\delta f)^2 \not> [-20 - 52 + 90 - 4\cdot77] \text{ dB Hz}^2 = (13\cdot23) \text{ dB Hz}^2. \quad (10.139)$$

$$\therefore \quad \delta f_{rms} = 4\cdot59 \text{ Hz in frequency band 250 Hz} - 1 \text{ kHz}.$$

and

$$\overline{\phi^2} \not> [-20 - 52 + 10 \log_{10} (3400 - 1000)] \text{ dB rads}^2$$

$$\therefore \quad \overline{\phi^2} = -38\cdot2 \text{ dB rads}^2$$

$$\therefore \quad \phi_{rms} = 12\cdot3 \text{ milliradians from } f_m = 1 \text{ kHz to } 3\cdot4 \text{ kHz} \quad (10.140)$$

The interested reader will note that the effect of companding is to reduce the required value of $(C/N_0)_T$ necessary to achieve the required baseband S/N ratio. It discriminates on a basis of signal level rather than offset frequency and thus has no other effect on the specification of permissible L.O. spurious angle modulation.

Knowing the permissible value of δf^2 or ϕ^2 it is not possible to calculate the phase noise density specification for the signal source unless the *law* of its variation with offset frequency, is known. (See Chapter 9).

10.4 FDM/FM Communication Systems

10.4.1 Introduction

In major communication systems carrying a very large number of telephone channels between two major telephone distribution centres it is usual to allocate 4 kHz bandwidth to each telephone channel and to arrange these telephone channels in ascending frequency order, by single sideband modulation of a series of carriers differing in frequency by 4 kHz. For example a group of 60 telephone channels might be arranged to occupy a baseband bandwidth of 240 kHz, extending from 12 kHz to 252 kHz. This arrangement is known as FDM (frequency division multiplex) as the combination and the ultimate reseparation of the individual channels is carried out on a basis of frequency.

The composite FDM baseband may then be used to frequency modulate a carrier with a centre frequency considerably higher than the maximum baseband frequency. The resultant signal is then known as an FDM/FM signal. Before the days of digital systems all point to point communication between large satellite earth terminals each carrying a large number of telephone channels, was carried out by this means. Even today most satellite borne telephone traffic is carried in this way.

10.4.2 The Required Value of C/N_0

A single conventional telephone channel, prior to combination in an FDM system, will be subject to variable power levels. The variations will be both with respect to time for a single speaker, as he talks more loudly or more softly, and also from speaker to speaker. A reference point is defined in each circuit so that the *average* power level is 0 dBm.

In an FDM/FM system each channel is allotted an rms Test Tone Deviation (δf) corresponding to 0 dBm at a circuit reference point. The multi channel loading will be statistical in nature due to speech gaps in individual channels as well as volume variations from speaker to speaker. The multi-channel rms deviation is calculated from:

$$\delta f_m = \delta f P_m \tag{10.141}$$

P_m is the multi-channel loading factor which, for not more than 240 channels, is:

$$P_m(\text{dB}) = (-1 + 4 \log_{10} N_c) \text{ dB} \tag{10.142}$$

where N_c is the number of channels.

$$\text{For 60 channels } P_m = 6 \cdot 11 \text{ dB} \tag{10.143}$$

For multi-channel FDM the waveform approximates to noise and the peak to mean ratio is normally taken to be 10 dB. This gives a fairly small statistical probability of overloading on peaks. Thus the peak multi-channel deviation ΔF is given by:

$$\Delta F \text{ (dB Hz)} = (20 \log_{10} \delta f + P_m + 10) \tag{10.144}$$

The signal bandwidth occupied is given approximately by Carson's Rule:

$$B_c = 2(\Delta F + f_m) \tag{10.145}$$

where B_c is the occupied bandwidth and f_m is the maximum baseband frequency.

The normalised baseband signal power output for a single channel is:

$$S = (\delta f)^2 \tag{10.146}$$

The corresponding baseband noise power output for (white) thermal noise see [(4.11)] is:

$$N = \overline{\phi_0^2} f_m^2 b$$

where b is the baseband bandwidth of one channel (3.1 kHz)

$$\therefore \quad \frac{S}{N} = \frac{\delta f^2}{\overline{\phi_0^2} f_m^2 b} \tag{10.147}$$

But

$$\overline{\phi_0^2} = \frac{2N_{op}}{C} = \frac{N_0}{C}$$

$$\therefore \quad \frac{S}{N} = \left(\frac{\delta f}{f_m}\right)^2 \frac{C}{N_0 b} \tag{10.148}$$

The reasons for the differences between (10.148) and (10.109) are as follows:

(a) The '2' in the denominator of (10.109) is due to the fact that Δf is a peak value.

(b) The '3' in the numerator of (10.109) disappears because f_m is assumed to be a constant over the relatively narrow signal bandwidth of a single channel in the FDM system to which (10.148) applies.

From (10.148)

$$\boxed{\frac{C}{N_0} = b \frac{S}{N} \left(\frac{f_m}{\delta f}\right)^2} \tag{10.149}$$

For a given system all the terms on the right hand side of (10.149) will have known numerical values for each channel; f_m being given the mean value for the channel being considered. Due to the triangular baseband noise distribution, the S/N ratio will be at its worst in the highest frequency telephone channel. To reduce the difference in performance between channels, pre/de-emphasis is normally employed. The baseband noise is also psophometrically weighted to allow for the frequency characteristics of the human ear. The combined effect of pre/de-emphasis and psophometric weighting is an improvement of about 5·5 dB in the top channel, which remains the most critical.

Thus the C/N_0 ratio required to give a specified TTNR may be calculated from (10.149) using f_m for the top channel, remembering that δf is the rms Test Tone deviation per channel and deducting 5·5 dB from S/N. Hence:

$$\left(\frac{C}{N_0}\right)_{\text{dB Hz}} = \left[\left(\frac{S}{N}\right)_{\text{dB}} - 5·5 + b \text{ (dB)} + 20 \log_{10}\left(\frac{f_m}{\delta f}\right)\right] \tag{10.150}$$

10 Table 1   *Some Intelsat IV Parameters 60 Channel FDM/FM*

Parameter	Global beam	Spot beam
b—baseband bandwidth	252 kHz	252 kHz
δf—rms test tone per channel	270 kHz	136 kHz
P_m—multi channel loading factor	6·11 dB	6·11 dB
Multi channel rms deviation	546 kHz	275 kHz
Peak to mean ratio	10 dB	10 dB
Δf—peak dev. multi-ch.	1·725 MHz	0·869 MHz
B_c—Carson bandwidth	3·954 MHz	2·242 MHz
Allocated bandwidth	5 MHz	2·5 MHz
C/N_0 for 50 dB S/N	78·7 dB Hz	84·7 dB Hz

As an example, assume 200 kHz Test Tone Deviation which is a fairly widely used value. Consider a 60 channel FDM system with a baseband extending from 12–252 kHz and a specified TTNR of 50 dB. Then:

$$\left(\frac{C}{N_0}\right)_{\text{dB Hz}} = \left[50 - 5\cdot5 + 10 \log_{10} 3100 + 20 \log_{10}\left(\frac{250}{200}\right)\right]$$

$$\therefore \left(\frac{C}{N_0}\right) = 81\cdot3 \text{ dB Hz} \tag{10.151}$$

As a more detailed example some parameters for the Intelsat IV communication satellite system are given in 10 Table 1 for the case of a 60 channel FDM/FM carrier.

It is usual for communication engineers to describe the noise level either in picowatts or in dBm at a circuit reference point where the signal is 1 mW (0 dBm = +90 dB w.r.t 1 pW). A S/N ratio of 50 dB is then achieved when the noise level is +40 dBp (= 10,000 picowatts).

For the widely used case where the rms Test Tone deviation = 200 kHz it is possible to relate noise power in dBm or in picowatts to the level of spurious frequency deviation at the channel offset frequency f_m.

10.4.3 Phase Noise Requirements

In Section 10.4.2 above it has been shown how the required value of C/N_0 thermal may be calculated. If we neglect any question of discrete spurious lines on the local oscillator output we may assume that $(N_{op}/C)_L$ is constant over the very small fractional bandwidth of one FDM telephone channel. The required value of $(N_{op}/C)_{f_m}$ then follows very simply once we have decided x [see (10.112)].

$$\left(\frac{N_{op}}{C}\right)_L = \frac{1}{x}\left(\frac{N_{op}}{C}\right)_{\text{Thermal}} = \frac{1}{2x}\left(\frac{N_0}{C}\right)_T \tag{10.152}$$

As discussed in Section 10.3 (see equation 10.117) we shall assume that for a satellite communication system $x = 100$ is normally a reasonable, if slightly conservative, choice.

The simple form of equation (10.152) justifies the statement made in Chapter 4 that it is more straightforward to specify the local oscillator spectral purity in terms of (N_{op}/C) rather than in terms of spurious frequency deviation. For those engineers who insist on the latter form:

$$\left(\frac{2N_{op}}{C}\right)_L = \overline{\phi_{0L}^2} = \frac{1}{x}\left(\frac{N_0}{C}\right)_T \tag{10.153}$$

$\therefore$ over a bandwidth b (3.1 kHz)

$$\overline{\phi_L^2} = \frac{b}{x}\left(\frac{N_0}{C}\right)_T = \frac{\overline{\delta f_L^2}}{f_m^2} \tag{10.154}$$

$$\therefore \quad \overline{\delta f_L^2} = f_m^2 \frac{b}{x}\left(\frac{N_0}{C}\right)_T \tag{10.155}$$

An Example

Consider the example from which equation (10.151) was derived. In this case:

$$\left(\frac{N_0}{C}\right)_T = -81\cdot3 \text{ dB/Hz}$$

Assume $x = 20$ dB

$\therefore$ from (10.152)

$$\left(\frac{N_{op}}{C}\right)_L = [-3 - 20 - 81\cdot3] \text{ dB/Hz}$$

$$\therefore \quad \left(\frac{N_{op}}{C}\right)_L = -104\cdot3 \text{ dB/Hz} \tag{10.156}$$

This value of phase noise density to carrier ratio applies strictly when the offset frequency f_m is that of the top channel. It may be slightly relaxed for the low channels where the thermally limited S/N ratio will be better.

Alternatively, using (10.155) for the same example:

$$\overline{\delta f^2} = (2\cdot5 \times 10^5)^2 \times \frac{3100}{100} \times \left(\frac{N_0}{C}\right)_T$$

$$= 14264 \text{ Hz}^2$$

$$\therefore \quad \delta f_{rms} = 119\cdot4 \text{ Hz.} \tag{10.157}$$

It is true that the engineer who prefers to think in terms of frequency deviation may argue that his approach is in fact the more simple. He may say that if 200 kHz Test Tone deviation gives a thermally limited signal/noise ratio of 50 dB, then for $x = 20$ dB, the random frequency deviation imparted to the signal by the local oscillator should be 70 dB below 200 kHz rms. This is 63·2 Hz rms.

The disadvantages of this approach are threefold:

(a) It neglects the de-emphasis and psophometric weighting advantages, which may or may not be relevant in a particular system. Hence the 5·5 dB difference between this value and equation (10.157).

(b) It is possible to carry out such a calculation knowing little about the system other than the Test Tone Deviation. To specify the local oscillator performance for a system which is only partially understood, is always dangerous.

An engineer who is competent to choose the correct value for x, should be capable of, and encouraged to, work out all the important system parameters. Having calculated the thermal carrier/noise density requirements, the natural way to specify the local oscillator spectral purity is in terms of (N_{op}/C) using equation (10.152).

(c) A full analysis also makes it possible to calculate the performance for channels other than the top one. It may be important to know what relaxation of the local oscillator phase noise specification can be permitted for the bottom channel.

10.5 Communication Systems Using PSK

10.5.1 Introduction

Many modern communication systems code the information carried, into digital form before transmission. Amongst its other advantages, such as uniformity of transmission characteristics regardless of the diversity of the information types, time sharing of multiple channels etc, it has the special feature that, with one qualification, signals may be accurately reconstituted at various stages in the transmission path to correct for the effects of distortion and added noise. The qualification is that distortion or noise shall not have resulted in a wrong interpretation of a received digit or digits. Thus the transmission performance of a system carrying digital signals is measured in terms of the bit error rate (ber).

A digital bit stream may be modulated on to a carrier in a variety of ways. In the case of satellite communication the two primary ways are:

(a) **Phase shift keying of the carrier (PSK); which may be two phase, called Phase Reverse Keying (PRK) or Quadriphase (QPSK). For PRK the phase difference between '0' and '1' is π radians. For QPSK there are four different levels differing by $\pi/2$ radians.**

(b) Frequency shift keying of the carrier (FSK). The simplest case is where only two carrier frequencies are used.

The efficiency of a modulation/demodulation system (apart from the question of spectral occupancy) is judged in terms of the required value of E/N_0 for a given ber. Here E is the energy per bit of information. From this point of view PSK is the most efficient system and the theoretically required value of E/N_0 is the same for biphase and quadriphase systems for the same error rate.

This is naturally so, as a doubling of the information rate (4 phase levels) will require twice the power if the energy per bit is constant.

Due to the great cost of increasing satellite transmitter power, or alternatively the figure of merit (G/T) of all earth stations, a satellite communication system carrying a very large amount of digital traffic will usually adopt PSK and most frequently QPSK. Thus we shall confine our detailed considerations to PSK systems. There is a further sub-division of PSK systems between these which reconstitute the carrier and use fully coherent detection, and those which use differential detection. We shall consider both systems, as the impact on the required phase noise specification is rather different in the two cases.

For the sake of completeness it is worth mentioning that there is a modulation system known as MSK (Minimum Phase Shift Keying) which may be regarded either as a special PSK system or a special FSK system. Whilst theoretically less efficient than QPSK it has a major offsetting advantage in that it is ideally a constant amplitude system and hence highly efficient hard limiting repeaters may be used in the satellite.

In analysing the phase noise requirements for local oscillators (usually synthesisers) for use in satellite communication systems it is necessary first to decide what is to be considered an acceptable ratio of thermal noise to LO phase noise contributions, both integrated over the relevant offset frequency bands. The background to this decision was discussed in Section 10.3.2 where it was stated that the decision was ultimately an economic one and that it involved consideration of a complete communications link—transmitting ground station/satellite/receiving ground station and budgeting also to allow for the effect of phase noise contributions from transmitter high power amplifiers (HPAs) which might be either kystrons or travelling wave tubes as well as all local oscillators in the link. A reasonable, if slightly conservative, decision was given by equation (10.117) which is repeated for convenience.

$$x = 100 = 20 \text{ dB} \tag{10.117}$$

where

$$\int_a^b N_{op}(f) \, df \text{ thermal} = x \int_a^b N_{op}(f) \, df \text{ local oscillator} \tag{10.158}$$

and a and b are the relevant offset frequency limits.

We shall deal only with a PSK system using coherent demodulation at the following bit rates: 300 b/s, 1·2 kb/s, 2·4 kb/s, 4·8 kb/s, 16 kb/s, 64 kb/s and 1

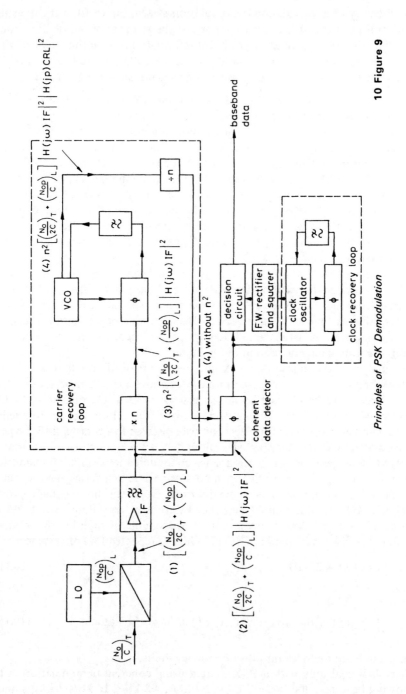

Principles of PSK Demodulation

10 Figure 9

Mb/s. It is thus a flexible system which may be switched to operate at different bit rates, although it may carry (quadrature) signals at two different bit rates simultaneously. Consideration of such a system illustrates clearly all the factors involved and is in fact more demanding in respect of *close to carrier* phase noise requirements than a high bit rate system carrying a single type of data stream at either 60 Mb/s or 120 Mb/s. High bit rate systems may be analysed using the same equations. We shall consider in turn the Carrier Recovery Loop, the signal ber and the Clock Recovery Loop.

Subsequently we shall discuss the phase noise requirements for differentially detected PSK and the effects of LO phase transients.

10.5.2 The Carrier Recovery Loop

In practice a coherent PSK demodulator may take one of many forms. A block diagram showing all those items which affect noise calculations is shown in 10 Figure 9. It is assumed that the IF input to the demodulator is at a sufficiently low frequency that the uncontrolled VCO may be designed to contribute negligibly to the phase noise budget.

The most critical item in such a demodulator is the carrier recovery loop.

This must reliably acquire the carrier in a specified maximum time at a given input noise density to carrier ratio

$$\left(\frac{N_0}{C}\right) = \left(\frac{2N_{op}}{C}\right).$$

The probability of subsequent loss of lock must be very small indeed. Loss of lock and the subsequent time taken for reacquisition will result in loss of signal information for this time. For a system working at 1 Mb/s an interruption as short as 1 second will result in the loss of 10^6 bits, thus having a catastrophic effect on the overall error rate.

The acceptable unlocking probability is a function of the system bit rate, the required ber and the relocking time. For a typical required ber of 10^{-5} at 1 Mb/s, a typical value for the acceptable unlocking probability might be 1×10^{-12}.

For many phase detectors the loop will unlock when the phase error, integrated over the 'relevant' offset frequency band, reaches $\pm \pi/2$ radians. For a simple sinusoidal error this represents the peak of the sinusoid: the loop will unlock when the rms value of the error reaches $\pm 1 \cdot 1$ radians and this must only occur with a probability of 1×10^{-12}. If σ is the acceptable phase error $1 \cdot 1$ radians is the $7 \cdot 15 \sigma$ value. Thus under these conditions σ (the long term rms value) must not exceed $0 \cdot 15$ radians and $\overline{\phi^2} = 0 \cdot 0225$ rads^2.

The 'relevant' offset frequency range extends from zero to an upper limit determined by the transfer function of the loop. At offset frequencies where the transfer function is negligibly small the VCO will not respond to an error and the loop will not therefore tend to unlock as a result of error components at such high offset frequencies.

Assume a high gain second order loop. Then the phase error which can contribute to unlocking is given approximately by:

$$\overline{\phi^2}_{\text{error}} = \int_0^{f_n} n^2 \left(\frac{2N_{op}}{C}\right)_f |1 - H(jp)|^2 \, df$$

$$+ \int_{f_n}^{\infty} n^2 \left(\frac{2N_{op}}{C}\right)_f |H(jp)|^2 \, df \qquad (10.159)$$

The error response and the transfer function of the loop may be approximated as follows:

$$|1 - H(jp)|^2 = \left(\frac{f}{f_n}\right)^4 \text{ for } f < f_n.$$
$$= 1 \text{ for } f > f_n. \qquad \text{(See 7 Figure 12)} \qquad (10.160)$$

$$|H(jp)|^2 = 1 \text{ for } f < Kf_n.$$
$$= \left(\frac{Kf_n}{f}\right)^2 \text{ for } f > Kf_n \qquad \text{(See Chapter 7)} \qquad (10.161)$$

where $K = 1 \cdot 5$ for $\zeta = 0 \cdot 707$
$K = 1 \cdot 75$ for $\zeta = 1$ (See Chapter 7)   (10.162)
$K = 2 \cdot 5$ for $\zeta = 1 \cdot 3 \rightarrow 1 \cdot 5$

$$\therefore \quad \overline{\phi^2}_{\text{error}} = \int_0^{f_n} n^2 \left(\frac{2N_{op}}{C}\right)_f \left(\frac{f}{f_n}\right)^4 df + \int_{f_n}^{Kf_n} n^2 \left(\frac{2N_{op}}{C}\right)_f df$$

$$+ \int_{Kf_n}^{\infty} n^2 \left(\frac{2N_{op}}{C}\right)_f \left(\frac{Kf_n}{f}\right)^2 df \qquad (10.163)$$

(a) White Thermal Noise

Consider an input of white thermal noise, due to the receiver noise temperature as follows:

$$\frac{N_0}{C} = \frac{2N_{op}}{C}$$

PLL Unlocking Sensitivity v/s f_m   **10 Figure 10**

After frequency multiplication this becomes

$$n^2 \left(\frac{2N_{op}}{C} \right)_{(DSB)} \tag{10.164}$$

Substitute in (10.163) and integrate:

$$\overline{\phi^2}_{error} = n^2 \left(\frac{2N_{op}}{C} \right) \left\{ \left[\frac{1}{5} \frac{f^5}{f_n^4} \right]_0^{f_n} + [f]_{f_n}^{Kf_n} - \left[\frac{K^2 f_n^2}{f} \right]_{Kf_n}^{\infty} \right\}$$

$$\therefore \quad \overline{\phi^2}_{error} = n^2 \left(\frac{N_0}{C} \right) (2K - \tfrac{4}{5}) f_n. \tag{10.165}$$

$\therefore$ required value of C/N_0 is:

$$\frac{C}{N_0} = \frac{n^2 (2K - 0.8) f_n}{\overline{\phi^2}_{error}} \tag{10.166}$$

But $\overline{\phi^2}$ error must not exceed 0.0225 rads^2

$$\therefore \frac{C}{N_0} = \frac{n^2 (2K - 0.8) f_n}{0.0225} \tag{10.167}$$

A damping factor greater than unity is often chosen for carrier recovery loops. 10 Figure 10 is a plot of the unlocking sensitivity versus offset frequency for $\zeta = 1.33$ and a white noise input.

This has been plotted using equations (10.159 and 10.162).

(b) A Specific Example
Assume a system with the following parameters:

$$f_n = 20 \text{ Hz}$$

$$n = 2 \text{ (a PRK System)} \tag{10.168}$$

$$\zeta = 1.33$$

Reliable acquisition at $C/N_0 = 42$ dB Hz*. Unlocking probability 1×10^{-12}.

Substituting into equation (10.162) to obtain K and then substituting all values, except the starred item, in equation (10.167).

$$\frac{C}{N_0} = \frac{4(5 - 0.8)20}{0.0225}$$

$$\therefore \quad \frac{C}{N_0} = 41.8 \text{ dB Hz say 42 dB Hz.} \tag{10.169}$$

This is the value of input carrier to noise density required to achieve an unlocking probability of 1×10^{-12} with this carrier recovery loop. In fact acquisition requires 42 dB C/N so that the system will operate at this level.

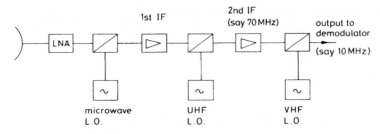

Simplified Block Diagram of a PSK Receiver **10 Figure 11**

(c) Local Oscillator Phase Noise Input
The phase noise contributions of all receiver local oscillators are additive. We will assume a microwave local oscillator, followed by a UHF local oscillator, followed by a VHF local oscillator to convert to the low IF input frequency required by the carrier recovery circuits. (See 10 Figure 11.) In a competent design the major phase noise contributions will be from the microwave LO. Having calculated the total permissible phase noise contributions for receiver local oscillators it is normally necessary to allocate a suitable proportion of this to each local oscillator in the phase noise budget.

If the combined local oscillator phase noise characteristics were white then their specification would be a very simple matter: it would consist merely of specifying the required number of dB below the phase noise contribution due to thermal noise.

From 10 Figure 10 it is apparent that the requirement is particularly critical at offset frequencies in the neighbourhood of f_n. If the *shape* of the (N_{op}/C) curve of the combined local oscillator contributions, is known over the important offset frequency range then the permissible values both of $(N_{op}/C)_f$ and the integrated phase jitter over critical parts of the band may be determined.

Fortunately, information as to the shape of the $(N_{op}/C)_f$ curve is relatively easy to establish in most cases. The majority of satellite ground stations derive all their local oscillators from a single stable (low close to carrier phase noise) 5 MHz Master Oscillator. Very close to the carrier both theory and measurement agree in predicting a $1/f^3$ law.

That is:

$$\left(\frac{N_{op}}{C}\right)_f = \left(\frac{N_{op}}{C}\right)_{f_c} \left(\frac{f_c}{f}\right)^3 \tag{10.170}$$

where f_c is the transition frequency, that is the highest offset frequency at which the $1/f^3$ law applies.

For very high quality MOs this law applies up to an offset frequency of 20 Hz (i.e. $f_c \simeq 20$ Hz) or very slightly greater. At higher offset frequencies one would expect a slope of $1/f^2$, $1/f$ or white phase noise, depending on details of the design. The most pessimistic assumption is that the phase noise is white above this transition frequency of approximately 20 Hz up to a considerable higher frequency, such as 100 Hz–200 Hz approximately, which represents the f_n of a phase lock loop of the synthesiser or PLO. Above this frequency the multiplied phase noise of the MO is not allowed to determine the overall phase noise performance of the local oscillator. Thus we shall make the following very slightly conservative assumptions in assessing the LO phase noise requirements as determined by the carrier recovery loop.

$$f_c = 20 \text{ Hz} = f_n$$

$$\left. \begin{array}{l} \left(\dfrac{N_{op}}{C}\right)_f = \left(\dfrac{N_{op}}{C}\right)_{f_n} \left(\dfrac{f_n}{f}\right)^3 : f < f_n \\[4mm] \left(\dfrac{N_{op}}{C}\right)_f = \left(\dfrac{N_{op}}{C}\right)_{f_n} : f > f_n \end{array} \right\} \tag{10.171}$$

Substituting (10.171) in (10.163) we get the 'error' of the carrier recovery loop which contributes towards unlocking and is due to LO noise alone:

$$\overline{\phi^2}_{\text{error LO}} = n^2 \int_0^{f_n} \left(\frac{2N_{op}}{C}\right)_{f_n} \left(\frac{f_n}{f}\right)^3 \left(\frac{f}{f_n}\right)^4 df$$

$$+ n^2 \int_{f_n}^{Kf_n} \left(\frac{2N_{op}}{C}\right)_{f_n} df + n^2 \int_{Kf_n}^{\infty} \left(\frac{2N_{op}}{C}\right)_{f_n} \left(\frac{Kf_n}{f}\right)^2 df \tag{10.172}$$

$$\therefore \quad \overline{\phi^2}_{\text{error LO}} = n^2 \left(\frac{2N_{op}}{C}\right)_{f_n} \left\{ \left[\frac{f^2}{2f_n}\right]_0^{f_n} + (K-1)f_n + \left[\frac{K^2 f_n^2}{f}\right]_{\infty}^{Kf_n} \right\} \tag{10.173}$$

$$\therefore \quad \overline{\phi^2}_{\text{error LO}} = n^2 \left(\frac{2N_{op}}{C}\right)_{f_n} \left[\frac{f_n}{2} + (K-1)f_n + Kf_n\right]$$

$$\overline{\phi^2}_{\text{error LO}} = n^2 \left(\frac{2N_{op}}{C}\right)_{f_n} (2K - 0{\cdot}5)f_n \tag{10.174}$$

If this is required to be $1/x$ times the contribution due to thermal noise (see equation 10.165) then:

$$\frac{n^2}{x}\left(\frac{N_0}{C}\right)(2K - 0{\cdot}8)f_n = n^2\left(\frac{2N_{op}}{C}\right)_{f_n}(2K - 0{\cdot}5)f_n \tag{10.175}$$

$$\therefore \quad \left(\frac{N_{op}}{C}\right)_{f_n} = \left(\frac{N_0}{C}\right)\frac{(2K - 0{\cdot}8)}{2x(2K - 0{\cdot}5)} \tag{10.176}$$

For $\zeta = 1\cdot33$: $K = 2\cdot5$

and $\quad\left(\dfrac{N_0}{C}\right) = -42$ dB/Hz

Then

$$\left(\frac{N_{op}}{C}\right)_{f_n} = [-42 - 3 + 10 \log 4\cdot2 - 10 \log 4\cdot5 - 10 \log x] \text{ dB/Hz} \quad (10.177)$$

$$= [-45\cdot3 - 10 \log x] \text{ dB/Hz} \quad (10.178)$$

If $10 \log x$ is set at 20 dB then:

$$\left(\frac{N_{op}}{C}\right)_{f_n} = -65\cdot3 \text{ dB/Hz} \quad (10.179)$$

It will be noted that $(x + 3)$ dB below N_0/C thermal is a very simple and apparently reasonable approximation to the requirement. For an equally simple but more accurate approximation see equation (10.180) below.

A graph of the required LO characteristics over the offset frequency range 1 Hz to 100 Hz is given as 10 Figure 12.

It should also be noted that the straight line approximations used for $|H(jp)|^2$ and $|1 - H(jp)|^2$ are about 1 dB optimistic for $\zeta > 0\cdot707$. For $\zeta < 0\cdot707$ they may be considerably more optimistic. Thus an approximate result with a greater degree of accuracy is:

$$\left(\frac{N_{op}}{C}\right)_{f_n} \text{(dB/Hz)} = \left(\frac{N_0}{C}\right) \text{dB/Hz} - (x + 4\cdot3) \text{ dB} \quad (10.180)$$

10.5.3 Signal Error Rate (Coherent Demodulation)

Perusal of 10 Figure 9 shows that the two DSB phase noise density to carrier ratios of inputs to the coherent phase detector are:

$$\overline{\phi_{01}^2}(f) = \left[\left(\frac{N_0}{C}\right)_T + \left(\frac{2N_{op}}{C}\right)^L\right] |H(jw)_{\text{IF}}|^2 \, |H(jp)_{\text{CRL}}|^2 \quad (10.181)$$

and

$$\overline{\phi_{02}^2}(f) = \left[\left(\frac{N_0}{C}\right)_T + \left(\frac{2N_{op}}{C}\right)^L\right] |H(jw)_{\text{IF}}|^2 \quad (10.182)$$

where $[H(jp)_{\text{CRL}}]$ is the transfer function of the carrier recovery loop.

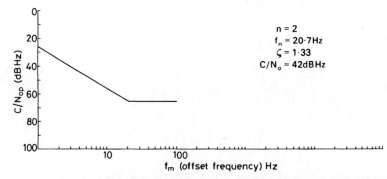

Local Oscillator Phase Noise Specification as **10 Figure 12**
Determined by a Carrier Recovery Circuit

The voltage output from the phase detector will be proportional to the phase difference (say ψ) and will be:

$$V_{ob} = K_d \psi$$

where K_d is the phase detector gain constant in volts per radian.

Thus the output voltage from the phase detector is given by:

$$\frac{V_{ob}}{K_d} = \left[\left(\frac{N_0}{C} \right)_T + \left(\frac{2N_{op}}{C} \right)_f^L \right]^{1/2} [H(jw)_{\text{IF}}] \, | \, 1 - H(jp)_{\text{CRL}} | \qquad (10.183)$$

Now over the significant signal bandwidth $| H(jw)_{\text{IF}} |$ is likely to be constant: without loss of generality we shall put it equal to unity.

$| 1 - H(jp)_{\text{CRL}} |$ is the modulus of the error response of the carrier recovery loop.

$$\therefore \quad \frac{V_{ob}^2}{K_d^2} = \left[\left(\frac{N_0}{C} \right)_T + \left(\frac{2N_{op}}{C} \right)_f^L \right] | 1 - H(jp)_{\text{CRL}} |^2 \qquad (10.184)$$

This will be subject to filtering by a baseband filter or its integrate and dump equivalent. We are interested in the relative outputs due to thermal noise and local oscillator noise and we shall assume a rectangular baseband filter of bandwidth b, where b is the bit rate. [In fact the 3 dB bandwidth would be $b/2$ and the noise bandwidth $(\pi/2)(b/2)$ if all the Nyquist shaping were concentrated here.]

The integrated output power/carrier ratio is given by:

$$\frac{V_b^2}{K_d^2} = \int_0^b \left[\left(\frac{N_0}{C} \right)_T + \left(\frac{2N_{op}}{C} \right)_f^L \right] | 1 - H(jp)_{\text{CRL}} |^2 \, df \qquad (10.185)$$

Now

$$|1 - H(jp)_{CRL}|^2 \simeq \left(\frac{f}{f_n}\right)^4 : f < f_n$$

$$\simeq 1 : f > f_n$$

$$\therefore \frac{V_b^2}{K_d^2} = \int_0^{f_n} \left[\left(\frac{N_0}{C}\right)_T + \left(\frac{2N_{op}}{C}\right)_f^L\right]\left(\frac{f}{f_n}\right)^4 df$$

$$+ \int_{f_n}^b \left[\left(\frac{N_0}{C}\right)_T + \left(\frac{2N_{op}}{C}\right)_f^L\right] df \qquad (10.186)$$

The output due to thermal noise, which is white over the band, is:

$$\frac{V_{bT}^2}{K_d^2} = \left(\frac{N_0}{C}\right)_T \left(\frac{f_n}{5} + b - f_n\right)$$

$$\frac{V_{bT}^2}{K_d^2} = \left(\frac{N_0}{C}\right)_T (b - \tfrac{4}{5}f_n) \qquad (10.187)$$

Thus we have reduced the noise bandwidth from b to $(b - \tfrac{4}{5}f_n)$. This reduction is due to the fact that the carrier reference used for coherent demodulation has similar noise characteristics to the input signal up to an offset frequency of approximately f_n.

Before assessing receiver local oscillator phase noise requirements we need to determine the value of $(N_0/C)_T$ in (10.187).

For coherent demodulation of PSK signals, assuming zero LO noise, the probability of error is given by:

$$P_e = \tfrac{1}{2} \operatorname{erfc} \left(\frac{E}{N_0}\right)^{1/2} \qquad (10.188)$$

where erfc is the Complementary Error Function, which is tabulated.

If a Texas TI 58/59 calculator is available this may be more conveniently written:

$$P_e = Q(x) \text{ where } x = \left(\frac{2E}{N_0}\right)^{1/2} \qquad (10.189)$$

Note: $(2E/N_0)^{1/2}$ is analogous to peak signal voltage to rms noise voltage density.

This may be evaluated using Module 1 Program 14. Assume a bit error rate of 1×10^{-5} to be required. Then

$$\frac{E}{N_0} = 9 \cdot 1 = 9 \cdot 6 \text{ dB} \qquad (10.190)$$

$$\left(\frac{C}{N_0}\right)_T = \frac{Eb}{N_0} \qquad (10.191)$$

As an example, consider a signal with a bit rate:

$$b = 2 \cdot 4 \text{ kb/s} = 33 \cdot 8 \text{ dB relative to 1 b/s.}$$

$$\therefore \quad \left(\frac{C}{N_0}\right)_T = (9 \cdot 6 + 33 \cdot 8) \text{ dB} = 43 \cdot 4 \text{ dB Hz} \tag{10.192}$$

This is a theoretical value. Due to imperfections in a real demodulator, such as, intersymbol interference, imperfect carrier recovery and imperfect clock recovery, there will be an implementation margin. Let the implementation margin be 2·6 dB.

Then required minimum value of

$$(C/N_0)_T = 46 \text{ dB Hz} \tag{10.193}$$

Substitute in (10.187), assuming $f_n = 20$ Hz:

$$\frac{V_{bT}^2}{K_d^2} = [-46 + 10 \log_{10}(2400 - 16)] \text{ dB rads}^2 \tag{10.194}$$

$$\therefore \quad \frac{V_{bT}^2}{K_d^2} = -12.2 \text{ dB rads}^2 \tag{10.195}$$

It is no coincidence that this is (actually approximately) equal to the practical value of $N_0/E = -(9 \cdot 6 + 2 \cdot 6)$ dB.

The difference from exact equality is the term $(-\frac{4}{3}f_n)$ in (10.187) which has been eliminated by rounding off errors.

Reverting to equation (10.186), the output power from the phase detector, with both thermal and LO phase noise contributions, is proportional to the variance of the relative phase jitter between the signal and the restored carrier used as local oscillator. The contribution due to the phase noise characteristics of local oscillators earlier in the receiver chain is given by:

$$\frac{V_{bL}^2}{K_d^2} = \int_0^{f_n} \left(\frac{2N_{op}}{C}\right)_f^L \left(\frac{f}{f_n}\right)^4 df + \int_{f_n}^b \left(\frac{2N_{op}}{C}\right)_f^L df \tag{10.196}$$

Assuming conservatively (see equation 10.171):

$$\left(\frac{N_{op}}{C}\right)_f = \left(\frac{N_{op}}{C}\right)_{f_n} \left(\frac{f_n}{f}\right)^3 \quad \text{for } f < f_n$$

and

$$\left(\frac{N_{op}}{C}\right)_f = \left(\frac{N_{op}}{C}\right)_{f_n} \quad \text{for } f > f_n$$

$$\frac{V_{bL}^2}{K_d^2} = \left(\frac{2N_{op}}{C}\right)_{f_n}^L \left[\int_0^{f_n} \left(\frac{f_n}{f}\right)^3 \left(\frac{f}{f_n}\right)^4 df + \int_{f_n}^b df\right] \tag{10.197}$$

$$\frac{V_{bL}^2}{K_d^2} = \left(\frac{2N_{op}}{C}\right)_{f_n}^L \left\{\left[\frac{f^2}{2f_n}\right]_0^{f_n} + b - f_n\right\}$$

$$\frac{V_{bL}^2}{K_d^2} = \left(\frac{2N_{op}}{C}\right)_{f_n}^L \left(b - \frac{f_n}{2}\right) \tag{10.198}$$

The ratio of the contributions due to thermal noise and to local oscillator phase noise, is obtained by dividing (10.187) by (10.198).

$$\therefore \quad \frac{V_{bT}^2}{V_{bL}^2} = \frac{\left(\frac{N_0}{C}\right)_T (b - \frac{4}{5}f_n)}{\left(\frac{2N_{op}}{C}\right)_{f_n}^L \left(b - \frac{f_n}{2}\right)} \tag{10.199}$$

Let it be specified that the local oscillator contributions be $10 \log x$ dB below the thermal contributions.

$$\therefore \quad \left(\frac{N_0}{C}\right)_T \left(\frac{(b - \frac{4}{5}f_n)}{b - \frac{f_n}{2}}\right) = 2x\left(\frac{N_{op}}{C}\right)_{f_n}^L \tag{10.200}$$

$$\therefore \quad \left(\frac{N_{op}}{C}\right)_{f_n}^L = \frac{1}{2x}\left(\frac{N_0}{C}\right)_T \left(\frac{b - \frac{4}{5}f_n}{b - \frac{f_n}{2}}\right) \tag{10.201}$$

If $f_n \ll b$

$$\left(\frac{N_{op}}{C}\right)_{f_n}^L \text{ dB} \simeq \left[\left(\frac{N_0}{C}\right)_T \text{ dB} - 10 \log 2x\right] \text{ dB} \tag{10.202}$$

It would appear that the local oscillator phase noise density must be maintained at this value or better over the offset frequency range f_n to b and it may rise as $1/f^3$ below f_n.

The required values of (N_0/C) thermal, adding a 2·6 dB implementation margin to the theoretical figure, are given in 10 Table 2 below, for various bit rates together with the required phase noise density to carrier ratios at f_n and above, assuming the LO contribution to be 20 dB below thermal. In calculating this table it is assumed that the required bit error rate is 1×10^{-5}.

It is immediately apparent that, for high bit rates, column 4 cannot be used to specify the requirement, which would probably not be capable of being achieved at f_n. The real requirement is that the local oscillator integrated phase jitter variance as determined by evaluating equation (10.196) shall be x dB below the similar variance due to white thermal noise as determined by equation (10.187). Putting $x = 20$ dB and evaluating equation (10.200) assuming $(N_{op}/C)_f$ is flat above f_n gives the permissible value for the integrated local oscillator phase jitter variance, which is detailed in column 6. It is over-stringent if it is interpreted on a phase noise density basis.

10 Table 2   *LO Phase Jitter v Bit Rate*

1	2	3	4	5	6	7
Bit rate	E/N_0 dB Hz	$(C/N_0)_T$ dB Hz	$(C/N_{op})_{f_n}$ dB Hz	Offset frequency range	$\overline{\phi^2} = \int_{f_n}^{b} \left(\frac{2N_{op}}{C}\right)_f \, df$	ϕ rms millirads
300 b/s	12·2	37	60	$f_n \to 300$ Hz	6×10^{-4}	24·6
1·2 kb/s	12·2	43	66	$f_n \to 1·2$ kHz	6×10^{-4}	24·6
2·4 kb/s	12·2	46	69	$f_n \to 2·4$ kHz	6×10^{-4}	24·6
4·8 kb/s	12·2	49	72	$f_n \to 4·8$ kHz	6×10^{-4}	24·6
16 kb/s	12·2	54·2	77·2	$f_n \to 16$ kHz	6×10^{-4}	24·6
64 kb/s	12·2	60·3	83·3	$f_n \to 64$ kHz	6×10^{-4}	24·6
1 Mb/s	12·2	72·2	95·2	$f_n \to 1$ MHz	6×10^{-4}	24·6

From the point of view of signal error rate, columns 6 and 7 express the real requirement in alternative ways. This is subject to simultaneously meeting the carrier recovery threshold requirements at offset frequencies close to f_n as discussed previously (see equation 10.180).

The highest bit rate determines the integrated phase jitter variance allowable up to an offset frequency equal to the bit rate. The permissible integrated local oscillator phase jitter variance will usually be partitioned over various sub bands.

The phase noise performance of a good microwave synthesiser designed for this type of application was discussed in Chapter 8 and plotted on 8 Figure 14. Using this as a basis as to what is readily achievable it is possible to calculate the integrated phase jitter over each of the offset frequency bands of column 5 of 10 Table 2 using the TI 58/59 program described in Sections 8.4 and 9.3. Knowing an achievable distribution of (N_{op}/C) over the offset frequency band makes it possible to overcome the difficulty described in Section 9.1 where it was stated that in the general case it is not possible to calculate density characteristics knowing only the integrated value.

Allowing some margins for synthesiser production tolerances an acceptable phase noise density/carrier ratio plot and integrated phase jitter specification, derived from 10 Table 2 and 8 Figure 14 assuming a maximum bit rate of 1 Mb/s is given in 10 Figure 13.

Satcom Link Budget Adjustment

Having derived a phase noise or phase jitter specification, such as that given in 10 Figure 13 or (over part of the band) in 10 Table 2, for each of the five major contributions to the phase noise budget listed in Section 10.3.2(a), it is necessary to include an appropriate allowance for their sum in the system Link Budget. Satcom system link budgets must include margins for weather attenuation and consequential noise temperature degradation, as well as polarisation losses, pointing error losses, interference, phase noise and several other factors. Some of these factors vary with time and, being uncorrelated, may be combined on a 'Root Sum Square' (RSS) basis. Phase noise degradation is closely constant from minute to minute or hour to hour and its rms value adds directly to the required margins.

10.5.4 Clock Recovery

An analysis of the phase noise requirements as decided by the clock recovery loop is relatively straightforward. It will have a 'carrier', that is VCO, frequency of half the bit rate. Phase variations in the clock oscillator at the transmitter will occur very slowly so that the natural frequency of the clock recovery loop can be very low. This will prevent VCO phase changes during a long sequence of digital 0s or digital 1s. In any event it must be very much less than $b/2$. In order to achieve accurate operation, the decision circuit (see 10

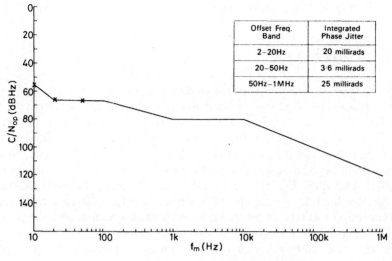

Offset Freq. Band	Integrated Phase Jitter
2–20Hz	20 millirads
20–50Hz	3·6 millirads
50Hz–1MHz	25 millirads

Required Phase Noise Performance　　**10 Figure 13**
(composite curve for all R/X LO contributions)

Figure 9), must operate near the centre of each signal bit, so that the rms phase error must not exceed say $\frac{1}{20}$ of a bit $= \pi/20$ radians at a frequency $b/2$.

The input signal to the clock recovery circuit, although it may contain a considerable amount of unfiltered high frequency noise, will already have been 'cleaned up' at low frequencies by the operation of the combined carrier recovery loop and demodulator.

As an example consider that f_n for the clock recovery PLL is chosen at 1 Hz and $\zeta = 1$. Using equations (10.166) and (10.162) we can find the permissible input noise density for a white noise input to the clock recovery circuits.

$$C/N_0 = \frac{2\cdot7}{\overline{\phi^2}} = \frac{2\cdot7}{\left(\dfrac{\pi}{20}\right)^2} = 20\cdot4 \text{ dB Hz} \tag{10.203}$$

and

$$\frac{C}{N_{op}} = 23\cdot4 \text{ dB Hz} \tag{10.204}$$

This value of C/N_{op} at the input to the clock recovery circuit is compatible with a very much poorer value at the input to the complete demodulator. This is because the carrier recovery loop and demodulator together attenuate phase noise below 20 Hz (the f_n of the carrier recovery loop) by 40 dB per decade.

Even if the phase noise input to the carrier recovery loop were rising by 30 dB per decade below 20 Hz, the resultant at the input to the clock recovery loop would still be falling by 10 dB per decade fall in frequency.

Thus it may be seen that the clock recovery loop phase noise requirements are entirely non-critical relative to those demanded in the low offset frequency region by the carrier recovery loop and in the higher offset frequency region by the coherent demodulator.

10.5.5 Differential Detection of PSK Signals

For the transmission of low bit rate signals, *coherent* demodulation would necessitate the use of a carrier recovery loop with a very low value of f_n. For example, for 50 b/s the natural frequency of the carrier recovery loop would have to be below approximately 2 Hz. This would impose very severe, and in fact unrealistically tight, requirements on the LO phase noise in the region of 2 Hz offset frequency. For this reason, differential demodulation is normally used for low bit rate signals and the sacrifice in basic efficiency consequent upon the use of this type of detection, is accepted as a reasonable price to pay, for the avoidance of an impossible phase noise specification.

The error rate of a PSK system using differential demodulation is given by:

$$P_e = \tfrac{1}{2}e^{-E/N_0} \tag{10.205}$$

for

$$P_e = 1 \times 10^{-5}$$

$$\frac{E}{N_0} = 10\cdot34 \text{ dB} \tag{10.206}$$

[Note the loss in efficiency relative to (10.190) for this error rate.] For a bit rate of 50 b/s:

$$\frac{C}{N_0} = \frac{Eb}{N_0} = 27\cdot3 \text{ dB Hz} \tag{10.207}$$

This is a theoretical figure.

Adding an implementation margin (say 2·6 dB) to allow for a non-matched filter and intersymbol interference we obtain the minimum required thermal carrier to noise density ratio.

$$\left(\frac{C}{N_0}\right)_{\text{thermal}} = 29\cdot9 \text{ dB Hz} \tag{10.208}$$

$$\therefore \quad \left(\frac{2N_{op}}{C}\right)_{\text{thermal}} = -29\cdot9 \text{ dB/Hz} \tag{10.209}$$

If we integrate this over the 'relevant' offset frequency band, we shall find the total permissible integrated phase jitter over this band.

A simplified block diagram of a differential demodulator is given in 10 Figure 14.

Assuming initially that the signal is accompanied by white noise, the effect of phase noise integrated over a given band will be dependent on the relative

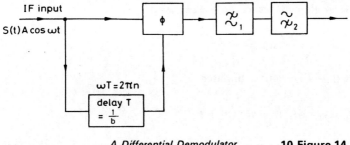

A Differential Demodulator **10 Figure 14**

sensitivity of the demodulator to phase noise at different offset frequencies. We must thus evaluate this sensitivity characteristic.

An an example assume that:

Bit rate $(b) = 50$ b/s

∴ T (the time for 1 bit) $= \frac{1}{50}$ sec

f_1 the basic baseband frequency $= 1/2T = 25$ Hz

Modulation—biphase $\left[s(t) = \pm \frac{\pi}{2} \right]$ (or 0 and π)

Input carrier angular frequency (IF) $= w$

offset angular frequency $= p$

It is likely that the baseband filter will be selected with a bandwidth b as a reasonable compromise between intersymbol interference and noise suppression.

Alternatively there is a possible implementation using integrate and dump filters at IF with a bandwidth of $\pm b$. (See Reference 44.) Thus as a fair approximation we may consider noise above 50 Hz to be attenuated by 20 dB per decade.

The noise in a 1 Hz bandwidth at an offset frequency f will be represented in the well known quadrature form (see Appendix III). We shall write an equation for the input signal plus the noise in a 1 Hz bandwidth. For reasonable signal to noise ratios the full signal plus noise may then be obtained by integrating the noise terms over the full offset frequency band, relying on the linearity applicable with low modulation indices ($\theta < 0.1$ radians).

∴ Consider the input at time t.

$$V_t = \pm \sqrt{2C} \cos wt + \sqrt{2N_0} \cos (pt + \psi) \cos wt$$
$$- \sqrt{2N_0} \sin (pt + \psi) \sin wt \qquad (10.210)$$

The phase angle with respect to the carrier is:

$$\theta_{ot} = \mp \sqrt{\frac{N_0}{C}} \, \sin \, (pt + \psi) \tag{10.211}$$

At a time $(t + T)$ exactly 1 bit later

$$\theta_{0(t+T)} = \mp \sqrt{\frac{N_0}{C}} \, \sin \, (pt + pT + \psi) \tag{10.212}$$

If the carrier phase angle is identical at $t = t$ and $t = t + T$ (that is if $w/2\pi$ is an exact multiple of $2f_1 = 1/T$) the phase difference between the two inputs to the phase detector is:

$$\Delta\theta_0 = \mp \sqrt{\frac{N_0}{C}} \, [\sin \, (pt + pT + \psi) - \sin \, (pt + \psi)] \tag{10.213}$$

[(10.213) gives two typical values out of four possibilities]. Now

$$\sin A - \sin B = 2 \cos \frac{A + B}{2} \, \sin \frac{A - B}{2}$$

$$\therefore \quad \Delta\theta_0 = \mp \sqrt{\frac{N_0}{C}} \left[2 \cos \frac{(2pt + 2\psi + pT)}{2} \, \sin \frac{pT}{2} \right] \tag{10.214}$$

Now the cosine term cannot exceed unity

$$\therefore \quad \Delta\theta_0 \not> \mp \sqrt{\frac{N_0}{C}} \, 2 \sin \frac{pT}{2} \tag{10.215}$$

The resultant sensitivity of such a differential phase detector to noise at different offset frequencies has been plotted in 10 Figure 15.

It will be noted that the most sensitive frequency is 25 Hz where there is a magnification of 6 dB, due to the factor '2' in equation (10.215).

$$\Delta\phi_0(f) = \mp \frac{1}{\sqrt{2}} \left(\frac{N_0}{C} \right)_f^{1/2} 2 \sin \frac{pT}{2} \tag{10.216}$$

The noise induced phase jitter is symmetrical for a signal of 0 or $-\pi$.

We may therefore drop the $\pm$ sign in determining the effects of noise and consider the effects on only the signal where the phase is 0.

$$\overline{\Delta\phi_0^2(f)} = 2 \left(\frac{N_0}{C} \right)_f \sin^2 \left(\frac{pT}{2} \right) \tag{10.217}$$

We may neglect noise above $2f_1$ as it will be rejected by filtering.

$$\overline{\Delta\phi^2} = \int_0^{2f_1} 2 \left(\frac{N_0}{C} \right)_f \sin^2 \, (\pi f T) \, df \tag{10.218}$$

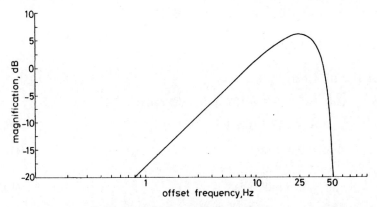

Differential PSK Detection Sensitivity to **10 Figure 15**
Phase Noise (bit rate = 50 b/s)

For white noise

$$\left(\frac{N_0}{C}\right)_f = \frac{N_0}{C}$$

For this case

$$\overline{\Delta\phi^2} = 2\left(\frac{N_0}{C}\right) \int_0^{2f_1} \sin^2\left(\pi f\,T\right)\,df \tag{10.219}$$

$$\int \sin^2 x\,dx = \left[\tfrac{1}{2}(x - \sin x \cos x)\right]$$

$$\therefore \quad \overline{\Delta\phi^2} = 2\left(\frac{N_0}{C}\right)\frac{1}{2\pi T}\left[\pi f\,T - \sin \pi f\,T \cos \pi f\,T\right]_0^{2f_1}$$

$$= \frac{N_0}{C}\frac{1}{\pi T}\left[2\pi f_1 T - \sin 2\pi f_1 T \cos 2\pi f_1 T\right]$$

Now $f_1 = 1/2T$

$$\therefore \quad \overline{\Delta\phi^2} = \frac{N_0}{C}\frac{1}{\pi T}\left[\pi - \sin \pi \cos \pi\right]$$

$$\therefore \quad \overline{\Delta\phi^2} = \frac{N_0}{C}\frac{1}{T} = \frac{N_0}{C}2f_1 \tag{10.220}$$

$$\therefore \quad \text{effective noise bandwidth} = 2f_1 = 50 \text{ Hz.}$$

Thus to a white noise input the overall integrated phase jitter variance over the band $0–2f_1$ may be calculated as *though* the noise sensitivity of the demodulator were flat from $0–2f_1$ although in fact it has a sine squared dependence on f.

Now for $P_e = 1 \times 10^{-5}$ (see equation 10.208)

$$\frac{N_0}{C} = -29\cdot9 \text{ dB/Hz}$$

$\therefore$ for 1×10^{-5} ber

$$\overline{\Delta\phi^2} = [-29\cdot9 + 10 \log 50] \text{ dB rads}^2$$

$\therefore$ $\overline{\Delta\phi^2} = -12\cdot9 \text{ dB rads}^2$ $\qquad\qquad$ (10.221)

$\therefore$ $\overline{\Delta\phi^2} = 0\cdot051 \text{ rads}^2$ $\qquad\qquad$ (10.222)

and

$$\Delta\phi = 226 \text{ millirads.} \qquad\qquad (10.223)$$

Consider now an input of non-white noise due to LO phase noise. For a typical master oscillator, phase noise density will rise as $(f_c/f)^3$ below 20 to 25 Hz. Assume, only a little conservatively, but very conveniently, that $f_c = 25$ Hz $= f_1 = 1/2T$.

Thus

$$\left(\frac{N_{op}}{C}\right)_f = \left(\frac{N_{op}}{C}\right)_{f_1}\left(\frac{f_1}{f}\right)^3 : f < f_1 \qquad\qquad (10.224)$$

assume, also a little conservatively, that

$$\left(\frac{N_{op}}{C}\right)_f = \left(\frac{N_{op}}{C}\right)_{f_1} : f > f_1 \qquad\qquad (10.225)$$

From (10.217) writing $N_0 = 2N_{op}$

$$\overline{\Delta\phi_0^2}(f) = 2\left(\frac{2N_{op}}{C}\right)_f \sin^2(\pi f T) \qquad\qquad (10.226)$$

$\therefore$ integrating in two separate segments from 0 to f_1 and f_1 to $2f_1$

$$\overline{\Delta\phi^2} = \int_0^{f_1} 4\left(\frac{N_{op}}{C}\right)_{f_1}\left(\frac{f_1}{f}\right)^3 \sin^2(\pi f T) \, df$$

$$+ \int_{f_1}^{2f_1} 4\left(\frac{N_{op}}{C}\right)_{f_1} \sin^2(\pi f T) \, df \qquad\qquad (10.227)$$

$$= 4\left(\frac{N_{op}}{C}\right)_{f_1}$$

$$\times \left\{\int_0^{f_1}\left(\frac{f_1}{f}\right)^3 \sin^2(\pi f T) \, df + \int_{f_1}^{2f_1} \sin^2(\pi f T) \, df\right\} \qquad (10.228)$$

The second integral (with integration limits of 0 to $2f_1$) has already been evaluated (see 10.218 and 10.220). It is symmetrical about f_1 and its value from

f_1 to $2f_1$ is half its value from 0 to $2f_1$.

$$\therefore \quad \overline{\Delta\phi^2}(f_1 \rightarrow 2f_1) = 2\left(\frac{N_{op}}{C}\right)_{f_1} f_1 \tag{10.229}$$

Using 10 Figure 15 we will approximate for the sine squared term in the first integral:

substituting $\left(\frac{f}{f_1}\right)^2$ $\tag{10.230}$

$$\overline{\Delta\phi^2}_{(0 \rightarrow f_1)} = 4\left(\frac{N_{op}}{C}\right)_{f_1}\left[\int_0^{f_1} \left(\frac{f_1}{f}\right)^3 \left(\frac{f}{f_1}\right)^2\right] df \tag{10.231}$$

$$\therefore \quad \overline{\Delta\phi^2}_{0 \rightarrow f_1)} = 4\left(\frac{N_{op}}{C}\right)_{f_1} f_1 [\log_e f]_0^{f_1} \tag{10.232}$$

Thus the integral (10.231) is not finite, if taken down to zero offset frequency. However the 'fundamental frequency' of the demodulated signal, assuming a regular sequence of 'one' $\rightarrow$ 'zero', 'one' $\rightarrow$ zero' etc, is 25 Hz. Even for a random sequence there will be negligible signal energy below (say) 2·5 Hz. Without loss in signal performance it is possible to incorporate a high pass filter (filter 2 in 10 Figure 14) in the output line of the demodulator. This filter might have a cut off frequency of (say) 2·5 Hz and might be of a multipole type resulting in a negligible transfer function below 1 Hz.

Hence a conservative approach is to evaluate equation (10.231) between 1 Hz and f_1.

$$\therefore \quad \overline{\Delta\phi^2}(1 \rightarrow 25) = 4\left(\frac{N_{op}}{C}\right)_{25} 25(3\cdot2 - 0) = 320\left(\frac{N_{op}}{C}\right)_{25} \tag{10.233}$$

$$\therefore \quad \overline{\Delta\phi^2}(1 \rightarrow 50) = 50\left(\frac{N_{op}}{C}\right)_{25} + 320\left(\frac{N_{op}}{C}\right)_{25} \tag{10.234}$$

$$\therefore \quad \left(\frac{N_{op}}{C}\right)_{25} = \frac{\overline{\Delta\phi^2}(1 \rightarrow 50)}{370} \tag{10.235}$$

For a ber of 1×10^{-5} [see (10.222)]

$$\overline{\Delta\phi^2}(1 \rightarrow 50) \simeq 0\cdot051 \text{ rads}^2 \tag{10.236}$$

$\therefore$ to be equal to thermal noise

$$\left(\frac{N_{op}}{C}\right)_{25} = 0\cdot00014 = -38\cdot6 \text{ dB/Hz} \tag{10.237}$$

$\therefore$ to be 20 dB below thermal

$$\left(\frac{N_{op}}{C}\right)_{25} = -58\cdot6 \text{ dB/Hz} \tag{10.238}$$

This is somewhat less severe than the carrier recovery loop requirement given by equation (10.179) for a system using coherent demodulation and a very much higher minimum bit rate (300 b/s). Coherent demodulation of a signal with a bit rate of 50 b/s would necessitate the choice of a very low natural frequency for the carrier recovery loop such as 3 Hz. The phase noise performance required of the local oscillators at *3 Hz* offset frequency might then be calculated from (10.180). For the same value of (N_0/C) this would give a requirement for $(N_{op}/C)_3$ numerically identical to $(N_{op}/C)_{20}$ given by equation (10.179) for a system operating at 300 b/s. As the local oscillator will almost certainly have a phase noise characteristic, in this offset frequency region, given by:

$$\left(\frac{N_{op}}{C}\right)_3 = \left(\frac{N_{op}}{C}\right)_{20}\left(\frac{20}{3}\right)^3 \qquad (10.239)$$

$$\therefore \quad \left(\frac{N_{op}}{C}\right)_3 = 296\left(\frac{N_{op}}{C}\right)_{20} \qquad (10.240)$$

Thus, for an LO to achieve the same phase noise performance at 3 Hz offset frequency as another LO at 20 Hz offset frequency implies that the former has a phase noise density characteristic 24·7 dB better than the latter. To achieve

$$\left(\frac{N_{op}}{C}\right)_3 = -65\cdot3 \text{ dB/Hz} \qquad (10.241)$$

would be exceedingly difficult (if not impossible) in the present state of the art, for a microwave local oscillator.

The great advantage of differential demodulation is that, at a price of less than 1 dB in required (C/N_0) thermal, it greatly relaxes signal source phase noise requirements for low bit rate digital communication systems.

10.5.6 The Effects of Phase Transients

In some synthesisers design constraints such as the required Q or the necessity to maintain a low oscillator tuning sensitivity K_0 (radians per sec per volt), result in a rather narrow frequency range being available if only voltage tuning is used on a UHF or SHF VCO. In such a case it may be necessary to supplement the voltage tuning range by the provision of mechanical tuning of the VCO cavity. If this is done using sliding contacts, random step function changes in frequency will occur (see Section 5.9.5).

Before considering the resultant deterioration in the ber of a digital communication system it is necessary to determine the nature of the effect as seen at the synthesiser output. Assume the faulty VCO to form part of a PLL as might be expected to be the case. For a frequency step input Δf, the phase error of a high gain second order loop with damping factor $\zeta < 1$ is given in

Reference 24 p. 33 as follows:

$$\theta_e = \frac{\Delta f}{f_n} \left(\frac{1}{\sqrt{1-\zeta^2}} \sin \sqrt{1-\zeta^2}\, w_n t \right) \exp - \zeta w_n t \tag{10.242}$$

for $\zeta = 0.707$: $\sqrt{1-\zeta^2} = 0.707$ and $1/\sqrt{1-\zeta^2} = 1.414$

$$\therefore \quad \theta_e = \frac{\Delta f}{f_n} [1.414 \sin (0.707 w_n t)] \exp - \zeta w_n t \tag{10.243}$$

This is a damped sinusoid with a frequency of $0.707 f_n$ and peak amplitude $1.414\, \Delta f/f_n$. If the VCO is followed by a well designed frequency multiplier

$$\theta'_e = 1.414 n \frac{\Delta f}{f_n} \tag{10.244}$$

where n is the frequency multiplication ratio.

Except in the case of a very unconventional synthesiser design, it is unlikely that the natural frequency of a PLL containing a UHF or SHF VCO will be less than a few kHz and it may be much greater.

If a UHF or microwave frequency multiplier incorporates poor pressure contacts anywhere in its RF high current path (including earth returns) then either or both of two effects may occur. Firstly there may be random phase jumps due to the multiplier itself. Secondly, and usually much more important, random changes in the phase of the load seen by the VCO feeding the multiplier will produce random steps in the VCO frequency. These frequency steps will be shaped by the PLL and then amplified by the multiplier (see equation 10.244). The cure in this case is to eliminate all pressure contacts and to provide an isolator between the VCO and the frequency multiplier.

The three potential deleterious effects on a digital communications system are as follows:

(a) Effects on the carrier recovery loop
(b) Effects on the clock recovery circuits. These effects may be either a tendency to unlocking or a tendency to follow the transient.
(c) Effects on the ber, either directly, or to choice of the 'wrong' sampling time due to clock phase jitter.

a. The Carrier Recovery Loop

For low bit rate PSK systems the carrier recovery loop will be designed to have a very low natural frequency. Perusal of 10 Figure 10 shows that even an undamped spurious phase modulation of the carrier at an offset frequency *much* greater than the natural frequency of the carrier recovery loop will be attenuated by 20 dB per decade above (say) $2.5 f_n$, as far as its tendency to produce loop unlocking is concerned. Thus except for high bit rate systems only very large phase transients will tend to produce unlocking of the carrier recovery loop. A conservative approach is to assume the phase transient to be an undamped sinusoid (of the measured or calculated amplitude) and to use the methods of Section 10.5.2 to assess its effect on the carrier recovery loop. If the effect is negligible, as will normally be the case unless the phase transients are very large, it will not be necessary to consider the further reduction due to the waveform damping factor [see the exponential term in equation (10.243)].

b. Effects on the Clock Recovery Circuits

The methods of analysis to be used follow those described in Section 10.5.4. If initially, and conservatively, we neglect the damping of the phase transient, we have two cases to consider, low bit rate systems and high bit rate systems. The natural frequency of the clock recovery PLL must be very low compared with the bit rate. Thus for low bit rate systems the transient will have to be very large indeed to affect the clock recovery loop in any way at all. Thus the clock oscillator should remain operative at its long term value, unaffected by the transient.

In the case of high bit rate systems the natural frequency of the clock recovery loop *might* be chosen at several kHz to facilitate initial locking, although this is unusual. The worst case would be when the natural frequency of this loop was approximately equal to the frequency of the transient. In this case a transient with a peak value of $\pi/2$ radians would tend to unlock the clock recovery loop. For smaller transients the clock would tend to follow the transient and thus clock phase jitter would result. In the vast majority of cases the clock will not be affected by the occurrence of a single transient.

c. Effects on the Bit Error Rate

Phase transients due to poor contacts in a synthesiser will normally be of greatest magnitude when they are due to (or affect) the highest frequency VCO. The PLL of which this VCO forms a part, will normally have a natural frequency not less than a few kHz and usually much greater. In this case the carrier and clock recovery circuits (having very low values of f_n) are unlikely to be affected.

Thus the major effect on ber is likely to be directly caused by the spurious phase shift imparted to a series of digits by the LO during each phase transient and of course the number of such transients which occur in the time over which the ber is measured.

For a QPSK system the phase difference between two information bits is $\pi/2$ radians. The decision threshold should therefore be set at $\pi/4$ radians $= 785$ millirads $= 45°$.

Consider the two extreme cases separately,

(i) When the digit duration is long compared with a half cycle of the LO imparted phase transient.

(ii) When the digit duration is very short compared with a half cycle of the phase transient.

In case (i) the unwanted phase transient is varying during a single digit and it is the mean power of the damped sinusoid over a digit period which constitutes interference. This power will obviously be less than the square of the *peak* value of the first half cycle of the damped sinusoid. The degradation of ber will therefore be less than in case (ii).

In case (ii), that of a high bit rate system, the greatest degradation will occur for a digit which coincides in time with a peak of the phase transient. During this time the phase shift due to the transient will add (algebraically) to that due to the signal. The degradation is thus a function of the phase shift (analogous to voltage) rather than to the power of the transient. Consider as an example that a single half cycle of the transient occupies 100 μS and that the digit rate is 10 Mb/s. There will be 1000 digits during the first half cycle of the transient. Only those near the peak of the transient will suffer the full degradation. It thus seems not too unreasonable to assume that the mean effect over the 100 μS period is not greater than that due to the power in the semi-sinusoid. However as a conservative procedure the square of the *peak* value of the transient will be added to the thermal noise power.

In the case of thermal noise [see (3.37)]

$$\overline{\phi_T^2} = \frac{N_o b}{C} \text{ rads}^2$$

But $C = Eb$, where E is the energy per bit and b is the bit rate.

$$\therefore \quad \overline{\phi_T^2} = \frac{N_o b}{Eb} = \frac{N_0}{E}$$

Assume that the system normally operates at a ber of 1×10^{-5}; C/N_0 being maintained at the required value even under the worst conditions of weather attenuation.

For ber $= 1 \times 10^{-5}$

$$\frac{E}{N_0} = 9 \cdot 6 \text{ dB} = 9 \cdot 1 \text{ (see 10.190)}$$

Allowing an implementation margin of 2·6 dB

$$\frac{N_0}{E} = -12 \cdot 2 \text{ dB}$$

$$\therefore \quad \overline{\phi_T^2} = -12 \cdot 2 \text{ dB rads}^2 = 0 \cdot 06 \text{ rads}^2.$$

Various values of θ_e (the peak value of the phase transient) are listed in 10 Table 3. To achieve a semi-worst case the square of this peak value is added to $\overline{\phi_T^2}$ to give a total value of $\overline{\phi^2}$ after adding thermal noise. The effective degradation (dB) is given in line 6. It is assumed that the implementation margin remains constant at 2·6 dB. The effective value of $\overline{\phi^2}$, allowing for this

10 Table 3   *Error Rate as Affected by Phase Transients*

1	$\theta_{e(\text{mrads})}$	200	250	300	331	389	400
2	$\theta_{e(\text{degrees})}$	11·5	14·3	17·2	19	22·3	22·9
3	$\overline{\theta_e^2}$ rads^2	0·04	0·0625	0·09	0·11	0·15	0·16
4	$\overline{\phi_T^2}$ rads^2	0·06	0·06	0·06	0·06	0·06	0·06
5	ϕ_{TOTAL}^2 rads^2	0·1	0·123	0·15	0·17	0·21	0·22
6	Degradation (dB)	2·2	3·11	4·0	4·52	5·46	5·64
7	Assuming Constant Implementation Margin of 2·6 dB						
8	$\overline{\phi^2}_{\text{effective}}$ (rads^2)	0·182	0·224	0·273	0·31	0·384	0·4
9	P_e	$4·6 \times 10^{-4}$	$1·4 \times 10^{-3}$	$3·4 \times 10^{-3}$	$5·5 \times 10^{-3}$	$1·1 \times 10^{-2}$	$1·27 \times 10^{-2}$

effect, is given in line 8. Finally the overall error probability, is listed in each case [for the method used see (10.189)].

Assuming phase transients to occur only rarely (a very doubtful assumption if they occur at all) a peak magnitude of 400 millirads would be likely to produce a very high error rate during the occurrence period.

Consider that the phase transient exceeds 400 millirads for x seconds and that it occurs y times per day.

Thus each day the total time with a high probability of errors $= xy$ seconds. For a bit rate b, errors per day $\ngtr bxy$

Data bits per day $\simeq 10^5 b$ (1 day $= 10^5$ secs very approximately).

$$\therefore \quad \text{max daily error rate} = \frac{bxy}{10^5 b} = \frac{xy}{10^5} \tag{10.245}$$

The analysis of the effects of phase transients on a PSK system could be carried out in much greater detail and with greater accuracy but this is not usually justified, as both the magnitude and the number of occurrences per day or per hour are likely to be very variable. It is sufficient to show, as has been done, that the effect is potentially very serious and that it should be eliminated by the use of non-contact tuning mechanisms and soldered contacts for all UHF and microwave connections, instead of making an attempt precisely to assess its effects.

10.6 Antenna Tracking

It is necessary that the antenna of a satellite ground station should point accurately in the direction of the satellite. In some cases where narrow satellite spot beams are used (e.g. direct satellite TV transmission to the domestic viewer) it is also necessary that satellite antenna pointing be maintained

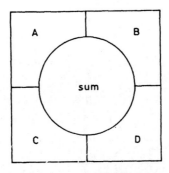

10 Figure 16
A 5-Horn Static Split Feed

correctly to considerably better than 0·1 degrees. For the highest pointing accuracy a position control servo system is used operating on a pointing error signal derived from the received signal or from a specially generated beacon signal.

We shall consider only one form of tracking system: that known as 'static split' or, in radar parlance as 'monopulse'. This system is capable of the most accurate tracking and would therefore be expected to impose the most severe phase noise requirements on any signal sources such as local oscillators.

Conceptually the simplest system to consider, and one with a good performance is a 5-horn static split system. The horn arrangement is illustrated in 10 Figure 16. This horn cluster might illuminate a sub-reflector, which itself illuminates the main reflector, thus forming a Cassegrain antenna.

The large centre horn is used for communication signals and for the 'sum' tracking channel. If microwave signals from the 'error' horns (A, B, C, D) are combined as follows, the error signals generated in the hybrid junctions will be:

$$(A + B) - (C + D) \rightarrow \text{Elevation error}$$

$$(A + C) - (B + D) \rightarrow \text{Azimuth error.}$$

The angular sensitivity of such a tracking antenna is expressed as volts of error signal per degree of pointing error per volt of sum signal. For small pointing errors the error signal will be small compared with the sum signal and its C/N ratio may also be disproportionately worse as the sum channel will be designed to have the minimum possible noise temperature.

Thus, for a good servo S/N ratio, it is usual to adopt coherent detection of the error signals, using the 'sum' signal as a relatively 'clean' reference. For perfect coherent detection the effective bandwidth of the error channel will be twice the post detection bandwidth. (Remember that a coherent detector is just a mixer with a zero frequency IF.) The post detection bandwidth will be the effective bandwidth of the servo system. Unless the antenna pointing position

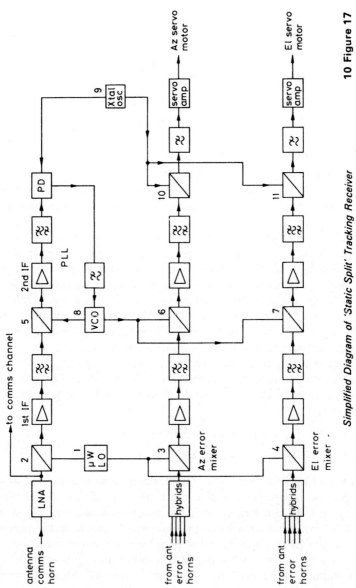

Simplified Diagram of 'Static Split' Tracking Receiver

10 Figure 17

is affected by vibration or by wind gusts, the servo bandwidth will be considerably less than 1 Hz and, even in the presence of wind effects, the trade-off between reduction of wind induced errors and thermal noise errors is likely to result in choosing a bandwidth no greater than 1 Hz.

As the servo bandwidth effectively rejects all noise components above (say) 1 Hz it might be thought that phase noise at lower offset frequencies than 1 Hz, due to any local oscillators in the receiver chain, would have to be held to a very low value if servo performance were not to be affected. A good phase noise performance at very low offset frequencies is always difficult to achieve, particularly in microwave local oscillators such as are required to drive the first mixers of down converters in a satellite ground station.

A simplified block diagram of a tracking receiver is given as 10 Figure 17 (Reference 45). It should be noted that the same microwave LO (1) is used to feed the 'sum' mixer (2) and both error mixers (3 and 4).

As long as there are no large relative delays in the three paths between the microwave LO and the three mixers, LO phase jitter at all three mixers, particularly at low offset frequencies, will be almost identical. The phase jitter added, at first IF, to the 'sum' signal and each of the two error signals, by the mixer action, will thus be practically identical.

Conversion from first to second IF is achieved in three mixers all driven by the same LO (VCO 8). This VCO forms part of a PLL in the sum channel which controls the phase jitter of the 2nd IF to be equal to that of a low noise crystal oscillator (9) operating at the 2nd IF frequency which is low enough (a few MHz) to achieve a very good close to carrier phase noise performance. Due to the use of the same VCO (8) to drive the two error channel mixers (6 and 7), the 2nd IF in both error channels will have the same LO induced phase jitter as the 'sum' channel, which is basically that of the 2nd IF crystal oscillator.

Coherent demodulation of each of the two error signals is carried out in coherent detectors 10 and 11 which are followed by low pass filtering and servo drive amplifiers. Assuming no large relative time delays the two inputs to mixer 10, for example, will have nearly identical local oscillator induced phase jitter, which will not therefore appear at the output. In any case the phases have been so controlled that coherent detectors (10 and 11) are primarily responsive to amplitude variations.

Thus, in spite of the low natural frequency of the servo system, a tracking system of this type does not impose severe close to carrier phase noise requirements on any of the signal sources used in the system and noise components at higher offset frequencies will have no effect, due to the narrow servo bandwidth.

Thus we may conclude that suitably designed tracking systems make less stringent demands on signal source phase noise than might initially be expected.

REFERENCES

1. ROBINS, W. P. 'Communication by geostationary satellites'. GEC Journal of Science & Technology, Vol. 40 No. 1 1973.
2. ROBINS, W. P. and SALTER, M. 'A Communications Satellite System for Europe'. Journal of the British Interplanetary Society, Vol. 26 No. 5 May 1973.
3. POPPER, K. R. 'The Logic of Scientific Discovery'. Hutchinson 1959.
4. AYER, A. J. 'The Central Questions of Philosophy'. p. 26. Pelican 1976.
5. ROBINS, W. P. 'New notation for phase noise and phase jitter'. IEE Proc. Vol. 128 pt. F. No. 3 June 1981.
6. HAFNER, E. 'The Effects of Noise in Oscillators.' Proc. I.E.E.E. Vol. 54. February 1966.
7. BEURLE, R. L. 'Comparison of noise and random frequency and amplitude fluctuations in different types of oscillators'. Proc. IEE, Part B Vol. 103 1956.
8. LEESON, D. B. 'Simple Model of a Feedback Oscillator Noise Spectrum'. Proc. I.E.E.E. Vol. 54. February 1966.
9. SMITH, W. L. and SPENCER, W. J. 'Quartz Crystal Controlled Oscillators'. Bell Telephone Laboratories Final Report, Contract DA 36–039 SC-85373, March 1963.
10. TERMAN, F. E. 'Electronic and Radio Engineering'. McGraw-Hill, Fourth Edition, 1955 pp. 106–108.
11. SARBACHER, R. J. and EDSON, W. A. 'Hyper and Ultra High Frequency Engineering'. Wiley 1946.
12. WHITWELL, A. L. and WILLIAMS, N. 'A New Microwave Technique for determining Noise Spectra at Frequencies Close to Carrier'. Microwave Journal November 1959.
13. WARNER, A. W. 'High Frequency Crystal Units for Primary Frequency Standards'. Proc. I.R.E. September 1952.
14. Austron Inc. 1915 Kramer Lane, Austin Texas (Austron 1120s).
15. Frequency and Time Systems Inc. Tech. Spec. FTS-1200 IPS 13th February 1980.
16. WHALE, A. V. 'A Klystron Oscillator having Low Frequency Modulation Sideband Noise'. 5th International Conference on Microwave Tubes : Paris : September 1964.
17. HINES, M. E. 'Microwave Power Sources Using Varactor Harmonic Generation'. Microwave Journal : April 1963.
18. RUPP, S. 'Phase Locked Signal Source Design'. Microwave Journal June 1974.
19. MAEDA, M., KIMURA, K., and KODERA, H. 'Design and Performance of X-Band Oscillators with GaAs Schottky Gate Field Effect Transistors'. I.E.E.E. Trans. MTT-23 No. 8 August 1975.
20. OHTOMO, M. 'Experimental Evaluation of Noise Parameters in Gunn and Avalanche Oscillators'. I.E.E.E. Trans. MTT 20 No. 7 July 1972.

21. BELL, R. L. 'Klystron Oscillator Noise Theory'. British Journal of Applied Physics Vol. 7 July 1956.
22. ROBINS, W. P. British Patent 1,017,236. 'Improvements in or relating to Doppler Radar Apparatus'.
23. PHYLIP-JONES, G. and ROBINS, W. P. British Patent 975,177. 'Improvements in or relating to Doppler Radar Apparatus'.
24. GARDNER, F. M. 'Phaselock techniques'. Wiley 1966.
25. Texas Instruments. 'The TTL Data Book for Design Engineers'. Second Edition pp. 7–65 and 5–34.
26. Mullard/Signetics TDA 1034 data sheet.
27. Texas Instruments. 'System 74 Designer's manual'. p. 16.
28. Rockland Systems Corporation Frequency Synthesiser Model 5600.
29. ZEPLER, E. E. 'The Technique of Radio Design'. Chapman & Hall 1943.
30. Anti Vibration Mount Characteristics. Barrymount Catalogues.
31. Active Vibration Cancellation. Barrymount Serva–Levl Isolation System.
32. CHAO, L. L. 'Statistics: Methods and Analysis'. McGraw-Hill 1974.
33. RUTMAN, J. 'Characterisation of Phase and Frequency Instabilities in Precision Frequency Sources: Fifteen Years of Progress'. Proc. IEEE Vol. 66 No. 9 September 1978.
34. SHOAF, J. H., HALFORD, D. and RISLEY, A. S. 'Frequency Stability Specification and Measurement'. National Bureau of Standards Technical Note 632.
35. SMITH, H. Marconi Space and Defence Systems Ltd; Technical Note L04/A/MS/05.
36. KENDALL, M. G. 'The Advanced Theory of Statistics'. Vol. 1. Griffin & Co.
37. GOLDMAN, S. 'Frequency Analysis Modulation and Noise'. McGraw-Hill 1948.
38. PEREGRINO, L. and RICCI, D. W. 'Phase Noise Measurement Using a High Resolution Counter with On-Line Data Processing'. Frequency Control Symposium Proceedings 1976 pp. 309–317.
39. RIDENOUR, L. N. 'Radar System Engineering'. Vol. 1 of the MIT Series. McGraw-Hill 1947.
40. WOODWARD, P. M. 'Probability and Information Theory with Applications to Radar'. Pergamon Press 1953–1960.
41. CLEGG, J. E. and THORNE, T. G. 'Doppler Navigation'. Proc. IEE March 1958, Paper No. 2568 R.
42. ISMAIL, M. A. W. 'A Precise New System of F.M. Radar'. Proc. IRE Vol. 44 September 1956.
43. CARLSON, A. B. 'Communication Systems' (Second Edition p. 236). McGraw-Hill 1975.
44. BENNETT, W. R. and DAVEY, J. R. 'Data transmission'. McGraw-Hill 1965.
45. PADLEY, J. S. and ROBINS, W. P. 'Skynet Types III & IV Stations: Technical Description:' IEE Conference Publication Number 72. 'Earth Station Technology'.
46. NYQUIST, H. 'Thermal Agitation of Electric Charge in Conductors'. Phys. Rev. 32, 1928.
47. RICE, S. O. 'Mathematical Analysis of Random Noise'. Bell System Technical Journal Vols. 23 and 24.
48. RIDLEY, B. K. and WATKINS, T. B. Proc. Phys. Soc. 78, 1961 p. 293.
49. GUNN, J. B. Solid State Comm. 1, 88, 1963.
50. GNERLICH, H. R. and ONDRIA, J. 'A New Look at Noise in Transferred Electron Oscillators'. IEEE Trans. MTT-25 No. 12 December 1977 pp. 977–984.

SUMMARY OF IMPORTANT FORMULAE

The symbols used are defined as they occur with the occasional exception of those given in the list which immediately precedes Chapter 1.

As far as possible the equation numbers given represent the expression occurring at the end of the first full (rather than simplified) derivation of each equation. It is hoped that this will facilitate the use of this summary for purposes of revision or reference.

1. For $T_s = 290°K$,

$$N_0 = -204 \text{ dBW/Hz} = -174 \text{ dBm/Hz} \quad (1.1)$$

2. Phase jitter due to superposed SSB noise.

$$\phi = \frac{\theta}{\sqrt{2}} = \sqrt{\frac{N_0}{2C}} \text{ rms radians} \quad (3.13)$$

3. The relationship between phase noise, AM noise and thermal noise due to DSB superposed white noise.

$$N_{op} = N_{oa} = \frac{N_0}{2} \quad (3.29)$$

4. The relationship between phase jitter variance and phase noise density.

$$\overline{\phi_0^2} = \frac{2N_{op}}{C} \quad (3.30.3)$$

5. Integrated phase jitter.

$$\overline{\phi^2} = \int_0^b \left(\frac{2N_{op}}{C}\right)_f df \text{ rads}^2 \quad (3.40)$$

$$\overline{\phi^2} = \int_{-b}^{+b} \left(\frac{N_0}{2C}\right)_f df \text{ rads}^2 \quad (3.49)$$

6. Relationship between frequency deviation and phase jitter.

$$\delta f = \phi f_m \quad (4.3)$$

where δf is the rms frequency deviation, and f_m is the offset frequency.

7. Integrated frequency deviation.

$$(\delta f)^2 = \int_{f_{m1}}^{f_{m2}} \left(\frac{2N_{op}}{C}\right)_{f_m} f_m^2 \, df_m \quad (4.8)$$

8. Phase noise performance of a linear oscillator.

$$\left(\frac{N_{op}}{C}\right)_{f_m} = \frac{FkT}{C} \frac{1}{8Q_L^2} \left(\frac{f_0}{f_m}\right)^2 \quad (5.27)$$

where F is the noise figure of the active device, Q_L is the loaded Q, C is the output power, f_0 is the carrier frequency and f_m is the offset frequency.

9. VCO phase noise due to noise voltage on varactor supply line.

$$\left(\frac{2N_{op}}{C}\right)_{f_m}^{1/2} = \frac{L(f_2 - f_1)}{V} \cdot \frac{V_m}{f_m} \quad (5.46)$$

where L is the ratio of maximum to minimum tuning slope, $(f_2 - f_1)$ is the oscillator tuning range corresponding to a control voltage range V, and V_m is the rms ripple voltage on the tuning supply.

10. The effect of frequency multiplication on phase noise.

$$\left(\frac{N_{op}}{C}\right)_2 = n^2 \left(\frac{N_{op}}{C}\right)_1 \quad (6.7)$$

where the suffix 2 refers to the output frequency and the suffix 1 to the input frequency. The offset frequency is supposed the same in both cases. Alternatively:

$$\left(\frac{N_{op}}{C}\right)_2 (dB) = 20 \log_{10} n + \left(\frac{N_{op}}{C}\right)_1 dB.$$

11. Phase detector sensitivities:
    (a) Analogue with signal inputs of $+10$ dBm and -10 dBm with 100 ohm load.

$$K_d = 0.07 \text{ V rms per rms radian} \quad (7.57)$$

(b) 'Exclusive Or' gate (TTL).

$$K_d = 0.64 \text{ V rms per rms radian}$$

(see Section 7.2.2b).

(c) Edge Triggered (TTL).

$$K_d = 0.32 \text{ V rms per rms radian}$$

(see Section 7.2.2c).
12. Phase Lock Loop Transfer Function:

$$H(s) = \frac{K_0 K_d F(s)}{s + K_0 K_d F(s)} \quad (7.14)$$

where $H(s)$ is the PLL transfer function, K_0 is the VCO tuning slope in radians per second per volt, K_d is the phase detector gain constant in volts per radian and $F(s)$ is the active filter transfer function.

13. Error Response of a PLL:

$$\theta_e(s) = \theta_i(s)[1 - H(s)] - \theta_n(s)[1 - H(s)] \quad (7.35)$$

where $\theta_e(s)$ is the phase error, $\theta_i(s)$ is the input or reference phase and $\theta_n(s)$ is the phase jitter of the uncontrolled VCO.

14. Straight line approximation to a PLL error response: 40 dB per decade below f_n, unity above f_n. See 7 Figure 12.

15. PLL Natural Frequency:

$$f_n = \frac{1}{2\pi} \left(\frac{K_0 K_d}{\tau_1 n} \right)^{1/2} \quad (7.52) \text{ [See also (7.79)]}$$

where f_n is the natural frequency, $K_0 K_d$ as in 12 above, τ_1 is the primary filter time constant ($R_1 C$ in 7 Figure 10) and n is the frequency division ratio inside the loop, which in many cases may be unity.

16. PLL Damping Factor:

$$\zeta = \frac{\tau_2}{2} \left(\frac{K_0 K_d}{\tau_1 n} \right)^{1/2} \quad (7.53) \text{ [See also (7.80)]}$$

where ζ is the damping factor, τ_2 is the secondary time constant $[(R_2 C)$ in 7 Figure 10] and all other parameters are as in 15 above.

17. Noise floors achievable by good design.
Analogue PD:

$$\left. \left(\frac{N_{op}}{C} \right)_{30 \text{ Hz}} = -141 \text{ dB/Hz} \\ \left(\frac{N_{op}}{C} \right)_{1 \text{ kHz}} = -144 \text{ dB/Hz} \right\} \quad (7.72)$$

Exclusive Or PD:

$$\left. \left(\frac{N_{op}}{C} \right)_{30 \text{ Hz}} = -160 \text{ dB/Hz} \\ \left(\frac{N_{op}}{C} \right)_{1 \text{ kHz}} = -163 \text{ dB/Hz} \right\} \quad (7.73)$$

Edge Triggered PD:

$$\left.\begin{array}{l}\left(\dfrac{N_{op}}{C}\right)_{30\text{ Hz}} = -154 \text{ dB/Hz} \\[2em] \left(\dfrac{N_{op}}{C}\right)_{1\text{ kHz}} = -157 \text{ dB/Hz}\end{array}\right\} \quad (7.74)$$

18. Noise Floor Degradation due to a frequency dividing PLL.

    Noise Floor Degradation = $(20 \log_{10} n)$ dB

    See Section 7.75.

19. Phase noise degradation of the Reference Oscillator in a frequency dividing PLL.

    Degradation = $(20 \log_{10} n)$ dB. See the end of Section 7.6.

20. Measurement of phase jitter using a phase detector:

$$\overline{\phi^2} = \frac{1}{2}\left(\frac{V}{V_c}\right)^2 \quad (8.41)$$

    where V is the rms jitter voltage and V_c is the rms calibration voltage measured when the two sources driving the PD are set to slightly *different* frequencies.

21. $\dfrac{N_{op}}{C} = \dfrac{1}{4}\left(\dfrac{V}{V_c}\right)^2 \dfrac{1}{b} \quad (8.43)$

    V and V_c as in 20 above.

22. $\langle(\Delta f_k)^2\rangle = 2\langle(\delta f_k)^2\rangle \quad (9.29)$

    where Δf_k is the difference between two successive frequency measurements and δf_k is the difference between one frequency measurement and the fixed mean frequency f_0.

23. Variance of the fractional frequency jitter.

$$\sigma_y^2 = \tfrac{1}{2}\langle(\overline{y}_{k+1} - \overline{y}_k)^2\rangle \quad (9.31)$$

    where

$$\overline{y}_k = \frac{\overline{f}_k}{f_0}.$$

24. Allan Variance.

$$\sigma_y^2(\tau) = \frac{1}{2(M-1)} \sum_{k=1}^{(M-1)} (\overline{y}_{k+1} - \overline{y}_k)^2 \quad (9.35)$$

    M is the number of individual frequency measurements, $(M-1)$ is the number of frequency differences, τ is the integration time of the frequency counter for each measurement and the dead space between measurements is assumed to be zero.

25. The Hadamard Variance.

$$\sigma_H^2(M, T, \tau) = \langle (\overline{y_1} - \overline{y_2} + \overline{y_3} \cdots \overline{y_M})^2 \rangle \quad (9.98)$$

where $\overline{y_k}$ is as defined in 23, M is the number of individual frequency measurements, τ is defined in 24 and $T = 1.5\tau$.

26. CW Radar—Modulation index of IF due to delayed echo.

AM: no change to transmitter modulation index,

$$\text{PM}: \theta'(\tau) = 2\theta \sin\left(\frac{p\tau}{2}\right) \quad (10.16)$$

where $\theta'(\tau)$ is the IF (relative) PM index, θ is the transmitter PM index at a modulation frequency p and τ is the echo delay.

27. CW Radar—Clutter to Signal Ratio.

$$\frac{I_c}{S} = \frac{1}{S} \sum_{k=1}^{l} C_{gk} \theta_0^2(p_k) b \sin^2 \frac{p_k \tau_k}{2} \quad (10.68)$$

For notation see text.

28. FM Radar—Minimum Frequency Deviation (ΔF) Required to give Range Discrimination δr.

$$\Delta F = \frac{nc}{2\delta r} \quad (10.100)$$

where the carrier is frequency modulated with a symmetrical linear sawtooth (see 10 Figure 7) with a peak to peak frequency deviation ΔF.

29. Signal/Noise Ratio of a simple FM communication system.

$$S/N = \frac{3M^2}{2b} \left(\frac{C}{N_0}\right)_T \quad (10.110)$$

where S/N is the baseband signal to noise ratio, M is the *peak* modulation index at the maximum baseband frequency, b is the baseband bandwidth and $(C/N_0)_T$ is the thermal carrier to noise density ratio.

30. Maximum Permissible Local Oscillator Phase Noise for a Simple FM System.

$$\int_{250 \text{ Hz}}^{1.1b} \left(\frac{2N_{op}}{C}\right)_{f_m} f_m^2 \, df_m \not> \frac{1}{x} \left(\frac{N_0}{C}\right)_T \frac{(1.1b)^3}{3} \quad (10.130)$$

where x should be selected on economic grounds, and for a single local oscillator in a satellite system x should be of the order of 200.

31. Signal to Noise Ratio of a Single Voice Channel in an FDM/FM System.

$$\frac{C}{N_0} = b\,\frac{S}{N}\left(\frac{f_m}{\delta f}\right)^2 \quad (10.149)$$

δf is the *rms* Test Tone Deviation.

32. L.O. Phase Noise Specification for an FDM/FM System.

$$\left(\frac{N_{op}}{C}\right)_L = \frac{1}{2x}\left(\frac{N_0}{C}\right) \quad (10.152)$$

33. Unlocking Sensitivity of a Carrier Recovery Loop used in a PSK System.

$$\overline{\phi^2} = \int_0^{f_n} n^2 \left(\frac{2N_{op}}{C}\right)_f |1 - H(jp)|^2 \, df$$

$$+ \int_{f_n}^{\infty} n^2 \left(\frac{2N_{op}}{C}\right)_f |H(jp)|^2 \, df \quad (10.159)$$

where n is the order of the phase modulation (4 for a quadriphase system) and $H(jp)$ is the transfer function of the carrier recovery loop.

34. Oscillator Phase Noise Requirements as determined by the Carrier Recovery Loop of a PSK System. (Simple approximation).

$$\left(\frac{N_{op}}{C}\right)_{f_n} \text{dB/Hz} = \left(\frac{N_0}{C}\right) \text{dB/Hz} - (x + 4\cdot3) \text{ dB} \quad (10.180)$$

x is of the order of 20 dB and is determined on economic grounds.

35. PSK System—Oscillator Phase Noise Specification as determined by Bit Error Rate Degradation.

$$\overline{\phi^2} = \int_{f_n}^{b} \left(\frac{2N_{op}}{C}\right)_f df \ngtr \frac{b}{x}\left(\frac{N_0}{C}\right)_T \quad \text{derived from (10.202)}$$

where $\overline{\phi^2}$ is the permissible integrated LO phase jitter variance over the baseband bandwidth f_n to b. f_n is the natural frequency of the carrier recovery loop and b is the bit rate; x is chosen on economic grounds.

NOISE FIGURE REVIEW

AII.1 Background

For many years it has been possible to build electronic amplifiers with an almost unlimited gain. If it were not for the effects of noise and interference it would be possible to amplify the weakest of signals to any required level. Communication and radar systems would use low power transmitters and would no longer be limited in range. In a sense communication and radar engineering would become trivial.

The Equi-Partition Theorem of Statistical Thermodynanics shows that, for example, a gas molecule possesses a mean energy of $\frac{1}{2} kT$ joules per degree of freedom (k is Boltzmann's Constant and T the temperature in degrees Kelvin). This principle applies even to electrons.

If, to avoid rather lengthy subtleties, we merely assume that an electronic circuit possesses two degrees of freedom per second per Hz of bandwidth, then the mean energy is kT joules per second per Hz. Note that in a 1 Hz bandwidth the energy is uncorrelated from one second to another. Thus the mean power is kT watts per Hz.

These ideas lead (Reference 46) to the conclusion that the open circuit noise voltage of a resistor due to thermal agitation is:

$$V_n = \sqrt{4kTBR} \qquad \text{(A.1)}$$

where V_n is the (long term) rms voltage, B is the equivalent rectangular bandwidth, R is the resistance in ohms.

A noise generator consisting of a resistor R_1 at temperature T will deliver maximum power to another resistor of the same value of resistance (Maximum Power Theorem).

The rms voltage across R_2 due to the noise voltage of R_1 (A Figure 1)

$$= \tfrac{1}{2}\sqrt{4kTBR}$$

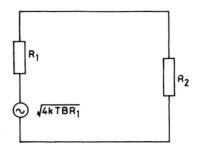

A Figure 1

Available Noise Power from a Resistor

The power dissipated in R_2 due to thermal agitation in R_1 is

$$N = \frac{kTBR_1}{R_1} = kT\,B \text{ watts.}$$

Then the noise density is

$$N_0 = kT \text{ watts per Hz.} \tag{A.2}$$

This is the maximum *available* noise power density from a resistor as we have assumed a matched load ($R_2 = R_1$).

In a state of thermal equilibrium R_2 will have an equal open circuit noise voltage and the power dissipated in R_1 due to thermal agitation in R_2 will be equal to that dissipated in R_2 due to thermal agitation in R_1. If R_2 produced no noise voltage of its own, then the power dissipated in R_2 due to R_1 would raise the temperature of R_2. If R_1 and R_2 started at the same temperature this would violate the Second Law of Thermodynamics. If R_2 were replaced by a pure reactance, which cannot dissipate power, it follows that a pure reactance cannot generate noise. If it did, this noise power would be dissipated in R_1, whereas the noise power from R_1 would be reflected from the reactance. Thus the temperature of R_1 would rise.

AII.2 Noise Temperature and Noise Figure

The *Noise Temperature* of a system at a selected reference point may be defined by (A.2) if N_0 at that point has been measured or may be calculated from other data.

Consider an amplifier or other network with an available power gain G. (If the network is an attenuator, G may be fractional). Assume the network to contribute no noise of its own, in this regard it is a perfect amplifier, and that its input is provided by a pure resistance source at a temperature of 290 degrees Kelvin. The available noise power density of the source is $290k$.

(Remember this k is Boltzmann's Constant $= 1{\cdot}38 \times 10^{-23}$ joules per °K). The available output noise power density

$$W_{01} = 290 \; kG \; \text{Watts/Hz} \tag{A.3}$$

and this is due to source noise alone.

For a real noisy network the available output noise power density is:

$$W_{02} = F \; 290 \; kG \; \text{Watts/Hz} \tag{A.4}$$

$$\therefore \quad \frac{W_{02}}{W_{01}} = F \tag{A.5}$$

F is thus the ratio of the output noise density due to source and network noise, divided by that due to the source alone, *when the source is at 290°K*. As G cancels, it is also the ratio of both components referred to the input of the network. F is known as the *Noise Figure* or *Noise Factor* of the network. Strictly this is the narrow band noise figure as we have derived it on a basis of noise density, that is for a 1 Hz bandwidth. To find the mean noise figure over a wider band it is necessary to integrate all noise density components over that band allowing for the frequency response of the real network.

The output noise power density due to the network alone, that is without the source contribution, is:

$$W_{02} - W_{01} = 290 \; kG \; (F - 1) \tag{A.6}$$

Referred to the input this is:

$$N_0(\text{network}) = 290 \; k \; (F - 1) \tag{A.7}$$

and the noise temperature contribution of the network alone is:

$$T_N = 290 \; (F - 1) \tag{A.8}$$

$$F = \left(\frac{T_N}{290} + 1 \right) \tag{A.9}$$

and

$$N_0(\text{network}) = k T_N \tag{A.10}$$

If we have two networks in cascade with noise temperatures T_{N1}, T_{N2} and available power gains G_1 and G_2, fed from a source of noise temperature T_S, and denote the output noise power density by W_{o12}, then:

$$W_{012} = G_1 G_2 k T_s + G_1 G_2 k T_{N_1} + G_2 k T_{N_2} \tag{A.11}$$

Moving the reference point to the input, by dividing by the total gain $G_1 G_2$ and calling the equivalent noise power density at the input N_{012}.

$$N_{012} = k \left(T_s + T_{N_1} + \frac{T_{N2}}{G_1} \right) \tag{A.12}$$

The effective noise temperature of source plus networks is then:

$$T_{(s+N)} = \left(T_s + T_{N_1} + \frac{T_{N_2}}{G_1} \right) \tag{A.13}$$

in which the individual contributions are immediately apparent.
Using (A.8) and (A.13)

$$T_{(s+N)} = \left[T_s + 290\,(F_1 - 1) + 290\frac{(F_2 - 1)}{G_1} \right] \tag{A.14}$$

where F_1 and F_2 are the noise figures of network 1 and network 2 respectively
and

$$F_{12} = \frac{N_{012}}{kT_s} = \left(1 + \frac{T_{N_1}}{T_s} + \frac{T_{N_2}}{G_1 T_s} \right) \tag{A.15}$$

(Noise figure (F) is *defined* for $Ts. = 290°K.$). Substituting (A.9) in (A.15)

$$F_{12} = F_1 + \left(\frac{F_2 - 1}{G_1} \right) \tag{A.16}$$

The argument may of course be extended similarly for n networks in cascade
to give:

$$F_{123-n} = \left[F_1 + \frac{F_2 - 1}{G_1} + \frac{F_3 - 1}{G_1 G_2} + \cdots \right] \tag{A.17}$$

AII.3 Waveguide Losses at Different Temperatures

In practice, for example in the waveguide run between a satellite ground
station antenna and the input to the low noise amplifier (LNA), we may have
sections of waveguide each with losses and in some cases at different physical
temperatures. To distinguish between physical temperatures and noise
temperatures we shall use t_1, t_2 etc., for physical temperatures and T_1, T_2 etc.,
for noise temperatures.

Consider first the simplest case of a single piece of waveguide at a uniform
physical temperature t and loss L where $L = 1/G$. The transmission coefficient
$a = G$. Power lost in transmission, which is the power dissipated in the
waveguide (WG) $= (1 - a) = (1 - G)$. Only the power dissipated will actually
add noise, as distinct from attenuating the signal. The loss coefficient, or
dissipation coefficient, l is:

$$l = (1 - G) = \left(1 - \frac{1}{L} \right) \tag{A.18}$$

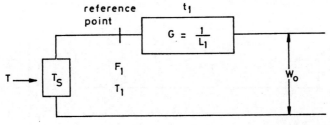

A Figure 2

Waveguide Loss at Physical Temperature t_1

Let the physical temperature of the WG, and therefore of the loss coefficient l, be t_1. The arrangement is illustrated in A Figure 2.

Available noise power density at output

$$W_0 = GkT_s + GkT_1$$

Also available noise power density at output

$$W_0 = GkT_s + t_1\,kl_1$$

$$\therefore \quad T_1 = \frac{t_1 l_1}{G} \tag{A.19}$$

From (A.18) and (A.19)

$$T_1 = \frac{t_1\left(1 - \dfrac{1}{L_1}\right)}{G} = t_1 L_1\left(1 - \frac{1}{L_1}\right) \tag{A.20}$$

Hence the overall noise temperature, including that of the source (T) at 290°K:

$$T = T_s + T_1$$

$$T = T_s + t_1 L_1\left(1 - \frac{1}{L_1}\right) \tag{A.21}$$

Consider two cascaded waveguide sections with losses L_1 and L_2 and physical temperatures t_1 and t_2 feeding a low noise amplifier (LNA) as shown in A Figure 3.

$$W_{03} = kT_s G_1 G_2 G_3 + kT_1 G_1 G_2 G_3 + kT_2 G_2 G_3 + T_3 G_3$$

Dividing by the total gain, the noise density at the reference point (N_0) is:

$$N_0 = k\left[T_s + T_1 + \frac{T_2}{G_1} + \frac{T_3}{G_1 G_2}\right] \tag{A.22}$$

Thus the overall noise temperature T is:

$$T = T_s + T_1 + L_1 T_2 + L_1 L_2 T_3 \tag{A.23}$$

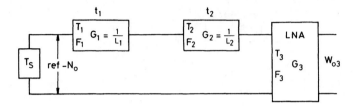

W.G. Losses at Different Temperatures   **A Figure 3**

Substituting for T_1 and T_2 using (A.20)

$$T = T_s + t_1 L_1 \left(1 - \frac{1}{L_1}\right) + L_1 L_2 t_2 \left(1 - \frac{1}{L_2}\right) + L_1 L_2 T_3$$

$$\therefore \quad T = T_s + t_1(L_1 - 1) + L_1 t_2(L_2 - 1) + L_1 L_2 T_3 \tag{A.24}$$

AII.4 The Reasons for a Mismatch at the Input to a Low Noise Amplifier

To assist physical appreciation we shall consider a low noise FET amplifier: the simple physical relationships would be somewhat obscured by the device complexity if we considered a parametric amplifier. A block diagram is shown in A Figure 4.

It is assumed that all reactances in the input circuit have been tuned out at the operating frequency. The losses, as represented by the parallel impedance of any tuned circuit, in parallel with the input resistance of the FET are included in R_{in}.

The shot noise of the FET, referred to the input (gate) is represented by a voltage noise generator $\sqrt{4kTR_n}$. Remember that R_n is not itself a resistor: it is merely a convenient method of representing a noise voltage. Even if our source impedance is fixed, it might be an antenna or a coaxial line, it may be changed by the use of an ideal transformer to produce any value of R_s at the FET input. Practical transformer design problems in a real case are irrelevant to this argument.

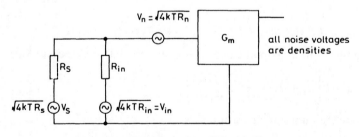

Noise Representation of an FET Amplifier   **A Figure 4**

Consider two extreme cases as follows:

(*a*) where R_s is very large.
(*b*) where R_s is small.

(*a*) R_s Large

Due to the potentiometer effect of R_s and R_{in}, the noise voltage at the gate of the FET due to R_s will be:

$$V'_s = \sqrt{4kTR_s}\frac{R_{in}}{R_{in} + R_s} \tag{A.25}$$

$$\therefore \quad \frac{V'_s}{V_n} = \sqrt{\frac{R_s}{R_n}}\frac{R_{in}}{R_{in} + R_s} \tag{A.26}$$

For a normal FET at medium frequencies

$$R_{in} \gg R_n \text{ and having chosen } R_s \gg R_n$$

$\dfrac{V'_s}{V_n}$ will be very large

The effect of a high value of source impedance is that the source impedance noise completely swamps the FET shot noise component.

However, a large value of R_s ensures that the noise voltage at the FET gate due to R_{in} (V'_{in}) is also large.

$$V'_{in} = \sqrt{4kTR_{in}}\frac{R_s}{R_s + R_{in}} \tag{A.27}$$

R_{in} will usually be orders larger than R_n so that even though $R_s/(R_s + R_{in})$ is fractional ($=\frac{1}{2}$ for the matched case)

$$V'_{in} \gg V_n \tag{A.28}$$

From (A.25) and (A.27)

$$\frac{V'_s}{V'_{in}} = \sqrt{\frac{R_s}{R_{in}}\frac{R_{in}}{R_{in} + R_s}} \cdot \frac{R_s + R_{in}}{R_s}$$

$$\frac{V'_s}{V'_{in}} = \sqrt{\frac{R_{in}}{R_s}} \tag{A.29}$$

Assuming V_n to be completely negligible in this case:

$$\left(\frac{V'_s}{V'_{in}}\right)^2 = \frac{R_{in}}{R_s} \tag{A.30}$$

From (A.30) it is apparent that an increase in an already large value of R_s will worsen the noise figure and make the noise due to R_{in} even more dominant.

(b) R_s *Small*
From (A.25) putting $R_s \ll R_{in}$

$$V'_s = \sqrt{4kTR_s} = V_s \tag{A.31}$$

From (A.27)

$$V'_{in} \ll V_{in} \tag{A.32}$$

$$\frac{V'_s}{V_n} = \frac{V_s}{V_n} \tag{A.33}$$

Thus a low value of source resistance R_s greatly reduces the noise contribution of the FET gate input circuit resistance R_{in} (by partially short circuiting its noise voltage) at the price of reducing the noise contribution of the source itself (and also the signal) relative to the FET shot noise. If the noise power density contribution of the source were completely dominant relative to all other contributions then, of course, the noise figure would approximate to unity (or 0 dB).

(c) *Optimum Source Resistance*
It is apparent from the foregoing discussion that there will be an optimum value of source resistance, which will produce a minimum noise figure, and that this will not be represented by an impedance match for which $R_s = R_{in}$.
In fact the optimum value of $R_s(R_s$ opt) is given by:

$$R_s(\text{opt}) = R_{in}\sqrt{\frac{R_n}{R_n + R_{in}}} \tag{A.34}$$

As R_n is defined with reference to 290°K (A.34) is derived assuming that R_{in} is also at 290°K.

AII.5 The Noise Figure of an FET Amplifier

The noise figure will be defined by the sum of the noise powers (assuming them all to be incoherent) at the gate input circuit divided by the noise power due to R_s. Due to the fact that the noise powers are compared at a common point in the network, they are all referred to a common impedance and the summing and division may therefore be carried out in terms of (volts)^2.

Now $V_k^2 = 4kTR_k = $ (say) DR_k \tag{A.35}

where V_k is the open circuit noise voltage per $\sqrt{\text{Hz}}$ of bandwidth of any one of the resistors. Thus in taking the ratio of the square of two noise voltages $4\,kT$ will cancel, and we shall be left with

$$\frac{V_{k1}^2}{V_{k2}^2} = \frac{R_{k1}}{R_{k2}} \frac{D}{D} \tag{A.36}$$

The sum of the squares of the noise voltages at the gate of the FET is:

$$V_0^2 = (V_s')^2 + (V_{in}')^2 + (V_n)^2 \tag{A.37}$$

Now (A.25)

$$(V_s')^2 = DR_s \left(\frac{R_{in}}{R_{in} + R_s}\right)^2 \tag{A.38}$$

and (A.27)

$$(V_{in}')^2 = DR_{in} \left(\frac{R_s}{R_{in} + R_s}\right)^2 \tag{A.39}$$

$$\therefore \quad V_0^2 = (V_s')^2 + D\left[R_{in}\left(\frac{R_s}{R_{in} + R_s}\right)^2 + R_n\right] \tag{A.40}$$

Now the noise figure is equal to the total noise power density divided by that due to the source, $(V_s')^2$.

$$\therefore \quad F = 1 + \frac{R_{in}}{R_s}\left(\frac{R_s}{R_{in}}\right)^2 + \frac{R_n}{R_s}\left(\frac{R_{in} + R_s}{R_{in}}\right)^2$$

$$\therefore \quad F = 1 + \frac{R_s}{R_{in}} + \frac{R_n}{R_s}\left(1 + \frac{R_s}{R_{in}}\right)^2 \tag{A.41}$$

AII.6 The Noise Bandwidth of a Single Tuned Circuit

The circuit is normalised so that at resonance a 1 ohm source is driving a tuned circuit with a parallel impedance at resonance of 1 ohm due to the final load R_L. All voltages are rms. The input noise density is considered to be derived from a high gain amplifier with an output impedance of 1 ohm. $N_0(f)$ is supposed to be very much greater than the thermal noise power density in R_L and to be white over a very large frequency band. Being independent of frequency it will be written N_0. (See A Figure 5.)

$$V_{On} = \frac{\sqrt{N_0}}{1 + 2jQ\dfrac{\delta f}{f_0}} \tag{A.42}$$

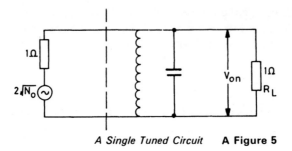

A Single Tuned Circuit   **A Figure 5**

The modulus of the denominator is:

$$\sqrt{1 + \left(\frac{2Q\delta f}{f_0}\right)^2}$$

Writing $f = \delta f$ to simplify the notation

$$|(V_{0n})^2| = \frac{N_0}{1 + \left(\dfrac{2Qf}{f_0}\right)^2}$$

Remembering that the input noise is white, the *DSB* output noise power is then:

$$N = 2N_0 \int_0^\infty \frac{1}{1 + \left(\dfrac{2Qf}{f_0}\right)^2} \, df$$

$$\therefore \quad N = N_0 \frac{f_0}{Q} \left[\tan^{-1} \frac{2Qf}{f_0} \right]_0^\infty$$

$$N = N_0 \frac{f_0}{Q} \frac{\pi}{2} \tag{A.43}$$

$$\therefore \quad \text{effective DSB noise bandwidth} = \frac{f_0}{Q} \frac{\pi}{2} \tag{A.44}$$

which is equal to $\pi/2$ times the 3 dB bandwidth. Section 5.2 makes use of this formula.

AII.7 Mixer Noise Figure, Noise Temperature and NTR

Due to shot noise in the semiconductor material of a mixer diode or diodes, whilst passing the rectified current due to LO drive, the noise temperature of the diode resistance (T_d) will normally be greater than its physical temperature.

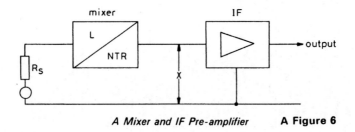

A Mixer and IF Pre-amplifier　　**A Figure 6**

It is usually assumed that its physical temperature is 290°K which we shall designate T_0. The Noise Temperature Ratio (NTR) is then:

$$\text{NTR} = \frac{T_d}{290} = \frac{T_d}{T_0} \tag{A.45}$$

In practice the NTR is usually defined to include any contribution from LO noise (and source) but the reader will readily understand this aspect (see Section 10.1.3 and also AII.8).

A mixer, driven from a signal source resistance R_s, feeding an IF amplifier is shown in A Figure 6.

Let: the noise temperature and the noise figure of the IF amplifier be T_{IF} and F_{IF}, the loss of the mixer be $L = 1/G$, the noise power density at the IF amplifier input, excluding the contribution of the mixer $= N_{01}$.

$$\therefore \quad N_{01} = kT_{IF} = kT_0(F_{IF} - 1) \tag{A.46}$$

If the IF amplifier were driven from a resistive source at temperature T_0 then the additional noise density contribution at X would be kT_0. Due to the noise temperature of the diode(s) resistance this contribution is kT_d.

From (A.45):

$$kT_d = kT_0(\text{NTR})$$

Thus total noise density at IF input $(N_{0\,IF})$ is:

$$N_{0\,IF} = kT_0(\text{NTR}) + kT_0(F_{IF} - 1)$$

$$N_{0\,IF} = kT_0(\text{NTR} + F_{IF} - 1) \tag{A.47}$$

The noise density at the input to the mixer is:

$$N_{OM} = kT_0\,L(\text{NTR} + F_{IF} - 1) \tag{A.48}$$

The available noise power density due to the signal source R_s is kT_0.
Thus the mixer/IF amplifier overall noise figure is

$$F_{\text{overall}} = L(\text{NTR} + F_{IF} - 1) \tag{A.49}$$

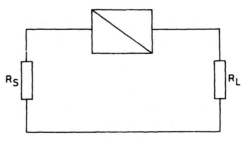

A Mixer Feeding a Load   **A Figure 7**

Hence the overall noise figure of a receiver with a mixer input stage, and zero losses between the signal source and the mixer input, is given by (A.49)

The overall noise temperature of the mixer/amplifier combination, neglecting the contribution of the signal source (R_s) is:

$$T_{(M+A)} = T_0(F_{overall} - 1) \tag{A.50}$$

$$\therefore \quad T_{(M+A)} = T_0 L(\text{NTR} + F_{IF} - 1) - T_0$$

$$= T_0 L\left(\text{NTR} + \frac{T_{IF}}{T_0}\right) - T_0 \text{ [using (A.9)]}$$

$$T_{(M+A)} = L(\text{NTR } T_0 + T_{IF}) - T_0 \tag{A.51}$$

The output noise temperature of a mixer feeding a resistive load is required when calculating the noise floor of an analogue phase detector (see Section 7.7.1 and equation 7.59). The arrangement is shown in A Figure 7. Adding T_0 to (A.51) (to allow for the noise contribution of R_s) and replacing T_{IF} with T_0, as a result of substituting R_L for the IF amplifier, gives the overall noise temperature (say T overall).

$$T_{overall} = L(\text{NTR } T_0 + T_0)$$

Multiplying by the gain $(1/L)$ gives the resultant noise temperature at the output; (say) T_m out.

$$\therefore \quad T_{m \text{ out}} = T_0(\text{NTR} + 1) \tag{A.52}$$

AII.8 The Suppression of Local Oscillator AM Noise by a Balanced Mixer

Let the signal be

$$e_s = \sqrt{2} E_s[(1 + S) \sin \omega t] \tag{A.53}$$

where S may be any complicated function of t, with a spectrum extending out to f_c; the local oscillator voltage be:

$$e_h = \sqrt{2} E_h\{[1 + V_n(p, t)] \sin ht\} \tag{A.54}$$

where

$$V_n(p, t)(\text{rms}) = \left[\int_{f_1}^{f_2} (N_{oa})_{f_m} \, df_m\right]^{1/2} \tag{A.55}$$

and f_1 and f_2 are the limits of the relevant offset frequency range. With the exception only of very poor oscillators

$$V_n(p, t) \ll 1 \text{ over any limited offset frequency range.} \tag{A.56}$$

The two signals are combined in a hybrid in such a way that half the power from e_h and e_s goes to each mixer diode. In one of the diodes either e_h or e_s is 180° different in phase from what it is in the other diode. We shall consider only the case where the phase of e_s is the same in both diodes and the phase of e_h differs by π radians at the two diodes. The other case where, the local oscillator is in the same phase and the signal is in antiphase at the two diodes, may be analysed in a similar manner and leads to the same value of local oscillator AM noise suppression.

(a) Mixer 1
Thus at one diode

$$E_1 = \frac{e_h}{\sqrt{2}} + \frac{e_s}{\sqrt{2}} \tag{A.57}$$

$$\therefore \quad E_1 = E_h\{[1 + V_n(p, t)] \sin ht\} + E_s[(1 + S) \sin \omega t]$$

Assume the local oscillator frequency to be above the signal frequency. Let the angular IF frequency be q.

$$h - \omega = q$$

$$\therefore \quad E_1 = E_h[1 + V_n(p, t)] \sin [(q + \omega)t] + E_s[(1 + S) \sin \omega t]$$

$$\therefore \quad E_1 = E_h[1 + V_n(p, t)][\sin qt \cos \omega t + \cos qt \sin \omega t]$$

$$\qquad + E_s[(1 + S) \sin \omega t]$$

$$\therefore \quad E_1 = E_h[1 + V_n(p, t)] \cos qt \sin \omega t + E_s(1 + S) \sin \omega t$$

$$\qquad + E_h[1 + V_n(p, t)] \sin qt \cos \omega t$$

$$\therefore \quad E_1 = \{E_h[1 + V_n(p, t)] \cos qt + E_s(1 + S)\} \sin \omega t$$

$$\qquad + \{E_h[1 + V_n(p, t)] \sin qt\} \cos \omega t$$

The square of the amplitude $(A_1)^2$ is given by the addition of the squares of the $(\sin \omega t)$ and $(\cos \omega t)$ terms.

$$\therefore \quad (A_1)^2 = E_h^2[1 + V_n(p, t)]^2 \cos^2 qt + E_s^2(1 + S)^2$$
$$\qquad + 2E_h E_s[1 + V_n(p, t)](1 + S) \cos qt \tag{A.58}$$
$$\qquad + E_h^2[1 + V_n(p, t)]^2 \sin^2 qt$$

But

$$E_h^2[1 + V_n(p, t)]^2[\cos^2 qt + \sin^2 qt] = E_h^2[1 + V_n(p, t)]^2$$

$$\therefore \quad (A_1)^2 = E_h^2[1 + V_n(p, t)]^2 + E_s^2(1 + S)^2$$
$$\left. + 2E_h E_s[1 + V_n(p, t)](1 + S) \cos qt \right\} \qquad (A.59)$$

Superficially it might seem that the only term to be passed by a following IF amplifier with an angular centre frequency q, would be the last term and that both the first and second terms of (A.59), being at baseband frequencies, would be rejected.

However in practice $E_h^2 V_n^2(p, t)$ will have components at an angular frequency q, and E_h^2 being very much larger than $2E_h E_s$, they may be of significant power relative to the wanted signal component.

Such terms will be the result of local oscillator AM noise sidebands at an offset frequency approximately equal to the IF.

Thus after dropping terms which will be rejected by the IF amplifier:

$$(A_1)^2 = E_h^2[1 + V_n(p, t)]^2 + 2E_h E_s[1 + V_n(p, t)](1 + S) \cos qt \qquad (A.60)$$

(b) Mixer 2

Consider now another mixer and assume the signal phase is the same as that in mixer 1 and that the local oscillator phase is shifted by 180° relative to its phase at mixer 1.

An analysis exactly similar to that carried out for mixer 1, after allowing for the rejection of baseband terms will give an equation similar to (A.60) except that the sign of the second term will be changed. The equation is:

$$(A_2)^2 = E_h^2[1 + V_n(p, t)]^2 - 2E_h E_s[1 + V_n(p, t)](1 + S) \cos qt \qquad (A.61)$$

From (A.60) and (A.61)

$$(A_1^2 - A_2^2) = 4E_h E_s[1 + V_n(p, t)](1 + S) \cos qt$$

Now $(A_1^2 - A_2^2) = (A_1 - A_2)(A_1 + A_2)$

$$\therefore \quad (A_1 - A_2) = \frac{4E_h E_s[1 + V_n(p, t)](1 + S) \cos qt}{(A_1 + A_2)} \qquad (A.62)$$

Now $E_h^2 \gg E_h E_s$: (A.60) and (A.61) may be approximated by:

$$A_1^2 = E_h^2[1 + V_n(p, t)]^2$$

$$\therefore \quad A_1 \simeq E_h[1 + V_n(p, t)] \qquad (A.63)$$

and

$$A_2 \simeq E_h[1 + V_n(p, t)] \qquad (A.64)$$

$$\therefore \quad A_1 + A_2 \simeq 2E_h[1 + V_n(p, t)] \qquad (A.65)$$

$$\therefore \quad (A_1 - A_2) = \frac{4E_h E_s[1 + V_n(p, t)](1 + S) \cos qt}{2E_h[1 + V_n(p, t)]}$$

$$\therefore \quad (A_1 - A_2) = 2E_s(1 + S) \cos qt. \tag{A.66}$$

It will be noted that $(A_1 - A_2)$ is the result of connecting the outputs of the two mixers in opposition and that, in doing so, the LO noise component has been cancelled. In practice the mixer balance will not be perfect but 20 dB to 30 dB rejection of local oscillator AM noise may readily be achieved in practice.

THE QUADRATURE REPRESENTATION OF NARROW BAND NOISE

In many communication or radar systems the bandwidth occupied by the signal, and hence the bandwidth in which we are interested, is only a very small percentage of the centre frequency. The output noise from, for example, an IF filter will be white over most of the filter passband. Thus narrow band white noise is both a simple and a practically interesting case. This is the case we shall consider although the analysis may readily be adapted to the case of coloured noise by breaking the frequency band into a number of sub-bands each with a different noise level.

Consider white noise of noise power density N_0 restricted to a narrow bandwidth $f_0 \pm b$. If we represented the noise by a large number of sinusoids of varying instantaneous amplitudes and phases but with frequencies within the range $f_0 \pm b$ we might expect that as a consequence of the Central Limit Theorem, the resultant would be a good representation of a Gaussian random process.

Approximate the noise by a number of discrete sinusoidal components separated from one another by Δf: the mean noise power of each component being $N_0 \Delta f$. The total mean noise power N in the bandwidth under consideration will be $N_0 = 2N_0 b$ and will be zero or irrelevant outside the frequency range $f_0 \pm b$. This is illustrated in A Figure 8.

Let the individual spectral lines be numbered $\pm k$ about 0 where the numeral '0' relates to the line at f_0 and the offset frequency of any line is $k\Delta f$. It is not essential but convenient and in accordance with the treatment in the rest of this book, to work in terms of density, and hence to put $\Delta f = 1$ Hz. The offset frequency $k\Delta f$ becomes equal to k and in a bandwidth of $\pm b$ Hz there will be a total number of lines $(2b + 1)$. The total noise power will then be given by:

$$N = \sum_{k=-b}^{+b} kN_0 \tag{A.67}$$

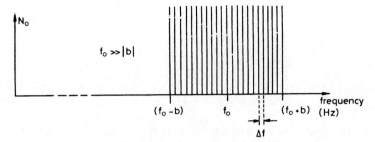

Representation of Narrow-Band Noise   **A Figure 8**

Putting different phase constants for each of the $(2b + 1)$ sinusoids an expression for the noise voltage is:

$$V_n(t) = \sum_{k=-b}^{+b} \sqrt{2N_0} \cos \left[(\omega_0 + 2\pi k)t + \psi_k\right] \tag{A.68}$$

Now $\cos (A + B) = \cos A \cos B - \sin A \sin B$

$$\therefore \quad V_n(t) = \left[\sum_{k=-b}^{+b} \sqrt{2N_0} \cos (2\pi kt + \psi_k)\right] \cos \omega_0 t$$

$$- \left[\sum_{k=-b}^{+b} \sqrt{2N_0} \sin (2\pi kt + \psi_k)\right] \sin \omega_0 t \tag{A.69}$$

This equation represents two RF carriers in quadrature with envelope modulation by a large number of sine waves with frequencies (k) ranging from $+b$ to $-b$, that is over the offset frequency range. As $f_0 \gg b$ the modulation frequencies are all very low compared with the carrier frequency. Equation (A.69) is the general expression for the pseudo-deterministic, but relatively accurate quadrature representation of narrow band noise. It is a more complete representation than that given in Chapter 3 as it represents the noise voltage for a finite bandwidth.

It is interesting to note that it bears a close relationship to a Fourier Series over a period of 1 second. In fact it may be derived by starting with a large number of random current pulses in the time domain of both polarities and with zero long term mean and taking the Fourier Series for the voltage developed across a resistive load. This approach is adopted in a classic paper by Rice (Reference 47).

Equation A.69 may be written:

$$V_n(t) = x(t) \cos w_0 t - y(t) \sin w_0 t \tag{A.70}$$

where

$$x(t) = \sum_{k=-b}^{+b} \sqrt{2N_0} \cos (2\pi kt + \psi_k) \qquad \text{(A.71.1)}$$

and

$$y(t) = \sum_{k=-b}^{+b} \sqrt{2N_0} \sin (2\pi kt + \psi_k) \qquad \text{(A.71.2)}$$

$x(t)$ and $y(t)$, which are themselves the sum of a large number of sinusoidal components each of frequency equal to the offset frequency $k = (p/2\pi)$, represent the instantaneous peak value of each of the two carrier frequency terms.

As the two RF carriers of equations (A.69) and (A.70) are in quadrature we may convert to the polar form as follows:

$$r^2(t) = x^2(t) + y^2(t) \qquad \text{(A.72.1)}$$

$$\beta = \tan^{-1} \frac{y(t)}{x(t)} \qquad \text{(A.72.2)}$$

where $r(t)$ is the resultant amplitude and β is the phase angle wrt $x(t)$. Consider the simple but important case when the noise bandwidth is 1 Hz, $k = 0$ and the bandwidth covers the range $f_0 \pm \frac{1}{2}$Hz. For this case:

$$x(t) = \sqrt{2N_0} \cos \psi_0 \qquad \text{(A.73.1)}$$

and

$$y(t) = \sqrt{2N_0} \sin \psi_0 \qquad \text{(A.73.2)}$$

$$\therefore \quad r^2(t) = 2N_0[\cos^2 \psi_0 + \sin^2 \psi_0]$$

$$\therefore \quad r^2(t) = 2N_0 \qquad \text{(A.74.1)}$$

$$\beta = \psi_0 \qquad \text{(A.74.2)}$$

Thus the polar form is

$$V_n(t) = \sqrt{2N_0} \cos (w_0 t + \psi_0) \qquad \text{(A.75)}$$

Firstly it will be noticed that in the polar representation the *carrier* peak amplitude given by (A.75), that is $\sqrt{2N_0}$ is identical to the peak amplitude of each of the two low frequency *modulating* sinusoids of the quadrature representation.

Consider the mean square values of $x(t)$, $y(t)$ and $V_n(t)$

$$\overline{x^2(t)} = N_0 : \overline{y^2(t)} = N_0 : \overline{V_n^2(t)} = N_0 \qquad \text{(A.76)}$$

Thus the long term mean power in *each* of the *modulating* waveforms of the two quadrature carriers is equal to the long term mean power of the single carrier used in the polar representation. (Remember that N_0 is *defined* in

Chapter 3 as a long term mean value). To eliminate any possible confusion consider the long term rms value of $x(t)$ and $y(t)$.

$$x(t)_{rms} = \sqrt{N_0} \text{ and } y(t)_{rms} = \sqrt{N_0}.$$

Hence when we are concerned purely with power relationships (and still for the case of a 1 Hz bandwidth) we may replace $x(t)$ and $y(t)$ by a constant value $\sqrt{2N_0}$. Substituting these values in (A.70).

$$V_n(t) = \sqrt{2N_0} \cos w_0 t - \sqrt{2N_0} \sin w_0 t \qquad (A.77)$$

$$\therefore \quad V_n^2(t) = [\sqrt{2N_0} \cos w_0 t - \sqrt{2N_0} \sin w_0 t]^2 \qquad (A.78)$$

As the two terms in the bracket on the RHS are orthogonal, the cross product term will be zero.

$$\therefore \quad V_n^2(t) = 2N_0 \cos^2 w_0 t + 2N_0 \sin^2 w_0 t \qquad (A.79)$$

$$\overline{V_n^2(t)} = N_0 \qquad (A.80)$$

Thus, in spite of appearances, the quadrature form (for a 1 Hz bandwidth):

$$V_n(t) = \sqrt{2N_0} \cos (pt + \psi_0) \cos w_0 t$$
$$- \sqrt{2N_0} \sin (pt + \psi_0) \sin w_0 t \qquad (A.81)$$

gives the same long term mean power as the polar form:

$$V_n(t) = \sqrt{2N_0} \cos [w_0 t + \psi_0] \qquad (A.75)$$

Equation (A.75) is equivalent to equation (3.4).

$$V_n(t) = \sqrt{2N_0} \sin [(w + p)t + \psi] \qquad (3.4)$$

as may be shown by putting $w_0 = (w + p)$ which does not affect the power relations as the noise has been assumed to be white over the range $f_0 \pm b$ (see A Figure 8).

The significant differences between equation (3.4) and equation (A.81) are as follows. In equation (3.4) the peak amplitude of the sinusoid is represented by its long term mean value ($\sqrt{2N_0}$) with the understanding that in reality the amplitude will vary from second to second and by a single phase term with the same understanding. In equation (A.81) the amplitude of each of the two *quadrature* terms is explicitly varying at a baseband rate: the resultant is a carrier *explicity* varying in both amplitude and phase at a baseband rate. Thus equation (A.81) is a more complete representation than equation (3.4); although, as we have seen throughout the book, the simple approach, carefully used, is adequate for most purposes.

THE Q OF VARACTOR TUNED OSCILLATORS

(a) *A Single Frequency 500 MHz Oscillator*

As discussed in Section 5.5 it is often true that the loaded Q of an oscillator is given approximately by:

$$Q_L = \frac{Q_u}{4} \tag{A.82}$$

where Q_L is the loaded Q and Q_u the unloaded resonator Q.

For UHF and SHF single frequency oscillators it is relatively easy to obtain high unloaded values of resonator Q by using coaxial or waveguide resonators.

Consider as an example the case of a 500 MHz single frequency oscillator with requirements which include a power output of $+10$ dBm and good phase noise. Choose a $\lambda/4$ coaxial resonator with an inside diameter of 1 cm and of ratio of inner diameter of outside conductor to outside diameter of inner conductor of 3·6 (the ratio giving highest Q). Equation 4.23a of Reference 10 gives

$$Q = 0.0839 \sqrt{f}\, bH \tag{A.83}$$

where f is the frequency in Hz, b is the inner *radius* of outer conductor in cms and

$$H = 1 \text{ for } \frac{b}{a} = 3·6$$

$$\therefore \quad Q_u = 938 \tag{A.84}$$

$$\therefore \quad Q_L = \frac{938}{4} = 234·5 \tag{A.85}$$

The total RF power generated, allowing for drive power and all losses will have to be approximately $+16$ dBm.

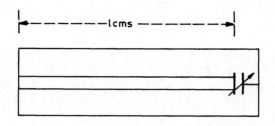

A Figure 9
A Capacitatively Tuned Coaxial Resonator

(*b*) *A Capacitatively Tuned Ocillator*
Assume that a tuning range of $\pm 5\%$ is required centred on 500 MHz. To ensure adequate frequency coverage allowing for design tolerances and temperature changes it might be wise to design for 500 MHz $\pm 6\%$; that is from 472 MHz to 530 MHz. Let the tuning range be achieved by using a variable (split stator) capacitor to load a coaxial line which' has a centre conductor with a physical length less than a quarter wavelength. See A Figure 9.

The reactance of a low loss short circuited transmission line is given by:

$$X_L = jZ_0 \tan \frac{2\pi l}{\lambda} \qquad (A.86)$$

where Z_0 is the characteristic impedance ($\simeq 75$ ohms for $b/a = 3\cdot6$) and λ is the free space wavelength. For resonance the capacitor will need to be set to have a reactance of equal magnitude (and opposite sign) at the desired resonant frequency. The required values of X_c and the required capacitor value C (in *pf*) have been calculated from equation (A.86) and are given in A. Table 1 for different values of l.

The equivalent circuit for such a resonator is given in A Figure 10 as viewed across the capacitor.

If the capacitor had very low losses the value of Q_u given by (A.84) might not be very greatly degraded: it might fall to perhaps 800 or so. Under

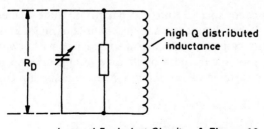

Lumped Equivalent Circuit   **A Figure 10**

A. Table 1 *Parameters of a Capacitatively Tuned Coaxial Resonator*

l cms	Parameter	472 MHz	500 MHz	530 MHz
13	X_c ohms	254	351	580
	C pf	1·32	0·906	0·518
12	X_c ohms	185	230	307
	C pf	1·82	1·38	0·978
11	X_c ohms	142	168	205
	C pf	2·36	1·89	1·46
10	X_c ohms	113	129	150
	C pf	2·96	2·45	1·99
9	X_c ohms	92	103	116
	C pf	3·65	3·09	2·58
8	X_c ohms	75	83	92
	C pf	4·45	3·83	3·26
7	X_c ohms	62	67	73
	C pf	5·43	4·72	4·08
6	X_c ohms	50	54	58
	C pf	6·68	5·85	5·1
5	X_c ohms	40	43	46
	C pf	8·35	7·36	6·46
4	X_c ohms	31	33	35
	C pf	10·78	9·54	8·43

operating conditions the rms voltage developed across the capacitor will be:

$$V = (Q_L X_c P)^{1/2} \tag{A.87}$$

Assuming $Q_L = Q_u/4$, reading X_c from A Table 1 and knowing P (the total power generated) V rms and hence V peak may be calculated for each value of *l*.

When using an air spaced tuning capacitor to tune a medium powered oscillator, this voltage will not normally be an area of major concern. However as we shall shortly consider replacing this capacitor with a varactor diode it will be necessary to know the peak oscillatory voltage developed across it. The results for the two extreme values of *l* in A Table 1, assuming $Q_L = 200$ are:

$$l = 13 \text{ cms}: V_{peak} = 96·4 \text{ volts} \tag{A.88}$$

$$l = 4 \text{ cms}: V_{peak} = 23·9 \text{ volts} \tag{A.89}$$

(c) A Varactor Tuned Oscillator

If a VHF, UHF or SHF oscillator forms part of a synthesiser it will normally be incorporated in a PLL and voltage controlled tuning will be essential. The substitution of a varactor diode for the split stator capacitor used in (*b*) above introduces two further limitations. Firstly the finite *Q* of the varactor will usually seriously degrade the undamped resonator *Q*. Secondly the peak signal voltage developed across the varactor must, on one hand not take the overall reverse bias of the varactor below (say) 4 volts, and on the other hand the peak signal voltage plus the standing bias must not reach the varactor breakdown voltage.

For a given value of *l* in A. Table 1 the capacitance change required to tune from 472–530 MHz must be achieved with a DC bias range on the varactor which, when added to the peak to peak RF voltage swing, does not violate the lower or upper limits. In addition there is often a further practical limit: the available output voltage swing from a low noise integrated circuit active filter amplifier may not exceed a specified value. For the TDA 1034, for example, this is 12 volts.

The *Q* of varactor diodes degrades as the reverse bias voltage falls. It is usually defined at a specific low bias voltage such as 4 volts. Over its useable frequency range the *Q* of a varactor is also inversely proportional to frequency.

$$Q_v = \frac{f_c}{f} \tag{A.90}$$

where f_c is the varactor cut-off frequency.

If we were not constrained by the peak to peak signal voltage, that is if we were designing a very low power oscillator or a receiver signal band pass circuit such as is the case in a varactor tuned UHF TV receiver tuner, then a high varactor *Q* and a large tuning range might be possible. Varactor diodes are available with cut-off frequencies as high as 600 GHz, yielding a *Q* of 600 at 1 GHz. or 1200 at 500 MHz. Unfortunately varactors with the highest cut off frequencies usually have a limited voltage rating.

Assume that it is necessary to restrict the tuning voltage range to 10 volts total due to active filter amplifier characteristics. Let V_0 min and V_0 max be the minimum and maximum tuning voltages respectively and V_{peak} the signal voltage. Then the minimum voltage across the varactor is:

$$V_{min} = (V_0 \text{ min} - V_{peak}) \tag{A.91}$$

and

$$V_{max} = (V_0 \text{ max} + V_{peak}) \tag{A.92}$$

V_{min} must not be less than the minimum voltage at which the required *Q* may be obtained and V_{max} must be below the varactor breakdown voltage.

The peak signal voltage across the varactor may be reduced in either of two ways. Firstly it is possible to couple the oscillator transistor and the output

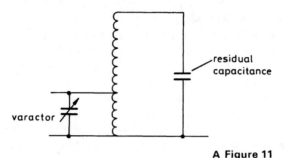

A Figure 11
Varactor Impedance Transformation

load to the resonator so that the loaded Q (Q_L) falls to the point where the signal voltage is acceptable (see A.87). This sacrifice in Q_L involves a direct sacrifice in phase noise performance as shown by (5.27) which is repeated below for convenience.

$$\frac{N_{op}}{C} = \frac{FkT}{C}\frac{1}{8Q_L^2}\left(\frac{f_0}{f_m}\right)^2 \tag{5.27}$$

If this approach were adopted for the case given by (A.89), and if it were necessary to reduce the signal voltage to 2·4 volts peak, then Q_L would have to be reduced from 200 to 2, giving a 40 dB degradation in phase noise performance.

An alternative approach is to provide an impedance transformation between the varactor and the resonator as illustrated in A Figure 11.

For a voltage transformation ratio t, the impedance transformation will be t^2. Thus the required varactor capacitance and the required varactor capacitance swing is greater by a factor t^2 than if it were connected directly to the high impedance point of the distributed inductance. In principle this may be achieved without sacrifice in Q_L and hence without sacrifice in phase noise performance. In practice a varactor with a large capacitance swing may well have a lower Q so that some (smaller) sacrifice in Q may be inevitable for medium power level oscillators, relative to that of low power oscillators.

It is also apparent that if the circuit of A Figure 11 is used, a narrowing of the required tuning range will permit the use of a varactor of smaller capacitance swing and in many cases higher Q. Use of the simple circuit of A Figure 10 will invariably result in a low working Q if it is used for a medium or high power oscillator: in many such instances Q_L may be as low as 10.

THE PHASE NOISE PERFORMANCE OF GUNN OSCILLATORS

Some few semiconductor materials, of which Gallium Arsenide is the best known example, have the property that the highest partly filled or filled energy band (at normal temperatures) is separated from the next higher permitted but empty energy band by a relatively small energy gap. In the case of GaAs this gap is 0·36 volts. If an electric field of more than 3·3 kV per cm is applied, some electrons will jump into this higher band. The mobility of electrons in this higher band is less than it is in the normally filled bands. Hence, as the voltage across a thin slice of such a material is increased, the current rises initially, followed by a fall in current with further voltage increase as electrons are transferred to the higher energy, lower mobility, band. Such a fall in current consequent upon a rising potential, constitutes a negative resistance which may be used as the active device in a microwave oscillator known as a Gunn oscillator or 'Transferred Electron' Oscillator. The 'Transferred Electron Effect' was predicted on theoretical grounds (Reference 48) and the prediction subsequently verified experimentally by Gunn (Reference 49).

A Gunn device is often referred to as a 'Gunn diode' although it does not incorporate any semiconductor junctions and is not a rectifier.

There are three distinct sources of noise in Gunn diodes as follows (Reference 50):

(a) Flicker noise from zero frequency up to perhaps as high as 10 kHz. This effect is common to most active devices including bipolar transistors, FETs and thermionic valves. Noise power varies as $1/f$.

(b) Generation/recombination noise which is a function of the trapping centres or impurities in the semi-conductor material. This also approximately follows a $1/f$ power law and is significant from low frequencies up to frequencies which may be as high as 20 MHz. It may be experimentally distinguished from flicker noise by the fact that it varies with temperature.

(c) Noise directly generated at RF, which is analogous to shot and thermal noise in other active devices.

If the oscillator were completely linear and the power output and frequency were independent of small variations in power supply voltage then only item (c) above would contribute to the RF noise output. Item (b), the generation/recombination noise, is effectively a noise current. The finite impedance of the power supply at each frequency from zero to (say) 20 MHz will convert this noise current into a power supply noise voltage at each frequency. This noise modulation of the power supply will amplitude modulate the oscillator carrier.

The mechanism of operation of a Gunn oscillator is such that changes in supply voltage affect the build up of high potential gradient domains and hence affect the susceptance of the diode as seen by the cavity. The resultant variation of frequency with supply voltage may be expressed as a pushing factor K_0' Hz/Volt. Noise voltage components, at absolute frequencies f_{m1} and f_{m2} superimposed on the power supply voltage will produce frequency modulation at rates f_{m1} and f_{m2} and (for small modulation indices) FM sidebands at offset frequencies f_{m1} and f_{m2}.

Thus both flicker noise and generation/recombination noise [items (a) and (b) above] will tend to produce both amplitude and frequency modulation of the oscillator.

The pushing factor of the oscillator will itself be dependent on the susceptance of the oscillator cavity at the diode interface and on the rate of change of diode susceptance with supply voltage. The resultant FM sidebands will be further filtered by Q_L and hence the frequency deviation resulting from low frequency noise superposed on the supply voltage will be greatly reduced by an increase in loaded Q.

For a full analysis see Reference 50 which also gives experimental results.

Careful design of the diode to reduce semiconductor trapping centres together with the use of a high loaded Q and a power supply with a low impedance up to video frequencies, make it possible to design Gunn oscillators with a relatively good phase noise performance (see for example curve (g) of 6 Figure 3).

INDEX

A

Accuracy
  frequency jitter measurements 193–195, 199
  frequency measurement 186
  linear approximation 13
  phase lock loop approximation 113
  phase noise measurements 168–171
Active filter 103, 125
Addition of sidebands
  power 14, 23, 24, 32
  voltage 14, 17, 23–24, 32, 33–36
Admittance
  matching 48, 57–59, 69
  mismatching for low noise 290–292
Allan Variance 184–185
Amplifier
  after frequency multiplication 81–85
  before frequency multiplication 81–85
  IF in PLL 133
  low noise 4, 290–292
  noise 80, 81–85
  noise figure 62, 70, 81–85, 290–293
  operational 68, 125, 162, 307
  saturating 128–130
Amplitude of carrier 8
Amplitude modulation (AM)
  after delay 209
  after frequency multiplication 75–76
  concept 1, 8, 22, 29
  local oscillator 2, 214–217
  measured performance 91
  noise 24, 310
  noise density 24, 29
  noise IF 209
  noise suppression 2, 216, 296–299
  oscillator noise 53, 91
  permissible in CW radar 211–217
  sidebands 8, 23, 24, 34–36, 207, 214
  theory 9, 12
  transmitter noise 207, 211–217, 229
Amplitudes, quadrature representation
  301–303
Analogue PDs 97, 122–127
Angle modulation 1, 9–16
Antenna
  gain 2
  polar diagram 213
  radar 213, 218
  tracking 272
Approximations
  linear 13, 40, 77
  to PLL performance 112–113
Arbitrary second 21, 27, 168–169
Available
  gain 7, 286
  power 286
Austron Inc 64, 90, 91, 277
Ayer A. J. 36, 277

B

Backward wave oscillator 92
Bandpass amplifier 6, 179, 255
Bandwidth
  baseband viii, 39, 45, 241
  equivalent filter of 199
  FDM 241–243
  FM 236–242
  noise 51, 52, 74, 238, 293–294
  offset frequency viii, 38–39
  Nyquist 255

positive feedback 50, 51, 56
time product 169
Barry Corporation 164, 278
Barrymount 164, 278
Baseband filter
  FM 233, 237, 239
  PSK 263
Baseband
  frequency 233
  phase relationships 28
  power 17, 32, 233, 240
  signal 1, 241
  sinusoid 17
  S/N ratio 234, 243
Beat frequency 226–229
Beurle R. L. 52, 277
Bell R. L. 91, 278
Bennett W. R. 263, 278
Bessell functions
  approximations 11
  general 10–11
  series 11
Bit error rates 254–260, 270–272
Bit rates 2, 247–249
Boltzmann's Constant 4, 285, 287
Breakthrough transmitter 207, 211–213,
  217–218
Broadcasting FM 232

C

Calculator TI 58/59
  program theory 153, 173–174
  program instructions 174–175
  program listing 175–177
Calibration
  correction 164–168
  phase detector 164–168
Cancellation
  DSB signals 222–224
  sidebands 23–24, 35
Capacitor
  losses 307–308
  split stator 72, 305
  varactor 57, 64, 72, 307–308
Carrier
  amplitude 8
  clutter 214, 218–221
  frequency viii, 50–53, 76, 77, 177, 181,
    185, 187
  power/modulation power 17
  recovery 249–254, 269–270
Carlson A. B. 236, 278

Carson's Rule 232, 236, 242
Cassegrain antenna 273
Cavity resonator 57, 72, 304–305
Cavity gap 92
Chao L. L. 169, 181, 182, 183, 194, 278
Characteristic impedance 305
Chi squared 194–196
Chirp radar 203
Clegg J. E. 224, 278
Clock oscillator 186
Clock recovery 260–262, 270
Close to carrier phase noise 2, 54, 64, 89, 93,
  249–254, 268
Clutter-ground, 204, 207, 213–214, 218–222
Coaxial resonator 57, 72, 304–305
Coherence time 7
Coherent
  demodulation 16–17, 248–260, 273–275
  local oscillator 206, 208, 214–216
  PSK 247–262
  radar 204–232
Comb filter 198
Communication systems 1, 203, 204, 232–275
Companding 232, 236, 241
Concepts
  phase noise 24, 36, 37
  potential sidebands 23–24, 33–36
Confidence level 169–171, 194–196
Conformable
  definition 14–15
  sidebands 14–15, 17, 19, 25, 37, 39, 80
  signals 154–156, 173–174
Confusion 13, 37
Contact
  RF 72, 269
  Spring fingers 72
Conventional symbols 3, 37, 185, 200
Convergence
  mathematical 65, 114–115, 177–178, 189,
    191–193, 267
Conversion-frequency
  by division 115–122
  by mixing 1–3, 131–133, 294–299
  by multiplication 75–92
Correlation 16
Counter 180, 186
  accuracy 186
  reciprocal 186
Court N. R. 91
Crystal
  oscillator 53, 54, 61, 62–64, 85–91, 134, 141,
    149–150, 206, 277

Q 61
  quartz 57, 61, 277
  rugged 57, 70
  vibration 57, 70
CW radar 204–224

D

Damped sinusoid 269
Damping factor 107–108, 119–120, 133
Data channels 2
Davey J. R. 263, 278
Dead space 182, 184–185, 195–197
De-emphasis 233, 236, 239–241, 243–246
Degradation
  communications system 2, 73
  communications system FM 234–241,
    244–246
  communications system PSK 253–254,
    258–260
  due to frequency multiplication 77–80
  due to frequency division 119–121, 130–131
  due to phase transients 268–272
  economic impact 2, 203, 236
  link budgets 2, 203, 235–236, 260
  oscillator performance 65–73
  radar systems 203, 211–222, 229–231
Degrees of freedom
  statistical 170, 194
  electrical 285
Delayed signals
  radar 207–211, 219, 225
  test equipment 211
Delta function 1
Delta modulation 204
Demodulation-coherent
  AM 17
  PM 16–17
  PSK 248–254
  tracking 273–275
'Derived' local oscillator 206, 208, 214–217
Density-frequency jitter 41, 44, 172–173,
    179–180, 185–186
Density-noise
  AM 24, 29
  PM 24, 26, 29
  thermal 4–5, 19–20, 285–286
Density-phase jitter 26, 30
Design-engineering 162–164
Deterministic waveform 18
Differential detection 247, 262–268
Differential equation 107
Digital comms. systems 73, 247, 268

Diode
  Gunn 309
  susceptance 310
  varactor 57, 64, 72, 307–308
Dividers-frequency
  need for 115
  programmable 116–119
  within synthesiser 136–139, 141–145,
    149–152
  effects on PLL 119–121, 130–131
Domain
  frequency 2, 172–202
  time 2, 172–202
Doppler
  frequency 203, 204, 218–221
  radar 204–224
Double sideband (DSB)
  superposed white noise-simplified 24–26
  superposed white noise-full treatment
    26–36
  suppression 222–224

E

Earth loops 68, 163
Economic factors 2, 203, 235–236
Edson W. A. 57, 277
Electron gun 92
Empiricism-extreme 36
Engineering design 162–164
Enhancement of spurious signals 156–161
Envelope modulation 301
Equi-partition theorem 285
Error horns 273
Error rate, 1, 73, 254–260, 262, 270–272
Error rate-formulae 256, 262
Errors of 3 dB 37
Expectation (statistical) 193
Experimental Results 61, 62, 85–87, 90–92,
    145–147, 152

F

FDM/FM 241–246
Feedback Q 52, 55, 56
FET 59, 290–292
Ferranti Ltd. 73, 91, 92
Ferranti klystron 73, 91, 92, 206, 213, 217, 222
Figure of Merit (G/T) 2
Filter
  active 102–104, 125, 159–161
  bandpass 36, 73
  bandstop 161, 206
  bank 206

baseband 233, 255, 263
comb 198
equivalent 178–179, 188–189, 199, 201
cut-off frequency 188–189
IF 179, 263
loop 100–104
low-pass 189, 192, 193, 201
passive 100–102
Financial aspects 2, 203, 235–236
First principles 3
Flicker noise 47, 55, 65, 309
Flip flops 116–118
F.M.-broadcasting 232
F.M. communications 232–246
Formulae 3
Fourier
    series 195–199, 301
    transformations 5
Frequency
    conversion 1, 2, 3, 131–133, 215, 216,
        294–299
    counter 180, 186
    deviation 41–46
    discrimination 219
    dividers 116–119, 136–139, 141–145,
        149–152
    division 115, 130–133
    domain 2, 172–202
    increment 135–138, 140, 149
    jitter 41–46, 172–202
    jitter density 41–45, 173, 186–189, 195–199
    jitter integrated 41–45, 173–174, 180–186
    multiplication 2, 35, 71, 75–92
    negative 5, 6
    offset viii, 53
    pulling 53
    range 135
    relevant 177–179
    selection 1
    spectrum 1
    stability/phase noise 172–202
    step changes 72
Frequency and Time Systems Inc. 70, 277
Frequency West Inc. 91
F.S.K. 2, 246

G

Gallium Arsenide 309
Gardner F. M. 96, 104, 109, 115, 278
Gaussian noise 6, 25
Generalisation of $N_{op/c}$ 171
Generation/recombination noise 309–310

Gnerlich H. R. 278, 309
Goldman S. 197, 278
Graphs
    UHF synthesiser 143–146
    SHF synthesiser 149–152
Graphical smoothing 171
Gravity 70
Ground clutter 204, 207, 213–214, 218–222
Ground stations 2
G/T 247
Gunn
    diodes 309
    effect 309
    oscillator 91, 309–310
Gunn J. B. 278, 309

H

Hadamard Variance 200
Hafner E. 48, 277
Halford D. 184, 278
Hines M. E. 86, 277
Harmonic content 69, 92, 161
Harmonic responses 199
Hybrid 273

I

I.C. amplifier 68, 125, 162, 307
IF modulation index 207–211
    AM 208–209
    PM 209–210
IF signal 219
Impatt oscillator 91
Impedance
    level 17,19
    transformation 48–51, 57, 58, 290–292, 308
Imperfections in signal sources 1
Incoherent sidebands 31
Independent measurements 168–171
Information-baseband 1
Instantaneous
    amplitude 21, 27
    frequency 77
    phase 21, 27
Intelsat IV 244
Integrate and dump 263
Integrated jitter
    frequency 45, 173–195
    phase 140, 146, 149, 152
Integration
    by calculator program 152–154, 173–177
    frequency deviation 45
    in FDM 45
    of output from PLL 114–115

oscillator noise 65
phase noise 38–39
pure noise 37–39
Inter-symbol interference 262
Intermediate calculations 3
Ismail M. A. 224, 278
Isolation-transmit/receive 211–213

J

Jitter (frequency) 41–46, 172–202
  Allan Variance 184–185
  calculation 173–177
  density 41–45, 173, 186–189, 195–199
  measurement 180–185, 195–199
  normalised 181
  practical concepts 179–180, 184–185, 200
  theoretical concept 173–177, 179–180,
    185–186
  transfer function 186–189
Jitter (phase) 18–40
  calculation 152–154, 173–177
  definition 26, 38
  density 26, 30, 171
  measurement 164–168
  oscillator 65
  synthesisers 140, 146, 149, 152

K

Kendall M. G. 194, 278
Kimura K. 44, 91, 277
Klystron
  low noise 73, 91, 92, 206, 213, 217, 222
  noise theory 91
Kodera H. 44, 91, 277

L

L-band 138–139
Lag-lead 103
Laplace Transformation 96
Leeson D. B. 53, 277
Limiter 35, 36
Limitations of simple sources 87
Linear
  approximation 13, 40, 77
  FM 224–229
  oscillator 47–53
Link
  budget adjustment 260
  degradation 235–236
Local oscillator
  AM noise 214–217
  AM suppression 2, 216, 296–299

derived 206, 208, 214–216
PM noise 232, 234, 237–239, 244–246, 254,
  261, 267–268
PM transfer to IF 2, 209–211
Logical Positivist 24, 36
Long term mean power 19–20, 31, 185–186,
  303
Loop filter 100–104
Loop gain
  oscillator 51, 64
  PLL 104–105
Losses-effect on noise temperature 288–290
Low noise amplifier 4, 290–293
Low pass
  analogue 16
  filters 189, 192, 193, 201
Lower AM sidebands 34–36
Lower PM sidebands 34–36

M

Maeda M. 44, 91, 277
Magnetrons 92
Marconi Space & Defence Systems Limited i
Maritime satellite 241
Master oscillator (MO)
  degradation in PLL 121
  enhancement 156–161
  frequency stability 201
  phase noise performance 62, 63, 141
  spurious outputs 156
Matching (admittance/impedance) 59
Mathematical
  manipulation 3
  sophistication 3
Maximum Power Theorem 285
Mean power
  instantaneous 19–21, 27–28, 31, 33–36
  long term 19–26, 29–31, 185–186, 303
Mean
  square 7
  statistical 169, 181–184
Meaningless 36, 178
Measured performance 61–63, 86–87, 90–91,
  146, 152
Measurement
  frequency jitter 180–195
  frequency jitter density 195–200
  phase noise 164–171
  phase noise density 164–171
  spectral purity 4
Mechanism 3
Merit (satellite ground station) 2

Microwave Associates i, 141, 142, 147, 148
Mismatch
  of oscillator 69
  for low noise 290–292
Misunderstandings 74, 168
Mixer
  as phase detector 97, 122–127
  balanced (AM suppression) 2, 216, 296–299
  in PLL 131–133
  noise figure 294–295
  NTR 123, 126, 127, 294–296
Modulating waveforms 302, 303
Modulation
  AM 1, 9
  angle 1, 9–16
  envelope 301
  PM 1, 9–16
  indices 8–13, 22, 26, 29, 33
  indices after multiplication 75–77
  indices IF 207–211
  spurious 1, 2
  theory 9–17
Monopulse 273
Moss S. H. i
Mullard/Signetics 125, 278
Multi-channel loading factor 242
Multiplier harmonics 161
Multiplication-frequency 2, 35, 71, 75, 92
  effect on:
  AM mod. index 75–76
  AM noise 75–76
  PM mod. index 76–77
  PM noise 75–92
  thermal noise 78–80

N

Narrow band noise
  basic representation 6–7, 19–20
  display of 6–7
  figure 287
  quadrature representation 300–303
Negative frequencies 5, 6
Noise
  added 69
  amplifier 3, 290–293
  AM 24, 29, 310
  bandwidth 51, 52, 74, 238, 255, 265, 293–294
  coherence time 7
  density 4–5, 19–20, 285–286
  density/carrier ratio 26
  density-single sided 5
  density-two sided 5

DSB 5
figure 3, 7, 49, 60, 62, 64, 70, 75, 81–85,
    88–90, 285–293
figure definition 287
figure amplifiers 290–293
figure mixers 294–295
floors 122–131, 138
Gaussian 6, 7
generation/recombination 309–310
Gunn oscillators 90–91, 309–310
mixers 294–296
oscillators 47–74
phase 24, 29
shot 122, 290
sinusoidal 8
SSB 5, 20–24
statistical nature 7, 20–21
suppression 2, 216, 296–299
temperature 286–290, 294–296
thermal 4–5, 19–20, 285–286
triangular 243
vector 8
voltage 50, 66–69, 285, 301, 310
white 24, 26, 38, 40, 202
Non-linear oscillator 47–48, 68
Normal distribution 169
Normalised
  frequency 181–185
  power 17
Notation 37
Novice 3
NTR 123, 126, 127, 294–296
Nyquist H. 278, 285

O

Observation of N.B. noise 6
Offset frequency range 177–180
Ohtomo M. 44, 91, 277
Ondria J. 278, 309
Operational amplifiers 68, 125, 162, 307
Optimum source resistance 292
Orthogonal sidebands 19
Oscillator/Multiplier 85–92
Oscillator Noise
  analysis 48–53
  degradation 65–73
  dual resonator 73
  examples 61–65, 90–91, 141–151
  figure 48–53, 81–90
Oscillator power supplies 66–68
Oscillators
  frequency jumps 72–73

frequency stability 201
Gunn 91, 309–310
manually tuned 72–73
master (MO) 62–63, 141
measured performance 61–63, 85–91,
    141–151
quartz crystal 53, 54, 61, 62–64, 85–91, 134,
    141, 149–150, 206, 277
reference 62–63, 141
SHF 90–91, 268
UHF 64, 143, 151, 268, 304–308
under vibration 57, 70
varactor tuned 304–308
Oscilloscope 6

P

Padley J. S. 275, 278
Parameters of oscillators 60–61
Parasitic oscillation 86, 92
P.D.s. (Phase detectors)
    analogue 97, 122–127
    as coherent detectors 16–17
    differential 247, 262–268
    edge-triggered 99–100, 128
    exclusive-or 97–99, 127–128
    gain of 97–100
    noise floors of 122–131, 138
    operating frequency of 97, 136–138
    other types 100
    phase error of 97–100
    signal breakthrough from 157–161
Peak/rms ratio 19, 185–186
Peak value of
    modulation index 8
    noise sinusoid 7
Pen recorder 171
Peregrino L. 199, 278
Perfect source 165
Phase
    angles 12, 19–20, 32
    constants 31
    detectors see PDs
    difference 25
Phase jitter
    after delay 209–211
    after frequency multiplication 76–85
    basic concept 7–8
    due to SSB noise 22
    due to DSB noise
        simplified treatment 26
        full treatment 30
    'density' 26, 30

due to discrete signals 154–156
integrated (measured and predicted) 146,
    152
integration
    calculator program 174–177
    of discrete spurious 155–156
    mathematical convergence 65, 191–192
    theory 152–154
measurement methods 164–168
intermediate frequency 217
PLL output 115
relationship to noise density 26, 30
relationship to frequency deviation 41–46
relevant offset frequency range 177–180
sinusoidal 18
specification for
    antenna tracking 275
    coherent PSK 259–261
    differential PSK 267
    FDM/FM 244–246
    FM 236–237
    radar 204, 207, 217–222, 229–231
variance 26, 30
Phase lock loops (PLL)
    approximation to response 112–113
    closed loop gain 105
    configuration 94–109
    damping factor 107–108, 119–120, 133
    error response 109–114
    error with frequency step input 269
    frequency division within 119–121, 130–133
    incorporating mixer 131–133
    natural frequency 105–107, 119–120, 133
    open loop gain 104–105
    reference oscillator in 93–133
    transfer function 109
Phase modulation 1, 9–11, 78
Phase modulation theory 9–11
Phase noise
    after frequency multiplication 76–78
    approximations 11, 22
    basic concepts 24, 81
    control 3
    density 24, 71
    experimental results 62–63, 85–91, 146, 152
    /frequency deviation 41–46, 172–174
    IF 214–217
    law of oscillator 54–55, 114–115
    local oscillator of 214–217, 232, 234,
        237–239, 244–245
    measurement of 164–168
    mechanism 3

recommendations (also see specification) 235–236, 260
reduction 3, 112–115
relation to frequency stability 41–46, 172–180
relation to phase jitter 26, 30
specification for
  antenna tracking 275
  bit error rate 259–261
  carrier recovery loop 254
  clock recovery loop 260–262
  differential PSK 267
  FDM/FM 244–246
  FM 236
  radar transmitter 204, 207, 217–222, 229–231
Phase shift keying (see PSK)
Phase signals of 14
Phase transients 268–272
Phases random 28
Phasor diagrams 7, 13–14, 27, 28, 31, 32, 33
Philosophic difficulties 36
Phylip Jones G. 92, 278
Physical temperatures 288–290
Physical understanding 3, 77
Picovolts 67
PM index 7–8, 10
Polar form 302–303
Popper K. R. 36, 277
Positive feedback 52, 55–56
Potential gradient 310
Potential sidebands 23–24, 33–36
Power
  density
    AM 24
    PM 24
  DSB 26
  losses 53
  normalised 17
  oscillator of 51–53
  output 70
  radar 206
  satcomms 2
  sideband 36
  SSB 23
  supply 66–69
  supply ripple 68, 161–164, 310
  transmitter 206
PRK (phase reverse keying) 246
Predicted performance 61–63, 81–87, 141–146, 149–152
Pre-emphasis 233, 236, 239–241, 243–244

Pre-scalers 119
Pressure contacts 72, 269
Program-calculator 153, 173–177
Psophometric weighting 243, 246
PSK 2, 246–272
Pump 1

Q

Q
  coaxial resonator of 304
  crystal of 61, 63, 277
  degradation 59
  filter of 73
  load, changes, effect on 70
  loaded 52, 55–60, 63, 69, 70, 304–308, 310
  positive feedback 52, 55, 56
  time domain equivalent filter of 199
  unloaded 59, 60, 304
  varactor tuned oscillator 304–308
  VCO 64
Quadrature carriers 301–303
Quadrature phase shift keying (QPSK) 246
Quadrature phasors 31
Quantitative contributions 3
Quartz crystal
  oscillators 53, 54, 61–64, 70–71, 85–91, 141–150, 206, 277
  parameters 61
  Q 61, 63, 277
  rugged 57, 277
  vibration 70, 71

R

Radar
  chirp 203
  coherent 204–232
  continuous wave (CW) 204–224
  delay 208–210, 225–229
  FM 224–232
  moving target indication MTI 204
  range 210, 225–229
  types of 203–204
Radio altimeter 226
Random phases 28
Range
  increment 226
  non-ambiguous 227
  resolution 226–229
Rate of change
  amplitude 6
  phase 6
Rayleigh distribution 20
Reactance-transmission line of 305
Receiver 2

Reciprocal densities 173
Reference oscillator
  degradation in PLL 120-121, 138-139
  enhancement 156-161
  typical performance of 62-64, 90, 121,
    140-141, 149
Reference point (noise temp.) 289
Relevant offset frequency range 177-180
Representation of NB noise 6, 19-20, 300-303
Resonance 52
Resonator
  capacitatively tuned 304-308
  coaxial 57, 72, 304-305
  dual 73
  manually tuned 72-73
RF
  contacts 72, 269
  noise 47, 48, 310
  power 2
Ribbon mounted quartz 57
Ricci D. W. 199, 278
Rice S. O. 278, 301
Ridenour L. N. 204, 278
Ridley, B. K. 278, 309
Ripple power supply of 66-69, 162-163, 310
Risley A. S. 184, 278
rms
  long term mean value 19-20, 185-186
  modulation indices 8, 22
  value for specific second 19-20
  value of noise 7
  voltmeter 168-169
Robins D. i
Robins W. P. 2, 37, 92, 222, 275, 277, 278
Rockland Systems Corp. 135, 278
RRE (now RSRE) 91, 92
Rupp S. 91, 277
Rutman J. 184, 185, 200, 278

S

SCPC 2, 232
SHF
  oscillator 72, 304
  resonator 304
  satcomms 235
  signal sources 90-93
SSB
  . concept 5-6
  superposed noise 20-24
  signals 222-224
Salter M. 2, 277
Samples independent 168-171

Sarbacher R. J. 57, 277
Satellite Communications
  economic factors 2, 235-236
  ground stations 2
  link budget adjustment 260
  link degradation 235-241, 260
  phase noise requirement 235-241, 244-246,
    249-268
  coherent PSK 259-261
  differential PSK 267
  FDM/FM 244-246
  FM 236
  systems 2, 203, 232, 235
Saturating amplifier 128-130
Screening 68, 163-164
Script '$\mathscr{L}$' 37
Second arbitrary 21, 27-35, 168-169
Semi-conductor trapping centres 310
'Serva-levl' system 164, 278
Servo system 103, 275
Shielding 68, 163-164
Shoaf J. H. 184, 278
Shot noise 122, 290, 310
Sideband
  cancellation 23, 35
  interaction 24, 35-36
  power 18-19, 23-24, 29, 36
Sidebands
  AM 8, 23, 34-36
  conformable 14-16, 37, 53, 80, 154-156
  FM 11-12
  lower 23, 34-36, 78-79
  potential 23-24, 34-36
  upper 23, 34-36, 78-79
Signal
  clutter ratio 213-214, 218-221
  noise ratio 128, 234, 241, 243
  processing 2
  sources 1, 3, 85-92, 134-152
  voltage 16-17, 307
Signetics 125, 278
Simple harmonic motion 106
Single channel per carrier (SCPC) 2, 232
Single sideband (see SSB)
Single sided density 5
Sinusoidal
  bandwidth 6
  representation 6, 19-20, 300-303
Smith H. i, 189, 193, 278
Smith W. L. 57, 70, 277
Source
  performance 90-91, 140-152

resistance 290–293
Specification of spurious AM, PM
  coherent PSK 259–261
  differential PSK 267
  FDM/FM 244–246
  FM 236
  radar CW 204–222
  radar FM 229–231
Spectral
  line 1
  width (minimum) 229
Spectrum
  analyser 36, 81
  signal of 1
  spreading 207
Spencer W. J. 57, 70, 277
Split stator capacitor 72, 305
Spring fingers 72
Spurious
  AM for coherent radar 204, 207, 211–217
  modulation 1
  outputs 92, 154–161
  PM for coherent radar 204, 207, 217–222
Stabiliser (voltage) 68
Stability (frequency) 41–46, 172–202
Standard deviation 169–171, 181, 193
Static split 273
Station MO 55, 62, 63
Statistical
  aspects
    frequency jitter 193–195
    phase noise 168–171
  bias 169
  confidence levels 169–170, 194–195
  equivalence 181–184
  expectation 193
  mean 169, 181–184
  value 7, 20–21, 185–186
  variance 181–195
  variance of the variance 194
Stress physical 73
Strip line 57
"Students" t Distribution 170
Successive refinement 3
Sum signal 273
Summary of phase jitter/noise relationships
    38–40
Superposed noise
  DSB 24–36
  SSB 20–24
Superposition Theorem 13, 33
Susceptance diode of 310

Symbols viii, 3
Synthesisers 134–152
  direct 134–135
  measured performance 146, 152
  multi-loop 140–152
  non coherent 134
  phase transients in 268–272
  PLL 135–152
  UHF 140–147
  X-band 147–151

T

't' Distribution 170
Tangent, approximation to 22, 28
Telephone channels 2, 241–242
Temperature
  noise 4, 206, 286–288
  physical 288–290
Terman F. E. 57, 277, 304
Test tone
  deviation 242, 244
  to noise ratio (TTNR) 244
Texas
  calculator TI 58/59, 153, 173–177, 256
  Instruments 97, 99, 127, 278
Time of oscillator build-up 56
Time domain 2, 172–202
Thermal
  equilibrium 286
  noise 4, 24, 26, 53, 78–81, 286, 310
  noise in FM comms 233, 243
Thermodynamics 286
Thorne T. G. 224, 278
Tracking
  antenna 272–275
  receiver 274
Traffic capacity 2
Transfer function
  active filter 103–104
  frequency jitter to 177–180, 186–189
  PLL of 109
  phase jitter to 177–180, 189–193
Transferred electron effect 309
Transients-phase 268–272
Transistor
  capacitance 69
  oscillator 47–74, 75, 85–87
Transmission line reactance 305
Transmitter
  AM 204, 207, 211–217, 229
  breakthrough 207, 211–213, 217–218
  PM 204, 207, 217–222, 229–231

power 206
Triangular noise 243
Triodes disc seal 73
Trough line 57
Truth tables 97, 99
Tuneable filter 165
Tuning
  bellows 72
  capacitor 72, 305-306
  manual 72
  plunger 72
  range 57, 64, 305-308
  slope 66-68
Two sided power density 5

U

UHF
  oscillators 64, 143, 151, 268, 304-308
  resonators 304-308
Unbalanced pick up 68
Unbiassed estimate 169
Uniform phase distribution 28
Upper sidebands
  AM 34-36
  PM 34-36

V

VCO
  gain 66-68, 96
  applications 75-164
  L-band 138-139
  power supply ripple 66-68
  UHF 64, 141-145
VCXO (voltage controlled crystal osc) 57, 148
VSWR 69, 70
Varactor diode
  cut-off frequency 307

multiplier 85
Q 307
tuned oscillator 57, 64, 72, 307-308
  voltage constraints 307-309
Variance
  Allan 184-185
  max-min 194-196
  frequency jitter 181-195
  phase jitter 26, 168-171
Variances
  of noise components 29
  relationships 185-186
Velocity ambiguty 204
Verification Principle 24, 36
Vibration 57, 70, 71, 161, 164
Voltage
  addition 17, 23-24, 32, 33-36
  tuning 57

W

Waveguide losses 288-290
Warner A. W. 63, 277
Watkins T. B. 278, 309
Weighting Function 195-196
Whale A. V. 73, 91, 277
White noise 38-40, 202, 300
Whitwell A. L. 62, 86, 277
Williams N. 62, 86, 277
Woodward P. M. 204, 278

X

X-band sources
  design considerations 93-94, 136-140
  measured performance 62, 90-92, 152

Z

Zepler E. E. 163, 164, 278

Printed in the United Kingdom
by Lightning Source UK Ltd.
9551000001B